"Engaging, provocative and with great attention detail, Brown dares to pose the stark alternatives facing humanity regarding food, population and the environment: limit the size of the global population or limit the rate of food production. This is no simple choice however. Brown argues that it is the simple 'solution' of industrial agriculture that has gotten us into the vicious circle of increasing food production to feed an increasing population while destroying the environmental foundation of sustainable food systems. Essential reading to anyone who wonders where our next meal ought to be coming from."
— **Brewster Kneen**, author of *Invisible Giant* (on Cargill) and *Farmageddon* (on Food and Biotechnology), publisher of the monthly magazine *The Ram's Horn*, is Canada's foremost analyst and critic of the food system

"Although it sums up Duncan Brown's thesis, the title *Feed or Feedback* is regrettably unlikely to convey to the layperson the urgency of this book. Here we have nothing less than an historical reading of man's relationship with his most essential life-support system—his food supply and the ecology that sustains it—and the stresses to which such vital systems are subject. We must take Brown's warnings seriously and act upon them." — **Susan George**, Associate Director Transnational Institute, Vice-President ATTAC France and author of ten books (most recently *The Lugano Report*)

"Everybody concerned about the future of humanity needs to know the basic, ominous biochemistry of food production and population growth. After an excellent scientific introduction to what happens when a society literally cannot get its shit together, Brown lays out the alternatives before us: fundamental reform of our agriculture, or ultimate collapse." — **Ernest Callenbach**, author of *Ecotopia* and *Ecology: A Pocket Guide*

"In using intensive farming to obtain our food, we have taken a road that is not only cruel to animals, but also inefficient and destructive to our environment. We urgently need to think again about agriculture. Professor A. Duncan Brown's closely-argued book is a valuable contribution to that debate." — **Peter Singer**, DeCamp Professor of Bioethics, Princeton University, and author of *One World*

"In 1972, the report on the *Limits to Growth* by a group of young scientists captured the world's imagination in an unprecedented way. Within a year, the initiators of this study, the Club of Rome, became world famous. Millions of copies were sold in over 50 languages, numerous articles appeared, conferences, speeches, debates, denial, political action, and new research, everything one can imagine followed.

In *Limits to Growth* a relatively new approach was taken to understand the global system of population, production, resource use, food supply and pollution, in such a way that all interactions between those variables could be measured together, and calculated in their common interrelated behaviour. The so-called systems-dynamics model WORLD-III was based on the principle of feedback-loops, taken from the science of cybernetics, the teachings of balance or homeostasis. These tell us that most systems in nature, be they geo-chemical, biological or physical, tend to a certain equilibrium, by means of feedback mechanisms which occur naturally.

In his book on agriculture, population and the carrying capacity of the planet, Duncan Brown does it all over again, and in an excellent way. His analyses are illuminating, his examples challenge the imagination, and his conclusions are dramatic, just as those from "Limits". Agriculture as we know it lacks proper mechanisms to achieve balances between input and output, and most commentators on the global food perspective have it wrong, as they generally do not understand the cyclic functions of soils, nutrients and water in relation to population dynamics. Consequently, our agricultural technologies are linear systems, not cyclic, and hence the forecasts that the world will run out of sufficient food production, not in the distant future, but soon, within some decades. It is a message that cannot too often be repeated."
— **Wouter van Dieren**, Club of Rome

"Duncan Brown shows why the positive feedback interaction between population and food supply is leading us to a collapse. We must break away from this vicious circle." — **John Hontelez**, Chairman, Friends of the Earth International (1986-1996), currently Secretary-General of the European Environmental Bureau, Brussels (representing 134 organizations from 25 countries)

"Duncan Brown analyses many well-known themes, but he sets them in a refreshing context. Of particular interest is the plausibility of Brown's prognosis that current intensive agriculture as a food-production system will collapse during this century. The combined effects of very low nutrient input efficiency, soil loss, water shortage and reduced biodiversity are persuasive in this connection. Structurally overstretched, this 'throughput' system loses its ability to recover from such degradation, even if world population were to stabilize.

The author develops some lines of thought which at first sight seem exaggerated, yet on closer inspection they are alarming.

A truly sustainable agriculture requires – among others – intensive recycling of nutrients and organic material. Of these, phosphorus is an essential element for the growth of plants for which no substitute exists. It will be the first physically restrictive factor for further betting on 'high external input agriculture'. Phosphorus is mined as phosphate and the expected exhaustion of phosphate reserves, against affordable prices within a century, aggravates the urgency of the problem. The great task of the coming half-century will therefore be to fundamentally redesign our systems of production.

We should not expect business interests – selling their inputs – to anticipate the needed adaptation of the food production system: under free market forces exploitation of phosphates will continue until scarcity is reflected in the pricing. By that time it is far too late for redesigning measures for which at least some decades will be needed. Therefore social organisations (such as universities) and governments should take the lead, I hope soon in cooperation with UN-organisations.

Serious recycling of nutrients requires fundamental social change and a different city-planning. Brown's analysis of the re-use potentials of urban waste and redesigning sewage systems is challenging. He concludes that people would have to disperse again over productive rural areas. But why should city-planners change their habits or implement controversial measures if the need and urgency for such redesign are not publicly accepted?

Cont. on p. 432

A. Duncan Brown

FEED OR FEEDBACK

Agriculture, Population Dynamics
and the State of the Planet

International Books

© A.D. Brown, Tuross Head, NSW, Australia, 2003
ISBN 90 5727 048 X

Cover design: Karel Oosting, Amsterdam
Cover photograph: Karel Tomeï, Flying Camera, Eindhoven Airport
Typesetting: Hanneke Kossen, Amsterdam
Printing: De Boekentuin, Zwolle

International Books
A. Numankade 17, 3572 KP Utrecht, the Netherlands
Tel. +31 30 273 18 40, fax +31 30 273 36 14, e-mail: i-books@antenna.nl

Table of Contents

Acknowledgments

This book could not have been written without the help I received. First I am indebted to Professor Henry Nix and the administration of the Centre for Resource and Environmental Studies (CRES) at the Australian National University for offering me a Visiting Fellowship and the facilities that went with it. I am grateful to many people—too many to enumerate—who offered encouragement, drew my attention to important reference material and, in some cases, lent me books which I would not otherwise have seen.

At a more specific level, I want to thank Peter Bellwood, Jenny Clark, Mark Elvin and Bob Wasson for reading sections of the text and for their constructive comments. And in that same vein I offer special thanks to Steve Dovers and Chris Watson for their critical assessment of most of the text and for their continuing encouragement.

I also thank Jan van Arkel of International Books for his level-headed assessment of the manuscript, for bringing some additional references to my attention and generally for being very helpful.

I have not mentioned Stephen Boyden. Actually I am not sure whether I should thank him or not, since it was he who talked me into writing the thing in the first place. I admit that he did suggest to CRES that they offer me a Visiting Fellowship. He also continued to offer constructive suggestions and much encouragement throughout the whole exercise—but still...

Finally I want to thank Joan, my wife, for putting up with me (and continuing to make muffins) during my preoccupation with trying to explain to the world the underlying reasons why it is likely to go down the drain.

Symbols and Abbreviations

ABC   Australian Broadcasting Corporation
BP   before the present (time)
d   day: d^{-1}; per day
G   giga..., 10^9, billion
h   hour
ha   hectare, 10,000 (10^4) m^2, 0.01 km^2: ha^{-1}; per hectare
hd   head: hd^{-1}; per head, per capita
J   joule (unit of energy), 10^7 ergs, 0.239 calories
k   kilo..., 10^3, 1000
km   kilometre, 10^3 m: km^{-1}; per kilometre: km^2; square kilometre, 100 hectares, 10^6 m^2: km^{-2}; per square kilometre: km^3; cubic kilometre, 10^9 m^3, 10^{12} l: km^{-3}; per cubic kilometre
l   litre, 1000 ml: l^{-1}; per litre
m   in isolation, metre: followed by another symbol, milli..., 10^{-3}
m/v   mass per volume
M   mega..., 10^6, million
P   peta..., 10^{12}, also phosphorus
ppm   parts per million
s   second: s^{-1}; per second
t   tonne, metric ton, 1000 kg
td   doubling time
W   watt, $J.s^{-1}$
y   year: y^{-1}; per year
μ   micro..., 10^{-6}, one millionth

Terminology

The terms 'agriculture', 'farm' and 'farming' are used loosely to include both cultivation and pastoralism. When distinctions are necessary other terms are used. The term 'Man' (with 'M' in upper case) is used as a common collective noun for the species *Homo sapiens*.

Nine 'Laws' of Ecological Bloodymindedness

The First Law   For every action on a complex, interactive, dynamic system, there are unintended and unexpected consequences. In general, the unintended consequences are recognised later than those that are intended.

The Second Law   Any system in a state of positive feedback will destroy itself unless a limit is placed on the flow of energy through that system.

The Third Law   Any sedentary community, by virtue of its sedentism, will encounter problems of sanitation. The manner in which sanitation is managed will affect the manner in which supporting agriculture is managed.

The Fourth Law   For every increment in the agricultural surplus there is a corresponding increment in the volume of urban sewage.

The Fifth Law   Stability or resilience in ecosystems requires that all essential reactions within the system function within ranges of rates that are mutually compatible.

The Sixth Law   The long-term survival of any species of organism requires that all processes essential for the viability of that species function at rates that are compatible with the overall functioning of the ecosystem of which that species is a part.

The Seventh Law   If any species of animal should develop the mental and physical capacity consciously to manage the ecosystem of which it is a part, and proceeds to do so, then the long-term survival of that species will require, as a minimum, that it understands the rate limits of all processes essential to the functioning of that ecosystem and that it operates within those limits.

The Eighth Law   Long-term stability or 'sustainability' in ecosystems (including agricultural systems) is dependent in part upon the recycling of nutrient elements wholly within the system or upon their replenishment from a renewable source, provided such replenishment is not itself dependent upon a finite source of energy.

The Ninth Law   If a population continues to grow exponentially it will eventually consume essential resources faster than they can be replenished. The provision of or access to additional resources will extend the 'life' of such resources, and hence the duration of growth of the population, only to a very small extent.

Preface

There is a widespread and growing awareness that the biosphere is facing a number of very serious man-made threats. Arguments and active attempts to confront these threats commonly focus on fairly specific issues such as deforestation, endangered species, pesticides, pollution, the 'greenhouse effect' and global warming, soil degradation and, of course, a large and growing human population.

All of these matters are important and should be taken very seriously. In a sense, however, they are symptoms rather than the primary causes of our problems. None of them could have achieved anything like their present significance without the stimulus provided, since the advent of agriculture, by (a) the nature of the dynamic interaction between the human population and its food supply and (b) the essentially 'linear' and irreversible (as distinct from cyclic) flow of nutrient elements in the current system of commercial agriculture. Unless we address those two basic phenomena the global ecosystem will certainly collapse—a process leading to the extinction of very many species, quite possibly including our own.

This book arose out of a submission I made in response to an Australian Federal Government discussion paper on 'Ecologically Sustainable Development'. My submission, in the context of (b), was to the effect that commercial agriculture, as currently practised, will eventually exhaust phosphate reserves. A friend (friend?) suggested I expand my submission into a book. I could see no point in that because it had been said before. On reflection, however, it seemed that, if the terms of reference were extended to include (a) and deal with agriculture and human population dynamics as components of a complex system, there could be some merit in the idea. So that is what I have tried to do.

The project obliged me to confront a number of topics in which I had little or no direct experience. Multiple authorship could have overcome that problem but I decided against it for two reasons. One was that, since my broad aim was to produce an accessible appraisal of a very complex situation,

I should first make sure that I could explain that situation to myself. The second was that the various specific topics would be treated, not as ends in themselves, but as components of a large and expanding dynamic system.

Perhaps the most basic 'message' of the book can be summarised by saying that, since the advent of agriculture, the species *Homo sapiens* has not met the requirements of the 7th 'Law of Ecological Bloodymindedness':

"If any species of animal should develop the mental and physical capacity consciously to manage the ecosystem of which it is a part, and proceeds to do so, then the long-term survival of that species will require, as a minimum, that it understands the rate limits of all processes essential to the functioning of that ecosystem and that it operates within those limits."

In their hunter-gatherer days, human communities responded spontaneously or empirically, so we must assume, to the challenges and rewards of the ecosystem of which they were a part. With agriculture and increasing sophistication, however, came an ability to manipulate the environment in ways and to degrees that were not previously possible. The problem that arose here was that understanding did not keep pace with the physical ability to modify and to manage. To complicate the process, empirical understanding was replaced to varying degrees by dogma from the clergy, from 'experts', from economists. All too often such people declare that the functioning of complex systems shall be assessed by one or two criteria that they, the specialists, have devised. This general process has usually been accompanied by a reluctance or inability to look objectively at the evidence—at the wider results and implications of practices decreed by the conventional wisdom of the time. So, to condense and simplify an enormously complex phenomenon, I consider this particular shortcoming to lie at the heart of the most fundamental long-term threat to the survival of *H. sapiens*.

The following chapters seek to address some of the more important pathways by which this general situation came about, some quantitative indicators of the current state of play, and some minimal changes that would seem to be necessary if the most serious consequences of our past and present activities are to be avoided. Up to that point the theme is essentially ecological. The final chapter is somewhat different. It considers, albeit in general terms, barriers in the way of the remedies suggested in Chapter 14. It suggests that the most fundamental of those barriers might lie in some peculiarities of human behaviour and, therefore, if we are serious about protecting the future of our species, a good way to start might be to take a hard look at ourselves.

PART I

CHAPTER I

The Beginning

There have been two biological events that, in my view, have overshadowed all others in terms of their effects on the global environment and ecology. By 'biological', I mean brought about by established organisms—I am not talking about the origins of life. The first, both chronologically and in order of importance, was the accumulation of elementary oxygen in the atmosphere. On present evidence that process began perhaps 3.5 billion years ago with the advent of cyanobacteria (blue-green algae).[1]

Until then the planet had supported only anaerobes—organisms that grew without oxygen, without any need of it and, more often than not, were poisoned by it. Oxygen is potentially a very dangerous substance. Part of the energy metabolism of modern aerobes, organisms that depend on oxygen, is devoted to mitigating its toxic effects. Some modern anaerobes can passively tolerate oxygen, but most cannot. We can safely assume that most forms of life that had evolved before this new gas accumulated were highly sensitive to it. But not only that; aerobic metabolism produces more energy from a given amount of substrate and does so much faster than anaerobic metabolism. The upshot was that gaseous oxygen became perhaps the most potent biologically selective molecule that the Earth has ever seen. Not only did this gas poison organisms that had existed until then, but it gave the new-fangled aerobes a great advantage in growth rate over any anaerobes that managed to survive either by hiding or by developing some defence against its toxic effects. Oxygen shaped the course of all subsequent evolution—but it produced a biosphere of great diversity and beauty.

The second event was the development of agriculture by Man. The transition from hunting and foraging brought about a series of ecological changes so profound and so extensive as to place in question the survival of our own species—and to place beyond question the fate of the myriad species that have already been driven to extinction or are on its brink.

Many of these ecological changes are widely recognised, sometimes with alarm, sometimes with satisfaction—according to one's predisposition or time and place in history. Others are so much a part of our lives that we take them for granted, consider them to be part of the essential nature of things. It needs only reflection and a rudimentary understanding of our past to appreciate that without farming, there would be no cities, no libraries, no hospitals, no computers, no nuclear weapons, no musical instruments of any sophistication, no motor cars—not even bicycles. There would have been little concept of wealth and probably no accountants or economists. Without farming there would have been no European invasion of America or Australia. There would have been no population explosion of Man—or sheep, or cattle, or rabbits.

Agriculture eventually placed into human hands the means to breach the defences of every ecosystem on the planet. Primary reliance on agriculture changed the relation between the human population and its food supply from one of negative to positive feedback, and it redirected the flow of nutrient elements in the biosphere. This fairly simple statement identifies, or perhaps disguises, trouble.

Before agriculture, Man's place in nature was not very different from that of any other animal species, insofar as his impact on the ecology was concerned. To be sure he had tools and weapons, which made him a capable, dangerous predator and, of course, he used fire which altered his habitat in a way not open to other species. This was especially true of Australia. But once the habitat was changed by fire, Man had to adapt to it as did the other species that survived the change. Certainly a number of species, notably large mammals and flightless birds, did become extinct when people arrived in the neighbourhood and, of course, the changes in animal numbers and types exerted their own effects on vegetation. Although each of these qualifications gives a foretaste of what was to come, they did not amount to anything different in principle from what other effective predators can do in a new environment.

When a new species arrives in an established ecosystem it is faced with two quite distinct possibilities. Either it will not obtain a foothold and will die out or, alternatively, it will adapt to the new habitat and become established. In the latter case it will necessarily disturb the balance of the preexisting species, frequently causing the extinction of one or more of them. This last has been experienced all too often when rats or cats have found their way into island communities, for example. The apparent extinction of megafauna at the hands of early Man is not necessarily any different in principle from extinctions caused by any other new arrival.

Of more immediate relevance is the impact of human communities on the distribution of nutrient elements within their habitats. Biogeochemical cycles are discussed in Chapter 2, but a few preliminary comments should not be out of place. Fig. 1.1 is a grossly oversimplified conceptual model of an eco-system.

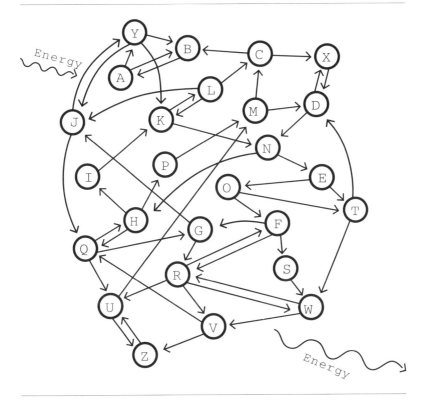

FIG. I.I *A grossly simplified conceptual representation of an ecosystem or, for that matter, virtually any complex dynamic system.*

It contains 26 units whereas an actual mature system contains millions. Each unit could itself be an entire ecosystem such as soil, a lake, an animal community, or the intestinal flora of an animal. The arrows denote the flow of something, a nutrient element for example; but in the end, they denote the flow of energy.

Except for the gross flux of energy through the system, provided in most cases directly and in nearly every other case ultimately by the sun, mature ecosystems are essentially self-sufficient. Nutrient elements pass through every component of the system, and every component may be said to affect every other. Another characteristic—the First 'Law'—is that, for every action taken on such a system, there is at least one unintended—and unexpected—consequence. It is very difficult, if not impossible, to predict accurately and precisely how ecosystems will respond to a perturbation. They cannot be modelled comprehensively because it is effectively impossible to collect enough data. All that can be done in that regard is to lump whole communities or subsystems into 'boxes' such as the sea, plants, animals, the soil etc—and hope for the best.[2] Inasmuch as social and economic systems can be depicted in the same way, similar generalisations also apply to them—which is one reason why I feel that the only thing I have in common with economists is that I don't understand economics.

Complex dynamic systems such as that of Fig. 1.1 contain within them the ingredients of all the ecological mechanisms that have been identified in the field—mechanisms such as competition, cooperation, parasitism, predation, symbiosis and so on. All of these processes contribute to the resilience of real ecosystems and to the checks that are maintained, in the wild, on the populations of individual species. Plagues are not a feature of mature complex ecosystems. In this type of environment early Man, the hunter-gatherer, consumed plants and animals and redistributed their nutrient elements in his excreta, in the food trimmings that he threw away, and in the burial of his dead. There is no reason to suppose that, in this respect, he differed in any significant way from the other animals whose habitat he shared. There is no reason to suppose that, for much of the time, that redistribution was anything other than random or that it amounted to anything less than complete recycling.

But not for all the time. There were some special circumstances when hunting societies did change the pattern of nutrient flow. These exceptions occurred when communities settled for a while in one spot and proceeded to exploit a migrating animal population. The animals might be bogong moths in the Snowy Mountains of South East Australia up to about a century ago[3] or mammoths 15,000 years ago in what is now the Ukraine. In this latter example, a palaeolithic community sustained a largely sedentary existence by the mass slaughter of migrating mammoths, a mammal that has now been extinct for 10,000 years. This occurred towards the end of the last ice age; the cold winters would certainly have permitted the extended storage of meat. Not to miss an opportunity, the people involved in this exercise used the mammoth bones to construct their dwellings.[4]

Migrating animals feed over a wide area. If they are killed and consumed in a restricted one, then the nutrient elements that they assimilated during their growth will, after passage through the human predators, raise the concentration of those elements in the soil of that small area. In other words, the migrating animals act as a catchment for the transfer of nutrient elements from a large area to a small one. It is a process that was also characteristic of fishing communities. Moreover, the phenomenon is not confined to Man. Other species can achieve the same thing if they forage over a wide area and camp or

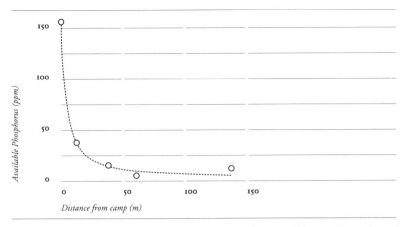

FIG. 1.2 *The content of soluble phosphorus in soil as a function of distance from a 'camp' occupied nightly by a flock of sheep. Plotted from analyses cited by Blair (1983).*

nest regularly in a small one. For example, Fig. 1.2 shows phosphorus in soil at various distances from a 'camp' where a flock of sheep normally spent the night.

A similar, but more dramatic, harvesting and concentration of nutrient elements occurs at the dormitories of bats and the nesting sites of marine birds that maintain the same communal areas over long periods of time. In such cases guano deposits can accumulate to commercially attractive levels.

— The First Farmers —

A great deal has been written about the origins of agriculture[5] and I shall therefore summarise only the barest essentials of the process.

Like any other study of the distant past, this one gives birth to and nurtures differences of opinion; interpretations are constantly changing in the face of new evidence. Nevertheless, there is probably a broad agreement about the essentials, as long as the definition of 'essentials' is sufficiently flexible.

The assignment of dates and, in many cases, recognition of an indigenous origin, will depend on one's definition of agriculture. Should it mean, for example, cultivating one or two native species at a seasonal campsite, should it mean small but intensively cultivated plots in a more or less permanent settlement for which much of the food supply was obtained by hunting and foraging, or should it mean what McNeish[6] calls 'village agriculture'—full-time farming based on a settlement with a central administration, supplementary industries (*e.g.* pottery) and religious ceremonies? McNeish excludes shifting agriculture.

If McNeish's definition is used, then the following 'hearths' of agriculture were established by the times shown (millennia BP): South West Asia, 8-9; China, 6-7; Mesoamerica, 3; South America (Andes), 4. If the definition is relaxed to admit small intensively cultivated plots, shifting agriculture and a loosening of the requirements for a highly structured village community, then the list can be amended to: South West Asia, 10; China, 8; Mesoamerica and South America, 5-6; New Guinea, 6-9. By relaxing the definition even more, the dates can be pushed back further. As they stand, however, the dates are sufficient to show that the change from hunting and gathering to farming did not happen overnight; there was a transition beginning with the small-scale cultivation of native plants and culminating in full scale village agriculture. The transition lasted several thousand years.

Given the different interpretations of archaeological evidence, it should be no great surprise to learn that there are also different opinions about the factors that propelled Man from a hunting and foraging existence to one in which he was largely sedentary and grew his own food. More evidence will doubtless resolve many remaining technical questions of where and how agriculture started. There is no such prospect for the question of 'why'; the answer to that will forever remain a matter of opinion. Embedded in that question, however, is the relation between the human population and its food supply—and that is crucial to my thesis. There are two very distinct schools of thought about it and, of course, various shades of qualification in between.

One of the two 'extreme' positions, if I may call them that, contends that once the essential biology of plant propagation was understood, tools available for its implementation and the climate favourable, people saw the advantages of farming and converted to it. Thereafter the increased food supply stimulated growth of the population. This mechanism can be encapsulated in the following dialogue:

"Hey! Come and look at this emmer!"

"Don't call me Emma! My name's Charlene—and just who is Emma anyway?"

"I didn't call you Emma! I'm talking about this emmer wheat."

"Oh yeah? What's so special about it?"

"I planted it! That's what special about it! I took the seeds of some wild wheat last season and planted them—right here. Don't you see what that means? We can give up all this hunting and gathering stuff, settle down, grow our own food, build a house, have lots of kids and... Where are you going?"

"I'm going a'gathering, Sir, she said—to see if I can find any more nuts."

In essence the other 'extreme' position is that growth of the human population eventually generated such pressure on natural food supplies that hunter-gatherer bands were forced to resort to the cultivation of plants and the domestication of animals to meet their needs. According to this view it is nonsense to suggest that hunter-gatherers of the late Pleistocene had to wait until they learned the essential biology of plant propagation before they could give effect to some innate urge to farm. As Cohen emphasises, whenever foraging communities have been appropriately observed in recent times, they have been found to have a detailed knowledge of the ecology of their environment with names for all the important plants and animals, as well as a clear understanding of the biological significance of seeds.[7] They could have turned to some form of agriculture at pretty well any time they chose. On this interpretation, the 'failure' of indigenous Australians to adopt agriculture should be seen as a disinclination because there was no need to make the change.[8]

The arguments that have been advanced in support of these two viewpoints are extensive and complex; they cannot receive justice in a few paragraphs. I personally find the second argument the more persuasive, while acknowledging that it would indeed be strange if there were not a spectrum of intermediate effects operating to various degrees at different times and in different places.[9] When the distinction between them is linked to human population dynamics, however, the questions lose their significance. 'Which came first, the technology or the population pressure?' is a chicken-and-egg question. The critical thing, as I have already noted, is that the transition to agriculture changed the relation between food supply and population from one of negative to positive feedback. And that, to coin a phrase, is dynamite.

— Feedback Systems —

All living organisms, most communities of organisms and many of our household and industrial gadgets are controlled or regulated in such a way that they keep some essential property(ies) relatively constant. Two of the most common household gadgets, at least among the relatively affluent, are a thermostatically controlled electric oven and a refrigerator. In each of these examples the output, either heat or the removal of heat, once it exceeds the preset level, will switch off the supply of electricity. This is a very simple mechanism. The power supply is either on or off and the temperature oscillates within a degree or so on either side of the desired value.

It is possible, of course, to construct regulators that are much more sensitive than this. Electronic circuits can be highly sensitive, but there are also some simple mechanical devices that respond progressively to a signal rather than simply switching on or off in the manner I have just described. For example, a thermostatically controlled gas oven regulates the *rate of flow* of the gas. The old-fashioned centrifugal governors on steam or diesel engines progressively restrict the rate of flow of steam or fuel as the engine speed approaches the set maximum. These are all simple engineering examples of negative feedback systems, the essence of which is that the output, whatever that might be, 'feeds back' and shuts off or slows down the input, whatever that might be.

Healthy biological systems have a pronounced tendency to be homeostatic. That is to say, they seek to maintain a relatively constant set of characteristics in a particular environment. For example, a cell, be it a free-living microorganism or part of a multicellular plant or animal, keeps its chemical composition, its size and its turgor pressure within fairly definite limits overall, and within very definite limits if the environment is constant. Homeostasis is achieved through a range of control mechanisms, all of which embody the principles of negative feedback. There are many metabolic pathways, such as the synthesis of a number of amino acids for example, in which accumulation of the ultimate product feeds back and slows the first reaction of the sequence. The sequence can be represented roughly like this:

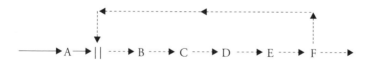

The body temperature of homeothermic (the so-called 'warm-blooded') animals is an example of a property that is easily measured and is regulated by negative feedback through a number of different processes. These include varying the food intake, sweating (and drinking), panting, shivering, varying the amount of exercise, moving into or out of the sun, varying the amount of clothing worn etc.

Fig. 1.1 contains the basic ingredients for negative feedback mechanisms to operate in an ecosystem which, if it is mature and natural, is also largely homeostatic within what might be called its normal operating range. Quite obviously, in any 'natural' environment, there must ultimately be such a relation between the supply of food on the one hand, and the population dynamics and biomass of any species of plant, animal or microorganism on the other. Many—probably most—multicellular animals have physiological and behavioural responses that can be brought into play to limit population growth rates in response to a food shortage. These responses amount to slowing the rate of reproduction by any of a number of physiological changes ranging from some form of dormancy in invertebrates to the very sophisticated suspension of embryonic development by female kangaroos when food is short. Killing or deserting the newborn is not unknown in higher animals, human or otherwise.

Man is unusual in some aspects of reproductive physiology; when all aspects are taken into account, we are unique. In most other mammals ovulation is advertised by a very specific and obvious period of oestrus during which the size and appearance of the external female genitalia change, and pheromones might be produced. If males are present, copulation will take place and the chances of conception from a single insemination are normally better than 75%. Copulation does not commonly happen outside the periods of oestrus; oestrus occurs with frequencies of a few days in small rodents, a month in cows and horses and a year in wolves and deer.

The human situation is quite different. Some individual women claim to know when they are ovulating but the claim might not always be reliable; it is certainly not universal and probably not widespread. More importantly, a man does not know when a woman is ovulating since, the menstrual period aside, she is potentially receptive at any time. Copulation therefore occurs more or less at random, or at least under circumstances decreed by factors other than ovulation. The chance of a single insemination leading to conception is probably no more than 4-8%. A successful breeding strategy therefore demands a lot of copulating, and it is sensitive to circumstances that reduce the frequency of intercourse. A reduced food supply, as in a drought, and the greater distances that need to be covered to obtain food under those conditions have the potential to do just that—for any of several possible reasons.

There are, however, more basic physiological responses to deprivation. Lactation has the potential to reduce fertility, especially in a physically active population when food is in short supply. Lactation requires the diversion of energy to milk production. When that milk is the only source of food for an infant, the mother continues to provide all the energy and nutrients for the growth of a child who is now bigger and, in absolute terms, growing faster than *in utero*. More precisely, lactation requires energy in the range of 2100-4200 kJ.d^{-1} in addition to that needed by a woman in the course of her normal activities.[10] In an active sedentary situation[11] this can amount to some

10000-12500 kJ.d^{-1}, which is about the energy consumption of the contemporary Hadza of Tanzania and rather more than foraging societies in less productive environments.[12] In a modern situation, a demand of this magnitude would be equated with heavy work for an eight hour day.[13] To emphasise the point, women in nomadic bands could need even more in a bad season when a lot of ground must be covered to obtain an adequate supply of food. A nursing nomadic mother could therefore need more than 17000 kJ.d^{-1} to sustain her physical activity and feed her child. If these energy needs were not met by food, she would use her own reserves of fat, in particular subcutaneous fat.

Human fertility is evidently dependent on a woman's subcutaneous fat's exceeding a critical level of about 20-25% of body mass. A moderate fall (10-15%) below that level can produce what Frisch refers to as a 'silent' suppression of ovulation, without a corresponding suppression of menstruation; it can also cause menstrual iregularities.[10] Amenorrhoea, the complete suppression of the menstrual cycle, will follow a slightly larger drop in fat reserves. All of these conditions are encountered in modern women, notably in athletes and those with anorexia. The phenomenon is ever present in those regions of the world exposed to chronic starvation. It would be fairly easy for a food shortage, and the extended foraging that it necessitated, to lower subcutaneous fat to a point at which fertility was either seriously impaired or lost.

The point of this short digression is to identify some of the physiological avenues by which the birth rates of hunter-gatherers could have responded to changes in the availability of food. On the whole, the evidence from recent hunting and gathering communities, arguments of the type advanced above and plausible assumptions suggest that a Pleistocene woman produced, on average, perhaps slightly more than four live births in her lifetime. Infant mortality is thought to have been close to 50%.[7] The whole process qualifies as negative feedback.

None of this should be taken to imply that hunter-gatherers always had a lean time. On the contrary, some of them clearly did very well for much of the time and were able to sustain a sedentary existence for long total periods, as attested, for example, by the wide distribution of shell middens, by the enormous size of some of them and by the construction of durable dwellings such as the stone huts constructed by the indigenous people in South East Australia.[14]

The proportion of the Pleistocene population that inhabited caves for any appreciable length of time is not known and is probably less than is often assumed.[15] Nevertheless, caves were certainly used and have provided a lot of important archaeological information. The 'Venus of Willendorf' is well known. It is a 10 cm statuette found in an Austrian cave and dated to 32000 BP. It depicts a very fat, probably pregnant 'human' female and is commonly regarded as a fertility symbol. That might or might not be so, but a familiarity of the sculptor with obesity is beyond dispute. We can conclude with some confidence that the sculptor's community was largely sedentary and had access to a substantial and reliable food supply, predominantly meat since a vegetarian diet would be unlikely to produce obesity of such severity. In any case, the lady was carved at a time of glaciation when, at that latitude, a reliable source of plant foods would be unlikely.

But what of positive feedback? Here the output, whatever that might be, feeds back to the beginning of the sequence and augments it. It can be depicted diagrammatically as:

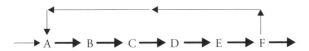

A simple demonstration of positive feedback can be made by placing in front of a loudspeaker a microphone that is appropriately connected through the amplifier. The microphone will pick up any hiss from the speakers or any background noise and feed it back through the system. The amplified sound is again picked up by the microphone, recirculated, reamplified—and so on. If the volume control is turned up and the amplifier is powerful, the resulting howl will not only be painful but might well do some damage.

Systems in a state of positive feedback are commonly known as vicious circles. They contain more than just the seeds of their own destruction; *unless a limit is placed on the flux of energy through it, a system in a state of positive feedback will destroy itself.*[16] That is certain and, even without any other considerations, it is enough to make human agriculture ecologically so very dangerous.

— The Transition —

Estimating populations that can be sustained by a hunting and foraging existence is not easy; estimates that have been made contain both inconsistencies and uncertainties. Some arise from the lack of a clear understanding of the land area that should be used for the calculation of density; this is perhaps especially true of fishing communities. Others derive from questionable observations. Table 1.1 illustrates some of those inconsistencies. Nevertheless, the comparison of hunter-gatherer population densities with contemporary densities supported by agriculture is stark.

Despite the enormous difficulties of estimating prehistoric population sizes and dynamics, especially of nomadic communities, the attempt has been made by a number of anthropologists—to the following effect. Growth rates of human populations in the late Pleistocene[7] have been put within the range 0.0007-0.003%.y^{-1}; this corresponds to doubling times (t_d) of 99000-23000 years.[17] Under the circumstances, a fourfold range is surprisingly good—and the rates are very different from everything that followed. By the time of the transition to agriculture, Man inhabited every continent on Earth, except Antarctica. Human migration from (presumably) Africa to the rest of the globe may well have been driven by many factors, including curiosity, but underlying all of it was almost certainly the pressure generated by a population that was slowly but inexorably growing. It was this same pressure that, according to the second theory already outlined, provided the impetus to change from foraging to cultivation, from hunting to pastoralism.

The global population in the late Pleistocene is thought to have been around 15 million which, averaged over the total global land surface amounts to a density of about 0.1.km^{-2}. This is the level considered by some to be the maximum that is sustainable for hunter-gatherers (Table 1.1), and it is an average that includes some very inhospitable land. It constitutes an important part of the argument for the population pressure model for the transition to agriculture.[7] It is also as well to remember that the transition began in the final stages of a glaciation. Not only was the change in climate fundamental to everything that followed, but the change in the area of usable land must also have had an impact. On the one hand, the melting of ice sheets exposed land that could be used for cultivation and grazing. On the other, rising sea levels inundated coastal areas—some of which would have been inhabited—and thereby contributed to the pressures of population.

The greatest manageable population density also set the size of nomadic bands because it specified the distance that must be covered to obtain enough food. For example, a band of 50 individuals living in the (supposedly) most favourable environment (Table 1.1, 2nd row), would require an area of 125 km^2 to sustain it. In other words, it would need to forage over a radius of about 6 km from a base camp—if it had such a thing. Halving the food supply (or doubling the population) would double the foraging area. This is the type of argument used earlier in relation to human fertility and it embodies an important, albeit obvious principle. That principle is that the more widely a resource is dispersed, the more energy is needed to harvest it. If the resource is food, then clearly circumstances can arise when an animal, human or otherwise, must use more energy in acquiring that food than it receives from it. If that happens, the animal starves.[18] Before that critical stage, however, the splitting of foraging groups into smaller bands followed by their spatial separation must have been highly likely, if not inevitable. And, assuming as we are that the population was growing despite these exigencies, that is the stuff of migration—and ultimately of population pressure.

TABLE 1.1 *Population Densities under Various Conditions*

	Density (individuals.km^{-2})	Ref.
Hunter-gatherers		
General	≤ 4	(a)
Maximum possible	c 0.4	(b)
Tropical rainforest	0.005 -0.12	(c)
Aboriginal Australians	0.01-c 0.8	(d)
Agriculture (20th century)		
New Guinea Highlands (swiddening)	c 30	(e)
India (cropland)	500	(f)
Indonesia (cropland)	870	(f)
China (dyke-pond system)	1700	(g)
Global		
(cropland)	360	(f)
(total land area)	40	(f)

(a) Boserup (1965): (b) Cohen (1977): (c) Bellwood (1985): (d) Harrison *et al.* (1977); Rose (1987): (e) Clarke (1971): (f) World Resources Institute (1992): (g) Ruddle and Zhong (1988). See also Table 6.1.

As we have already noted, the change to agriculture did not happen overnight. On a time scale appropriate to modern society it was a leisurely business taking, for example, about 5000 years to spread from its appropriate 'hearth' in the Eastern Mediterranean—South West Asian region to Scandinavia. Nor should we assume that it was an all-or-nothing affair since there were mixed 'economies' involving the cultivation of crops and the hunting of wild animals.[19] Indeed, such hybrids have persisted in various forms up to the present as has trading in foodstuffs between cultivators and hunters. From our immediate point of view, however, the important changes lay in the underlying dynamics of food production and population growth.

Agriculture simplifies an ecosystem. In effect, it pushes the system back to an early stage of 'succession' (see Chapter 2) in which there are relatively few species and nett growth rates are relatively high. The total biomass, and hence the total energy, is less than in a mature ecosystem, but a higher proportion of that biomass is available as human food—and as food for competing species. A human community can obtain more energy from a specified area of culti- vated land than from the same area in its pristine state—but there is less vari- ety, which is an important reason why nutrition commonly deteriorated with the transition to agriculture.

The change in population dynamics that accompanied farming ensured that the extra food that was now produced by a particular area of land was, in a very important sense, used differently by the people who ate it. Although these new farmers probably spent more time labouring to produce their food than did their hunting and gathering antecedents,[12, 20] they did not have to cover great distances, carrying small children, in order to do so. Not only were children no longer a burden in this very literal sense, but they did not have to be very old before they could be made to work the farm. Throughout the entire farming period up to the advent of modern mechanisation, a peas- ant farmer has needed a wife and children for the most businesslike and least romantic of reasons—he could not have run his farm without them.[21] Be- cause of poor sanitation and disease, child mortality in sedentary societies was evidently even higher than in hunting and foraging communities;[12, 22] it was important to space births as closely as possible. There was now an opportu- nity to wean children earlier than in nomadic bands. Sedentism and the rela- tively constant availability of grains facilitated the preparation of a soft food easily fed to infants; early weaning would shorten the period of lactation with its attendant potential for contraception.[23]

The upshot of all this was an increasing birth rate. Although more food en- ergy was available from a given area of land, it was used not to produce bigger fatter people but rather a greater number of people who were smaller than their hunting and foraging antecedents.[24] The population growth rate of the Neolithic period is put at about 0.1%.y^{-1} (t_d = 700 y). Although this could well be too high for reasons outlined by Cohen,[12] it is unlikely that the error would be greater than, say, a factor of 4, and that would still leave the rate 8 times as great as the fastest growth rate assigned to the late Pleistocene (see above).

A population growing at an average rate somewhere in the range of 8-140 times faster than its hunter-gatherer forebears was eventually going to run up against natural barriers to which it must respond in some way. The responses included migration, thereby bringing new land into cultivation or pasture, and improving technology to increase the yields of crops. Like the transition to agriculture, this took time—a long time by present standards, but a drop in the ocean of evolutionary time. The positive feedback was espressing itself. Overall there was now more food available and the population increased ac- cordingly. And as that population increased it produced more food.

Needless to say, there is more than one explanation of the geographical spread of agriculture. The two main contenders are (i), the diffusion of the idea through contact between communities and (ii), the migration of farmers. The two processes are not mutually exclusive and it is scarcely conceivable that either did not occur. Both archaeological and genetic evidence imply that Neolithic cultures spread throughout central Europe and adjacent regions at an average rate of about 25 km per generation (1 km.y^{-1}) taking agriculture with them, in many cases displacing indigenous hunter-gatherer communities. The driving force behind this migration has been attributed to high rates of population growth, especially at the frontiers.[25]

The neolithic migration did not proceed at a uniform rate however, and partly for that reason, it is unlikely that population pressure was the only driving force behind it. Bogucki[25] has commented that "Linear Pottery communities appear to have spread across central Europe in several spurts" and suggests that factors other than population pressure were implicated in the timing of these spurts. These other factors could have been cultural or ecological—bad seasons or outbreaks of disease for example. Occasional epidemics, or even the normal dynamics of endemic diseases, could produce pulses in the growth rate of a community and provide an incentive to escape either the disease or the population pressure. It has been suggested that populations grew fairly rapidly immediately after the establishment of a settlement only to slow, or even be offset, later by an increase in mortality, especially among infants.[22]

I should emphasise here an earlier brief comment. Throughout their existence, human communities have been confronted intermittently by bad seasons—droughts, floods, hard winters—and consequent food shortages, sometimes of great severity. We can confidently assume that the response of Palaeolithic hunter-gatherers to these exigencies included extending the range of their territory when the shortage was moderate and moving to other territories when the shortage was severe. It is unlikely that prehistoric nomadic bands were confronted with outright famine in the sense that we have come to understand that word. They were mobile and could escape the worst of its clutches. Moreover, unless they lived in an extreme environment, such as the high arctic or a desert, they had access to a range of types of food, foods that would not necessarily be affected equally by a climatic disaster.

The situation was quite different for agrarian communities. They cultivated a relatively small number of types of food and they were far less mobile than nomadic foragers—or nomadic pastoralists. The 1 km.y^{-1} rate of migration referred to above was not a response to the acute pressure of famine; it was a response to the chronic pressure of a growing population and perhaps mild food shortages. It bore almost no relation to the mobility of hunter-gatherers.

The first farmers might not have seriously disturbed the natural cycles of nutrient elements. Indeed, effects of population size aside, a farming community in its early stages probably redirected nutrient flow to no greater extent than did sedentary hunting bands. The farmers obtained their food from a relatively small area, they constantly moved within that area and, in the course of their daily activities, they must have returned their food wastes and their excrement to some part of the farm.

The dynamics of nutrient cycling changed when settlements grew from farming villages to towns and cities. The inhabitants of cities had a range of skills and crafts other than farming, their societies had distinct social structures, they eventually had some sort of currency to facilitate trade—and they had serious problems of sanitation. But cities, as distinct from villages could not develop until the surrounding farms produced a surplus, that is to say, more food than was needed by the farmers. Reaching that stage needed time, and increased efficiency as measured by yield per area of land and, more importantly, by yield per man-hour of labour. By each of these criteria, increased efficiency required improved practices. By each criterion, but especially the second, increased efficiency eventually required more energy than could be provided by man power. That energy was eventually obtained from draught animals. Sometimes improved efficiency also required irrigation.

The production of a surplus not only enabled some people to go and live somewhere else and do something else, it undoubtedly encouraged—and ultimately obliged them to do so. The production of a surplus with its resultant separation of farmers and city dwellers must have consolidated and then accelerated the concept of property as something that could be precisely delineated—and inherited. This, in its turn, gave rise to the uneven distribution of wealth, the development of political and social hierarchies, and the concept of power. Civilisation set in train processes that led Man to concentrate his attention on his culture, his religion, his politics, his economics—and to forget his biology. It set in train events with their own internal momentum, their own positive feedback loops, events that began to change profoundly the ecology of the planet.

CHAPTER 2

An Ecological Detour—and Some Ground Rules

There are two ecological concepts that, perhaps more than any other, need to be recognised if the ecological significance of agriculture is to be appreciated. I refer to 'succession' and to the cycling of nutrient elements—to 'biogeochemical' cycling. They are interrelated but, in the interests of comprehension, I have treated them separately to a large extent. A separate subsection is given to the role of dung in the recycling process generally and as a farmyard manure. I have done this because dung, be it cowdung, human poo, silkworm 'droppings' or earthworm 'castings' (there is no shortage of terms) is crucial to nutrient recycling both in the wild and on the farm, and because I have gained the impression that there are some misconceptions about its role, especially on the farm.

— Succession —

From time to time a new potential habitat is formed for any of a number of possible reasons. It might be that an island is thrown up in an earthquake, a dry depression in the ground is filled with water or a lake dries out, a major bushfire destroys the biota in a particular area, someone goes off for a holiday leaving an uncovered jug of milk on the kitchen bench—or some ground is freshly ploughed.

The habitats are ripe for exploitation and very little time is lost before the invasion begins. There follows a succession of different organisms, a succession guided by a few simple but compelling principles. The organisms that are first to establish themselves will be selected by the history of the new habitat (in essence whether there are any inocula, that is to say seeds, spores, eggs or viable tissue remaining from a previous régime), by the distance between the new habitat and established sources of inocula, the means of transport of those inocula (wind, birds, feet etc) and the type of selection exerted by the physical and chemical characteristics of the new habitat.

Our concern is with the colonisation of land where the course of succession is rarely as stereotypical as outlined in ecology text books. The norm in this context is usually a succession starting with grasses and other small ground plants, through shrubs and ultimately to trees and mature forests—if the physical environment will allow the growth of a forest.[1] If there is much organic matter from a previous régime, the pioneer colonisers may well be heterotrophs—microorganisms, fungi and detritus feeders of various kinds. If left solely to their own devices, however, theirs is essentially the succession of a closed system like that of the bacteria and fungi that colonise that jug of milk. An ecosystem in this condition would run down towards death or dormancy unless it were 'opened' by plants to the flux of solar energy that is necessary to sustain it.

Once plants are established, the gross morphological features of the subsequent succession are easy to see, although in most cases a human life span is not long enough to enable one observer to follow the process all the way to maturity. In the early stages the plants are small and there are not many of them; as the system matures, the size and number of individual plants increase. This is the pattern regardless of whether or not the system is dominated by grasses in its early stages and trees in its later ones. It occurs, if for no other reason, because mature plants are apt to be bigger than seedlings. At first there is plenty of bare ground and the pioneer plants are not likely to suffer from shading by other plants or from serious competition for soil nutrients. Apart from any selection imposed by the soil itself and its own biota, it is a relatively easy time for new arrivals, at least in terms of competition. That changes, however. The number of individual plants increases and there is competition for nutrients as roots spread laterally and downwards, and bigger plants begin to shade the smaller ones. All of this is accompanied by a fall of litter and the spread of roots which, between them, change the chemistry and structure of the soil and provide organic matter to support a growing population of fauna and microflora. As well as that, of course, the establishment of a plant community progressively supports a population of herbivores which, albeit without much enthusiasm, do their bit to help the carnivores along.

The feeding patterns of animals, be they foragers or hunters, are neither uniform across species nor uniformly destructive. They are selective and the consumption, for example, of one species of plant by a herbivore assists other plant species directly by keeping a competitor in check and, indirectly, by distributing nutrient from the first plant, now in the form of excrement, throughout the system.

The upshot of all these processes is a progressive increase in the density of the biomass (the biomass per unit area of ground) and in the number of spe-

cies. There comes a time, some centuries later if the progression leads to a forest, when the system reaches its 'climax'—a term used here to denote a phase
of relative overall stability. In this phase there are many species and, although
there may well be dominant types or even dominant genera, no one species
will obviously dominate unless the environment is very selective in some
physical sense. The more complex the system, the greater is the number of
species and hence the smaller the proportion of the total biomass that is represented by any one of them. Inasmuch as the food available to an animal, including Man, is restricted to a limited number of species, it follows that the
greater the complexity of the environment, the smaller is the proportion of
the total biomass available to that animal. Thus tropical rainforests, the dens-
est and most diverse of all ecosystems on this planet, can sustain only a small
human population (Table 1.1). It is not surprising that preagricultural Man
did better in open temperate forests and grasslands than in the jungle. There
are no population explosions of rabbits, mice, grasshoppers—or of Man—in
an undisturbed jungle.

The mature system has a pronounced capacity for self regulation (see
Chapter 1), a consequence of which, indeed it is almost a synonym, is homeostasis. This is essentially the maintenance of a relatively constant species composition, biomass, energy flux, turnover rate of nutrients—and so on. Because of its relative stability, an ecosystem at climax is often described by ecologists as being in a state of 'equilibrium', or 'dynamic equilibrium'. Nothing
could be further from the truth. If the term 'equilibrium' is to mean anything
at all in a scientific context, then the meaning must have its roots in thermodynamics. A system in a state of thermodynamic equilibrium is at the lowest
energy level attainable under the prevailing conditions. An ecosystem at equilibrium would contain only dead and/or dormant organisms. The *relative*
constancy of a mature forest approximates most closely to a 'steady state', inasmuch as it has a high energy content and its dynamics are sustained by a
flux of (solar) energy.[2] There is another important difference between a reaction system at equilibrium and one in a steady state; the former is freely reversible, the latter is not. The irreversibility of complex biological processes is
often alluded to—in relation to the notorious second law of thermodynamics[3]—by such metaphors as one's inability to 'run the tape backwards' in
which fallen trees will stand up, emit visible light and carbon dioxide while
assimilating oxygen—etc etc.

A more homely analogy can be found in what, at one stage, some of my
students were inclined to call the 'reverse mouse experiment'. I trust I need
not argue very strongly to convince a reader that it is not possible to take
mouse faeces, shove it up a mouse's rectum, apply heat to the mouse's body

and force carbon dioxide into its lungs, and expect the mouse to exhale oxygen and in due course to regurgitate pristine cheese—or whatever it ate that produced the faeces.[4] As I have already intimated, an animal's digestive tract is a very complex ecosystem that, when the fluctuations between meals are ironed out, is essentially in a steady state.

It can also be misleading, of course, to place too much emphasis on the constancy. The quantitative species composition and the overall dynamics of a mature ecosystem necessarily respond to seasonal and climatic fluctuations, to long term changes in climate and in topography.[5] Nevertheless, even with these variations, a mature system is more stable and a great deal more resilient than the earlier stages of succession.

The growth of a living organism or of a biological community is not simply a matter of producing more living tissue from the available food supply. It is, in fact, an expression of the difference between 'anabolic', or biosynthetic reactions on the one hand, and 'catabolic' or degradative reactions on the other. The anabolic reactions embody all those processes that produce the molecules from which living tissue is constructed—fats, proteins, polysaccharides, nucleic acids etc. The catabolic reactions consist predominantly, especially in heterotrophs, of the breakdown of foodstuffs and reserve 'fuels' to provide both energy and the raw materials for the anabolic processes but, in addition, they also include the degradation of existing tissues. In other words, vital tissues are continually broken down and renewed—they continually 'turn over'. At first glance this might seem like a singularly fruitless exercise but, among other things, it has potential repair mechanisms built into it.[6]

Concurrence of these two sets of processes is, if possible, of even greater significance in a community of organisms, for here there is a collective turnover determined not only by the sum of individual turnovers, but also by the way in which members of the community interact with one another. Crowding among plants will slow their biosynthetic reactions, eating a plant or an animal will stimulate degradation of the consumed and the biosynthetic reactions of the consumer. All of this leads to a communal turnover of nutrients within the ecosystem, a topic that is considered specifically below. It also produces a collective energy metabolism that can be expressed quantitatively, occasionally with a high degree of precision, but more often as an approximation.

The energy metabolism of a heterogeneous community is reflected in the exchange of carbon dioxide and oxygen with the atmosphere. The fauna consume oxygen and produce carbon dioxide as do some of the soil fauna and microflora. This is the nett expression of respiration. The rest of the soil biota, the anaerobes, release carbon dioxide without consuming oxygen.

Some also release methane. During the day plants photosynthesise, the nett expression of which is the production of oxygen and uptake of carbon dioxide. At night the plants also respire and the balance of their gas exchange is reversed.

The quantitative relation between these two processes, photosynthesis and respiration, is not only of critical long-term importance for the composition of the atmosphere, but it is also a useful indicator of just what is going on. The nett rates at which the two gases, oxygen and carbon dioxide, are exchanged can be converted to rates of gross photosynthesis (P) and gross respiration (R).

In the early stages of colonising soil that is not already well populated with heterotrophs, P is greater than R because plants are the pioneers—and they are growing. As the system matures, however, and the total biomass increases, the proportional rate of growth of the biomass slows, gross photosynthesis declines somewhat and respiration tends to catch up so that eventually in the theoretically ideal situation, P = R, the quotient P/R = 1 or, in other words, there is no nett increase in oxygen evolution or carbon dioxide uptake.[7] The P/R quotient remains constantly at 1 only in the human imagination, however, although in mature tropical rainforests it is probably never far from unity. In other environments the quotient will change with circumstances. It will oscillate seasonally above and below unity in temperate forests, it will oscillate for intrinsic reasons in all ecosystems,[8] and it will fall below unity in an ecosystem that, for any reason, is in decline.

But what of agriculture? The biological aim of agriculture is to push an ecosystem back to an early stage of succession in which the P/R quotient is high, the *proportional* growth rate is high and a major part of the total, albeit small biomass is available as human food. There are, of course, some unintended consequences.

Because the system is at an early stage of succession, if it is left to its own devices it will progress towards the later stages—to greater diversity. This progression is just what the farmer is trying to prevent; the aspiring colonisers are usually classified as weeds or pests. Moreover, the food that the farmer grows for himself and his family is by no means attractive exclusively to Man. There are plenty of other animals, both vertebrate and invertebrate, who know a good thing when they see it and, given half a chance, will beat the farmer to the fruits of his labours without so much as a twinge of conscience. Their task is made a great deal easier by the simplicity of the system; there is more of the target food, it is easily accessible and there are fewer potential competitors or predators to get in the way. All of this is true not only of herbivorous animals and the farmer's crops, of predatory animals and the farmer's livestock, but it also applies to infectious diseases, whether viral, bacterial, fungal or parasitic.

These consequences are inevitable. Their effects can be mitigated—but only by constant management and vigilance. From the time Man elected to grow his own food he became a prisoner of the system that he created; he had to devote his life to keeping the ecosystem that is his farm in that early stage of succession. The quantitative relation between Man's biomass and that of his crops, especially the grasses, is essentially symbiotic; neither could exist at anything like its present size without the other. There was another way by which Man could have achieved his primary aim of increasing his food supply without the total commitment demanded by agriculture, and without the complete change to a positive feedback relation between population and food supply that was engendered by farming. This other way was practised widely before it was displaced by agriculture, but nowhere was it used to such effect as in Australia. I refer to the deliberate use of fire.

Fire is one of the natural phenomena that push an ecosystem back to an earlier stage of succession. Just how far back depends on several factors including, of course, the severity of the fire. It is rare for a forest fire to kill all plants *and* their seeds. There have been some very serious bushfires in Australia, for example, that have killed all the vegetative plants in a badly affected area but have stimulated the germination of acacia seeds. As a result, the first colonisers of the burnt area were not grasses or 'weeds', but wattles.

When fire is used as part of a modern management strategy, the aim is primarily to reduce the amount of fuel and thereby lessen the severity of a wild fire. There is good evidence that hunter-gatherer communities in many parts of the world had their own management strategies for burning vegetation, strategies that had other aims but which, by and large, also affected the patterns of wild fires. Deliberate burning was evidently practised in some unlikely places such as the British Isles, as well as less surprising regions such as Central Europe, America and, of course, Australia.[9] There is also evidence that the ecosystems of South-West Asia had been extensively modified by deliberate burning, perhaps for a couple of millennia, before those same areas were eventually cultivated.[10]

Because of climate and vegetation, the practice was probably more effective and seems to have continued without interruption for a longer period in Australia than anywhere else.[11] Burning woodland produced a number of results that must eventually have been transformed into conscious objectives. There was a general opening up of the forests, thereby improving human access into and mobility within the forest. The incidence of some desirable types of plant such as grasses (desirable for their seeds and as forage), nut bushes and small edible plants was increased. Improved forage was likely to promote an increase in numbers and total biomass of herbivores; the hunting of those animals was made easier by the reduction in cover.[11, 12] By reducing understorey, fires might also have made forests more accessible to larger game animals.

All of these changes have an underlying objective in common with agriculture, namely to simplify an ecosystem, push it back to an earlier stage of succession, and thereby increase the proportion and accessibility of biomass that is available as human food. It is not for nothing that the first Australians have been described as 'firestick farmers.'[13]

— Biogeochemical Cycles —

Substances in the surface layers of a sterile planet will move from one location to another in response to seismic activity. If the planet has an atmosphere there should also be movement caused by wind erosion. If there is any liquid it too might cause erosion; if there is enough of the liquid, there will be some accumulation of the eroded material in pools. The presence of living organisms, however, changes the situation profoundly and the movement of surface materials becomes a very complex biological phenomenon. The biota in their entirety introduce such a wide range of biochemical transformations into the whole process that it now becomes appropriate to speak of the movement of *nutrient elements* rather then of substances. Moreover, as the previous section has foreshadowed, those elements cycle rather than merely move, and the cycling is dependent on a flux of solar energy.

Ecology texts usually discuss biogeochemical cycles with specific reference to individual elements (the carbon/oxygen/hydrogen cycle, the nitrogen cycle, the phosphorus cycle etc); this is quite artificial but it can scarcely be avoided if the topic is to be considered in any detail. It is also a common practice to illustrate the cycles with some kind of a flow chart in which the various components are expected to share boxes with labels such as 'Plants', 'Animals', 'Soil Biota', 'Inorganic Deposits', 'Oceans' etc. An attempt might be made to quantify the cycles, in which case there will be numbers in the boxes to indicate their sizes. There might also be numbers on the arrows to indicate their rates. With suitable labelling, Fig. 1.1 can be interpreted in these terms if one feels so inclined.

Every element that is assimilated by a plant, animal or microorganism is recycled in some way. This is true of elements that are the principal constituents of biological molecules (see below), of common minerals (*e.g.* calcium, magnesium, sodium, potassium, iron etc), of trace elements (*e.g.* copper, molybdenum, cobalt, selenium, iodine etc), and of adventitious elements. It is true of elements that are transmitted rapidly and of those that, in the short term, might seem to be removed from a cycle and shunted into relatively inaccessible mineral deposits, but which nevertheless continue to recycle over a greatly expanded time scale. All nutrient elements share some common pathways of transmission. These include the direct ingestion of one organism by another, the deposition of excrement and the subsequent assimilation of its elements by other organisms, the decomposition of plant litter and animal cadavers by (predominantly) soil biota and eventual assimilation of the elements by higher plants. The principal agents of litter decomposition are different for each type of habitat, ranging from soft-bodied invertebrates—worms, slugs, snails etc where there is plenty of moisture, to ants and termites which, in an arid environment such as that of inland Australia, carry the main burden of responsibility.[14]

In addition, there are specific reactions and cycling mechanisms peculiar to some of those elements that constitute the principal components of biological molecules. These elements, in descending order of abundance in the earth's crust, are oxygen, carbon, nitrogen, hydrogen, phosphorus and sulphur.[15] A central part of the cycling of carbon and oxygen is atmospheric.[16] The cycling of hydrogen is intimately tied to that of carbon and oxygen and, because some water is always in the form of vapour, hydrogen also cycles partly through the atmosphere. Nitrogen need not cycle atmospherically since nitrogenous compounds can be directly ingested or transferred through solution—but nevertheless an important component of the nitrogen cycle is atmospheric. This is brought about by two broad groups of bacteria and by cyanobacteria. The denitrifying bacteria can use nitrate (NO_3^-) instead of oxygen in their respiration and, in so doing, produce elementary (gaseous) nitrogen that escapes to the atmosphere. If their activities were not balanced by nitrogen-fixing bacteria and cyanobacteria, it is doubtful if there would now be any life on earth. The nitrogen fixers reduce gaseous nitrogen in essence to ammonia which is incorporated into amino acids and thence into proteins and other nitrogenous biological molecules.[17]

Neither must sulphur cycle through the atmosphere but, nevertheless, some usually does.[18] Phosphorus, however, is different. It has few naturally occurring compounds that are volatile within the range of temperatures of the earth's surface and it therefore does not cycle to any significant extent atmospherically.[19] It is this lack of significant volatile metabolites that sets the cycling of phosphorus apart from the other five nutrient elements just discussed and has enabled agriculture to convert phosphorus essentially into a non-renewable resource.[20]

A critical property of any reaction sequence, of any cycling mechanism, is the *rate* at which it proceeds. Cycling rates in complex systems are difficult to measure but there are some useful indicators that can at least enable comparisons to be made between different types of environment. One such indicator is the rate at which litter falls to the forest floor. This is useful because the decomposition of litter is an important, sometimes the major contributor to the 'non-plant' respiration of an ecosystem. It has been reported, for example, that wet tropical rainforests produce a mean annual rate of litter deposition of about 1100 $g.m^{-2}$ compared with about 100 $g.m^{-2}$ for arctic forests. Between these extremes the rates of deposition lie almost on a straight line when plotted against latitude.[21] It should not be too far from the mark, therefore, to conclude that the turnover rate of nutrients in a tropical rain forests is about an order of magnitude greater than that of high latitude—or high altitude—forests. A comparison of this type might fail, however, when applied to an Australian dry sclerophyll forest where the litter can accumulate until a fire releases its nutrients.

Cycling rates have another significance of more immediate relevance to agriculture and the theme of this book. I refer to the state and availability of phosphorus in the soil. Soil chemists classify soil phosphorus into groups according to the severity of the method used to extract it for analysis. Broadly speaking, soil phosphorus can be 'bound' in a number of ways (insoluble inorganic compounds, organic compounds, tightly adsorbed onto charged soil particles), or it can be freely soluble in water or dilute sodium bicarbonate solution.[22] The soluble forms are readily available to plants but phosphorus that is bound so tightly that it can be extracted only by conditions as severe as digestion with hot perchloric acid or fusion with sodium carbonate is commonly classified as 'unavailable'.

I would argue that essentially *all* the phosphorus is available, but at very different rates for the soluble and bound forms. Bound phosphorus is not available to the fast growers, not because it is locked behind an impenetrable barrier, but because the plants cannot get it fast enough to sustain the level of metabolic activity necessary to keep them growing or alive. If plants are intrinsically able to sustain a very slow growth rate, however, and if the 'bound' phosphorus is all that is available, then they will use it by virtue of its slow mobilisation through microbial activity and abiotic chemical changes in the soil. Indeed many plants, including a high proportion of Australian native species, grow very slowly and are inhibited by soluble phosphorus at the concentrations required by crops. The apparent recovery of soil fertility induced by the traditional practice of fallowing, and in land that has been vacated in the course of a slash-and-burn cycle (see Chapter 3), can probably be attributed partly to the gradual mobilisation of bound phosphorus. There are other reasons too, of course, among which biological nitrogen fixation is one of the most important. Tables A1-A3 give numbers that can be attached to some of the general statements made above.

— A Brief Discourse on Dung —

The impression to which I referred at the beginning of this chapter is that there is a widespread belief that animal manure makes a nett overall addition to soil fertility rather than merely replacing something that had been removed. The distinction between those two interpretations depends partly on what is meant by 'fertility'. If it means the total capacity of soil to produce a certain crop yield or sustain a certain biomass, then the distinction is warranted and the belief in enhanced fertility justified because of the contribution that animal dung can make to soil structure, water retention, earthworm populations etc. If, however, fertility is intended to mean the content of nutrient elements in the soil, then the statement needs qualification.

The breakdown of seeds, plant or animal tissue into an organic compost is achieved faster by an initial passage through the digestive tract of an animal than by any other natural process. What a mammal can achieve in about a day, for example, takes a well managed domestic composting system several weeks, although the products are different in several respects. A digestive tract is a complete microbial ecosystem, differing in principle from a 'conventional' (for want of a better adjective) terrestrial ecosystem only in the state of the energy flux on which it depends. A 'conventional' system is driven by a flux of radiant energy; a digestive system is driven by a flux of chemical energy. In both cases the output is 'degraded'. The unused radiant energy from the terrestrial ecosystem is emitted mainly as heat, excess chemical energy from the digestive system is discharged as heat and as degraded chemical compounds. In both cases the system approximates a steady state when the oscillations are ironed out. In a digestive tract the oscillations are probably better described as pulses, especially for those carnivores that eat large meals separated by long intervals of time. Grazers approach much more closely to a true steady state—or what is known in a laboratory or industrial situation as a 'continuous culture'.

If the animal is vertebrate—and farm animals usually are—the raw material is mechanically fragmented by teeth or, a bit further down the line, in a gizzard; it then passes along a disassembly line superbly designed for the purpose. The foodstuffs encounter a series of digestive enzymes operating under conditions of water content, pH and temperature that are close to optimal for each type of enzyme. If the animal is a mammal or a bird—and again farm animals usually are—the temperature of around 37°C is not only well suited to the action of the home-made digestive enzymes, but it is about ideal for the growth and metabolism of the massive population of intestinal flora—predominantly bacteria. If the animal has a rumen there is a large protozoal population as well.

The result of the combined activity of the animal's digestive enzymes and those of the microbial population is that, within about a day for large animals and much less for small ones, the residue from the two processes of digestion and assimilation is discharged as faeces. What does it contain? In a nutshell (figuratively speaking, of course), it contains what the animal ate, less what it withdrew for its own use. I shall ignore for the moment the organic content and confine attention to the nutrient elements, regardless of their chemical state. During the course of its growth, a young animal eats to meet its energy needs, obtained from the hydrogen and carbon of its food, and to obtain the nutrient elements from which it synthesises new tissue and bone, and from which it also replaces tissue and bone that are broken down as a part of the turnover referred to earlier. The faeces contain the excess, that is those elements that have not been assimilated. The urine contains those that have been released in tissue turnover and any that were transferred in excess from the intestines into the bloodstream but not otherwise used. Carbon used in energy metabolism is mostly exhaled in the breath, the corresponding hydrogen finds its way into 'metabolic' water which, in the case of many insects, can be sufficient to meet their entire water requirement.

Some modern steers will reach a mass of 700 kg in about two years from birth or, in other words, a bit less than three years from conception. The phosphorus content of medium to large mammals is about 1%; in reaching that mass, the steer will take about 7 kg of phosphorus from the soil (this is equivalent to about 78 kg of the fertiliser, superphosphate). Its growth rate is by no means uniform over that period but, if for the sake of simplicity we assume that it is, then the animal will withdraw phosphorus from the soil at an average rate of about 2.5 kg.y^{-1}. These values include the phosphorus provided by the mother's blood supply during gestation and by the milk afterwards. A modern cow might produce about 23 l of milk a day (during the height of the lactation period—not an annual average), equivalent to 18-20 g phosphorus, but a modern cow does not provide all of that to her calf. Most of it, and the nutrient elements it contains, is removed from the farm. The steer, however, is still not fully grown at 2 years but thereafter the growth rate will slow and, depending on what we can euphemistically call its lifestyle, subsequent growth might or might not consist largely in depositing fat (see Chapter 13). The important point, however, is that if the animal goes elsewhere, no matter how much dung it has deposited in the course of its life, it has removed about 7 kg phosphorus (and proportionate amounts of nitrogen, calcium, potassium, iron, sulphur etc etc) from the soil. If it dies on the farm, essentially all of its nutrient elements (apart from those taken by scavengers) will be deposited onto and concentrated within an area of soil about equal to that of the cross-sectional area of the animal. The phosphorus in the bone, however, represents about 80% of the total and will be released very slowly —too slowly under normal circumstances to be of use to the farmer. If the animal is eaten by the farmer, the subsequent distribution of nutrients will depend on what the farmer does with the uneaten parts of the carcase, his own excrement and ultimately, what happens to the farmer's own body.

If the steer remains on the farm after it is fully grown it no longer directly removes anything—except energy which is provided by sunlight. Of course it still has a battery of complex nutritional requirements but these, as well as providing energy, are basically to enable its energy metabolism to function and to replace losses. But neither does it add anything. It simply redistributes nutrients that it ingested but either did not assimilate or else assimilated and later excreted (some qualification is needed here—see below). Just how these nutrients are redistributed depends on circumstances.

If cattle or sheep are free to roam within an enclosed area, dung and urine will be deposited discretely (but rarely discreetly) over the entire grazing area during the day but might be concentrated in a small area at night, as represented in Fig. 1.2. If the dung is removed from the grazing land and applied to arable land, then the nutrient content of the latter is indeed enhanced—but at the expense of the former. If cattle are confined in stalls the nutrient flow to and from the stall is entirely in the hands of the farmer since he must bring food to the animal and remove the excrement. Mature cattle deposit about 8 tonnes of wet dung per year, containing about 12 kg phosphorus (and 45 kg nitrogen).[23] In other words, the mature animal distributes annually a bit more phosphorus than its body contains. During the first two years of its life, therefore, our steer would have redistributed perhaps about 1½ times its 'body equivalent' of phosphorus; if it remains on the farm the difference between body phosphorus and redistributed phosphorus will increase by roughly 12 kg $P.y^{-1}$.

Comparisons are often made of the relative value as manure, for the pasture on which they are grazing, of the excrement from different species of animal—sheep and goats for example; usually the comparison is based on the content of a specific nutrient element. If an animal is fully grown and maintaining a constant mass, then its excrement, faeces plus urine, will contain the total amounts of nutrient elements of the food that it eats, no more, no less—regardless of the species of the animal. If the animal is an efficient digester, such as a ruminant, then a higher proportion of the ingested elements will be excreted in the urine and a lower proportion in the faeces than will be the case for an inefficient herbivore, but the total (faeces + urine) in both cases will equal the amount ingested. Of course, the distribution changes if the animal is not at constant weight. If it is growing, pregnant or lactating, the proportions of nutrient elements in the urine will be reduced relative to those in the faeces. If it is losing weight, the opposite will be true and the proportion of the elements in the urine will rise relative to those in the faeces.

This argument might be taken to lead to the startling conclusion that it would not be possible to 'overstock' grazing land with fully grown, non-pregnant, non-lactating animals which, because they maintain constant weight, do not cause any nett removal of nutrients from the soil. Apart from the practical improbability of assembling such a herd or flock in the first place, and apart from the effects of trampling and soil compression, the conclusion would be wrong—desperately wrong—because it takes no account of the rates of the many processes from the growth of grass to the decomposition of the dung and the reassimilation by grass of the nutrients released from the dung. All of the steps in the cycle are affected in one way or another by the season, by water availability, by temperature and, of course, by the number of grazing animals. To summarise, if the overall rate of consumption of grass *per area of land* does not exceed the corresponding rate of growth of the grass then, by that criterion, the system is 'sustainable'. If it does exceed it, the land is overstocked. Of course, different answers can be obtained to the basic question according to the time scale used. A period of ten years that included at least one drought should give a different answer from a question limited to six months that embrace a wet summer.

These are the bare bones of the situation; a little flesh should now be added. The nutrient elements that are deposited on the ground are more mobile, more susceptible to physical removal as by rain or wind, than any that are part of the tissues of a living organism. Despite that, they are useful to a plant only if they are in a form that can diffuse through moist soil and be transported across the root wall. The first requirement here is essentially that they are soluble in water, a criterion that is clearly met by those nutrients that are voided in urine. In faeces, however, the situation is not so simple.

Apart from water, the major constituent of plants is cellulose; the digestion of this polysaccharide is one of the biggest nutritional challenges confronting a herbivore. Considering that it comprises only glucose molecules, cellulose is a remarkably refractory substance. Were it not, there would be no old trees, nor old wooden buildings or boats. Professional herbivores digest cellulose with the indispensible help of symbiotic gut microflora. At the top of the list of skilled professionals must be termites at one end of the size scale, and ruminants at the other. There are other herbivores, such as rabbits, that are not as good at it in the first place, but what they lack in innate digestive ability they make up in determination. Rabbits produce two types of faeces. The first is the partial digest of the plant material they have recently eaten; this is deposited within the warren. They are coprophagic, that is to say they eat faeces—in this case their own.[24] Coprophagy has an important advantage in arid environments in that it conserves water. It is a common practice among desert animals and its capacity for water conservation was no doubt an additional benefit for those rabbits that plundered inland Australia. The point of the exercise in the present context, though, is to enable the rabbit to expose the now partly degraded cellulose to a second encounter with its digestive equipment. The faeces from this second passage are deposited outside the warren. I have to confess to some doubt about the rabbits' ability to avoid confusion about just which batch they happen to be processing at any particular time but, by and large, they seem to manage—all too well. Omnivores—amateur herbivores, including Man—digest cellulose at most to a very limited extent. In such cases, however, the cellulose provides bulk and contributes to bowel function and health.

In all animals the digestive process is anaerobic—it is a series of fermentations; it does not and cannot proceed as far as it would were oxygen available. The products of these fermentations, apart from some gases (carbon dioxide, methane, dimethyl sulphide etc) are themselves relatively complex and can be inhibitory to plants. Moreover, nutrient elements can still be bound in a way that makes them unavailable to plants. Once deposited, the faeces are exposed to oxygen which, with the mechanical assistance of many invertebrates and the biochemical agency of soil microflora, takes the breakdown process a lot further. The dung is consumed and dispersed by those same invertebrates (flies, beetles etc) and quite often by vertebrates as well. (Herbivore dung can be an important dietary supplement for some carnivores.) Although this second stage of consumption and digestion by animals is also anaerobic, it does take the process further. The upshot of all this is partly that the dung is dispersed and, perhaps more importantly, the residual complexes are further broken down with the release of nutrient elements in states that can be assimilated by plants.

Just how uniform or random is the nutrient distribution achieved in this way? There are some 'natural' conditions, already discussed, in which the distribution is anything but random. In a forest or on natural grassland grazed by animals free to migrate and to be consumed by predators, it is likely to be as close to random as one could reasonably expect. On a farm the situation is different—but that is considered in the next chapter.

Now, after all that, can animal excrement make a nett addition to the total nutrient content of soil? If it does to any significant extent, it can only be with nitrogen through the medium of the microbiological fixation of that element. One way in which that does sometimes happen is when urine or faeces relieves a deficiency of another element, in particular phosphorus, for a leguminous plant and its bacterial symbiont, *Rhizobium*. There are two other possibilities. The intestines of many, perhaps all, animals contain bacteria that, at least in the laboratory, can fix gaseous nitrogen. Indeed the enzyme nitrogenase and the capacity for nitrogen fixation are much more ubiquitous than was thought even a decade or so ago. It is far from certain, however, that these bacteria do actually fix nitrogen in the digestive tracts of those animals where they have been found. Perhaps the sheep is the farm animal best studied in this respect. Under a set of conditions that would seem to be appropriate, nitrogen fixation has never exceeded 2% of a sheep's total nitrogen intake, with most measurements indicating a proportion of less than 0.1%. On the other hand some insects, and most termites, apparently do fix significant amounts of gaseous nitrogen in their digestive tracts under normal conditions. In this way termites sustain their life cycles on a wood diet containing negligible amounts of chemically bound nitrogen.[25] The remaining possibility is that the organic content of dung stimulates non-symbiotic nitrogen fixation in the soil by providing energy for bacteria such as *Clostridium pasteurianum* and *Azotobacter*. The reduction of elementary nitrogen (N_2) to the level of ammonia (NH_3) or its organic equivalent ($-NH_2$) requires a lot of energy, whether the reaction is achieved biologically or chemically.

But with these exceptions—the stimulation of nitrogen fixation in legumes and the soil, and possibly small amounts of fixation in intestines or rumina—animal excreta do not and cannot add to the total content of nutrient elements in soil. The organic component of dung does enhance soil quality in various other ways but, in so doing, it facilitates processes that would have occurred anyway at the hands of the soil biota. Animals distribute nutrients horizontally and, when they feed on deep rooted plants, vertically; but vertical redistribution would also happen anyway from the fall of litter. It is the horizontal distribution that can change things. If it is essentially random a 'natural' cycle occurs. If it is not random, then some soil is enriched at the expense of other soil somewhere else. And that forms an important part of the story of agriculture.

CHAPTER 3

The Transition

Regardless of which came first, animal or plant domestication,[1] sedentary agriculture was ultimately and absolutely dependent on the cultivation of plants. When an aspiring farmer or gardener wanted to grow plants of his own choosing on land that had not previously been cultivated, his first task was—and is—to remove the native plants. It is unlikely that this could have been achieved without some loss of soil nutrients. Just how significant was the loss depended on many factors, including the scale of the operation. A few holes made with a digging stick in grassland would scarcely have been a major disaster. Complete or extensive clearing would be another matter. Although with a good understanding and a great deal of care and effort, forest and grassland perhaps could have been made arable without soil degradation, it is scarcely conceivable that this was achieved with any greater success in those early transitions from foraging to farming than it was by more recent pioneers. The Neolithic farmers began their new industry with an initial loss of capital.

Nutrients are removed from soil essentially in three ways: (i) by the removal of plant material; (ii) by leaching, that is by dissolving in water and running off in solution; (iii) by the physical removal of the soil itself, that is by erosion. Of course the three processes are interrelated but they can be considered separately since there are some dominant factors in each case. For the moment I want to focus on the first mechanism.

In an agricultural context, removal of plant material is brought about by direct human consumption, by the physical removal of parts of plants, by consumption of herbivores or by the transfer of those herbivores somewhere else. Loss of nutrients from soil can be compensated to some extent by natural processes, as discussed in Chapter 2 and elsewhere. The effectiveness of such compensation, however, depends on the relative rates of removal and restoration. It is likely to be most effective on a flood plane that is regularly inundated. That, of course, is a special case and, although it is a process that has helped sustain farming for a very long time on the Nile, the Tigris and

Euphrates, the Yangtze and the Yellow River for example, it was not an option available to everyone.

In the final analysis, however, the sustainability of agriculture, be it Neolithic or recent, depended fundamentally on management of the farm and the way in which nutrients harvested from the farm were redistributed. Attempting to assess this redistribution boils down largely to theorising and guess-work, with some help from assorted pieces of evidence, much of it indirect. The question needs to be considered first for subsistence farms and then, with the production of surpluses, the various forms of commercial agriculture. The latter are addressed mainly in later chapters.

With subsistence farming in its strictest sense there might be an exchange of types of produce with other farms but, in most situations, that exchange would approximate a balance. In other words, there would rarely be a significant nett removal of nutrient elements from the 'system'. The fate of such farms would depend largely on their internal management, an important part of which was the distribution of manures.

The principal types of manures that a Neolithic farmer had at his disposal were foliage and plant cuttings, food wastes, his own excrement, the dung of his animals, the carcasses of any animals that died and were not eaten and, of course, his own body and those of his family after death (as long as 'at his disposal' is given sufficient flexibility of meaning).[2] In addition there was the likelihood that leguminous crops were used as specific fertilisers, using 'specific' to denote their significance for a particular element—in this case, nitrogen. Doubtless they were first used as a manure by chance—pulses were among the early crops in many regions. Legumes were used deliberately as fertilisers by the Romans,[3] so the conscious but empirical recognition of their value as a manure had occurred at least in the Mediterranean region by the second half of the first millennium BC.

Until pigs were domesticated it is doubtful if there was much redistribution of food scraps around the farm. Human societies seemingly have a tendency to establish garbage dumps of some kind as revealed by the ubiquity of middens and, in many early farming settlements, the use of waste pits.[4] Food scraps were therefore likely to have been concentrated into a relatively small area near the farmstead or the settlement.

Similarly human excrement was statistically more likely to have been disposed of close to a dwelling, or a settlement, than anywhere else around the farm. There is evidence in support of this although some amounts to an extrapolation from the historic period, and some is rather curious. For example:

"Human excrement and household refuse are commonly distributed by
    modern primitive communities in the garden plots around their settlements

and the distribution of such 'night soil' was the normal practice in medieval Europe. Spreads of potsherds and other domestic refuse around many pre-historic sites, particularly in the field systems adjacent to many late prehis-toric farmsteads in northern Europe, surely indicates such manuring."[3]

If the distribution of potsherds is evidence for the disposal of 'night soil', then I can only assume that those responsible were either given to a very unusual and difficult diet or, alternatively, they anticipated with a freshly filled cham-ber pot the much later ritual of tossing a freshly emptied champagne glass over the shoulder.

There is, however, better evidence in some regions for the disposal of food wastes and/or 'night soil' in and around settlements. For example, a survey undertaken in 1938 revealed anomalously high levels of soil phosphate in dis-crete zones of the Swedish island of Gotland. These zones were identified as the sites of stone-, bronze- and early iron-age villages.[5] A later soil analysis of a known iron-age farm site in southern Norway was interpreted to indicate de-liberate manuring.[6] In this case, however, the highest phosphate concentra-tions were found in a small area immediately surrounding the remains of one of the buildings. This might indicate the use of dungheaps, a practice that was common during much of the historic period. It is virtually axiomatic that sedentism *ipso facto* challenges hygiene.

A settlement of 2000 or more people, as at Tell Abu Hureyra (Chapter 1), would almost certainly have suffered from a chronic sanitation problem.[7] This is not to say that Neolithic farmers were necessarily predisposed to foul their own nests. It is quite probable that Man, together with a number of other species, has some innate sense of hygiene. In my school days, the con-ventional wisdom had it that Australian Aborigines had no sense of hygiene but, despite that, managed well enough in their native state because they stayed but briefly at a camp site and, by the time they returned, the site had been sterilised by the sun. The same conventional wisdom attributed the alarming increase in death rate when Aborigines were forced into settlements, as in missions, to this lack of any sense of hygiene. In this context the follow-ing quotation from James Dawson in 1881 is relevant and instructive.

"It is worthy of remark that nothing offensive is ever to be seen near the habitations of the aborigines, or in the neighbourhood of their camps; and although their sanitary laws are apparently attributable to superstition and prejudice, the principles of these laws must have been suggested by experi-ence of the dangers attendant on uncleanness in a warm climate, and more deeply impressed on their minds by faith in supernatural action and sor-cery. It is believed that if enemies get possession of anything that belonged to a person, they can by its means make him ill; hence every uncleanness belonging to adults and half-grown children is buried at a distance from their dwellings. For this purpose they use the muurong pole (yam stick), about six or seven feet long, with which every family is provided. With the sharpened end they remove a circular piece of turf, and dig a hole in the ground, which is immediately used and filled in with earth, and the sod so carefully replaced that no disturbance of the surface can be observed. Children under four or five years of age, not having the strength to comply with this wholesome practice, are not required to do so; and their excreta are deposited in one spot, and covered with a sheet of bark, and when they are dry they are burned."[8]

Shifting agriculture, slash-and-burn agriculture or swiddening—all synonyms—seems to have been practised in most parts of the world at some time and is still used in some areas, mainly in the tropics. Its essence is that forest or scrub is cleared by felling or ringbarking trees, slashing undergrowth and burning the cut plant material on the spot. Most of the ashes—and hence the mineral content of the plants—accumulates on the soil, but some is lost in smoke. The land so cleared is cultivated for a few years until yields begin to fall; it is then abandoned to fallow and the process started afresh in another area of forest. The sequence is repeated on several new areas and eventually, perhaps after a couple of decades, the cultivators return to the first plot and the whole cycle is restarted. Few contemporary shifting cultivators keep livestock; those that are kept are mainly pigs and poultry. Animal dung is rarely used as manure.[9] (See also Chapter 7.)

In the present context the significance of swiddening is that the need to move at intervals to a fresh patch of ground is occasioned by the impoverishment of the cultivated land, an impoverishment that, in the absence of significant erosion, would not occur were the nutrients harvested from it returned to their source. This is an integral characteristic of shifting agriculture and has been since its inception. The farmers clearly did not return all their food wastes and excrement to the arable land. The recovery of the land during its fallow period can be attributed to nitrogen fixation, to the extraction of nutrients from the sub-soil by deep-rooted plants, to the weathering of minerals from rock and clay, to the release of bound phosphate in a soluble form, and to the deposition of nutrient elements in rain, in animal droppings and in smoke from neighbouring fires. The quantities of nutrient elements in the subsoil, like their counterparts in the topsoil, are finite and subject to depletion; all these recovery mechanisms combined demonstrably do not equal the rate at which nutrients are removed while the land is actually farmed.

Fixed plot agriculture was either chosen from the outset or eventually dis-placed swiddening in most of the temperate world. As with swiddening, it de-pended on a form of rotation but in this case the variable was the crop rather than the plot. The forms of crop rotation that are known to have been prac-tised throughout history—and assumed to have been used before that—have been discussed at length by many authors.[10] The effects of crop rotation were similar in some respects to those of swiddening. Legumes could restore soil nitrogen, relatively deep-rooted crops such as lucerne could bring additional nutrients to the surface, plant pathogens could be separated from their hosts for a season or so. And perhaps soil moisture might be conserved during a pe-riod of bare fallow—albeit at the cost of degrading organic matter (see Chap-ter 10). The frequency of crop rotation was normally higher than that of plot rotation in shifting agriculture and the system lacked the means of reaching deeper nutrients that were accessible to the roots of trees.

We can reasonably assume that the value of animal dung and possibly also of legumes was recognised by many (but not all) early farmers before there was archaeological evidence for their use in that way. At the same time, there was probably inadequate appreciation that the improvement of those early crops by animal manure was achieved at the expense of their neighbouring pastures, if only because the nutrients of the manure were taken from a relatively large area of pasture and concentrated into a relatively small area of crops. This is a simple but important point with a counterpart in modern societies. If, for the sake of argument, farm animals graze on 10 hectares and their dung is used as manure on 1 hectare of crops, the effect of the manure will be about ten times as obvious on the arable land as will its removal from the pasture.[11]

The movement of produce within a framework of subsistence agriculture was not, of itself, likely to have had a major effect on average nutrient levels over a wide area of farmed land, although within that area the distribution al-most certainly became 'spotty'. There are two important qualifications, how-ever. The first, as already acknowledged, is that the effect could be completely masked by flooding. The second is where primary deforestation or subse-quent mismanagement led to soil erosion. An example of this is given by the collapse, in about 8000 BP, of a sedentary community in southern Levant (embracing modern Israel, Jordan and southern Syria). The collapse of this community, previously attributed to a climatic change, has now been laid at the door of severe soil erosion caused by extensive deforestation (in this case for the timber), exacerbated by the hilly nature of much of the terrain—and by goats which prevented regrowth of saplings and shrubs.[12] This was a couple of millennia before the first cities in Mesopotamia—before primary produce was removed from the land to meet the needs of a commercial market.

The essence of agriculture changed when towns were formed and food was brought to them from the countryside. The process took place over a long time in human terms, in many regions, and there is no *a priori* reason for assuming that the mechanism and its dynamics were the same in every case. The conventional view is that agriculture had to produce a surplus before towns were possible. Jacobs, on the other hand, has argued that 'cities' developed first and then stimulated the (further?) development of agriculture.[13] Like the earlier question of which came first, farming technology or population pressure (see Chapter 1), this one—which came first, the surplus or the city?—is also something of a chicken-and-egg question. From the point of view of the effects on the dynamics of nutrient flow, the question is of little significance. The important thing here is that, however they began, towns and cities on the one hand and agricultural production on the other eventually locked themselves into a mutually dependent association—another positive feedback loop. With that association came major changes in social structure and dynamics as well, of course, as in the dynamics of nutrient flow and the logistics of its management.

If all that it is meant by a surplus is simply the regular production of more food than is needed by the farmers who grew it, then it is likely to be a short-lived benefit. If the surplus merely allows the farming population to increase, as apparently happened in Tell Abu Hureyra, then the surplus is ephemeral. Another dimension is added if we take into consideration the time the people had for pursuing activities other than managing the farm. This is not so easy to pin down either. There is a prevailing opinion that, at least under favourable conditions, hunter-gatherers had more 'spare time' than did early or, for that matter, the majority of later farmers.[14] But both types of society did have spare time. Both types of society painted, carved in several media, made tools, ornaments and weapons. In due course, sedentary societies made extensive use of ceramics—and, of course, they built houses.

Does this amount to a division of labour? Only partly. It is not, in itself, sufficient to enable a division of labour to be used as an argument for a town and hence a surplus of food production. There has always been some division of labour in all human societies, be they hunter-gatherers or subsistence farmers. There has apparently always been some sort of division between the work done by men and by women—the women commonly carrying the heavier burden of responsibility for providing the more dependable components of the food supply. In addition there have always been specialist activities undertaken by individuals with particular skills in painting, pottery, making spears, carving out canoes, or putting the roof on a house. The fact that some people undertook some of these tasks for some of the time is not sufficient in itself to show that that was all they did—that they did not help on the farm at other times. Where there have been full-time craftsmen, the archaeological evidence for it is usually clear. To take an extreme example, it is not conceivable, even for a complete neophyte, that the Athenian Parthenon could have been designed and constructed by farmers in their spare time.

In the course of human cultural evolution, societies were presented, if you like, with two questions. If they answered 'Yes' to Question 1, they proceeded to Question 2. If they answered 'No' to the first question, that was the end of it—until, in the fullness of time, an answer was imposed by the descendants of those who had answered 'Yes' to both. The first question was, 'Do you want to convert from hunting and foraging to farming?'; the second was, 'Do you want to build cities?'. Those who answered 'Yes' to both grabbed the tail of the most implacable tiger the world has ever seen. With few exceptions, those who answered 'Yes' to Question 1 and 'No' to Question 2 lived in the wet tropics.[15]

Urbanisation required effective food storage, a challenge that was both more urgent and easier to achieve in dry temperate regions than in the wet tropics where seasonal fluctuations in crop yields were small, and high humidities made storage of any foodstuff difficult. It is more than likely that the centralised storage of grain, and the concentration of control of the grain into the hands of a few, stimulated a hierarchical rearrangement of social structure and, with it, a genuine division of labour. This process also has the ingredients of a positive feedback loop. Certainly the control of the granaries had far-reaching social and political implications.

But there is another factor in the transition from a village to a city that preceded control of the granaries and, in most cases, was a prerequisite of that control. The first cities arose in Mesopotamia and all depended on irrigated agriculture, as did their subsequent counterparts in Egypt and, much later, in Meso America.[16] Irrigation produced a gradient whereby productivity decreased with distance from the river. It has been argued[17] that those who had the good land—the land close to the river—became rich and powerful and, indeed, were the ones who got control of the granaries. This growing disparity between the rich and poor, as much as anything else, may well have produced the social and economic conditions necessary for major divisions of labour and the genesis of a city. Indeed such divisions of labour would almost certainly have been necessary to produce the sophisticated hydraulic engineering that distinguished virtually all early civilisations.

It is simply not conceivable that early agrarian communities did not respond to the inevitable bad seasons by a deliberate policy of accumulating and storing the excess from good seasons.[18] It is also tempting to think that better technology, that is to say, better farm implements led to improved yields per man-hour of labour. This happened eventually of course; the question is whether or not it was significant in the gestation of a city. Improved equipment is not likely to have had a dramatic effect until it enabled energy in addition to man-power to be applied to the problem. This happened conspicuously when draught animals were persuaded to haul a plough. The first ploughs, or more accurately 'ards', appeared in Mesopotamia in the 4th millennium BC (5000-6000 BP)[1]; the timing corresponds well with the evolution of Sumerian settlements into cities. On the other hand, the plough was not used in America until it was introduced by the Europeans, nor was it adopted in Sub-Saharan Africa until the 19th century,[17] but that was not a birthplace of cities.[19]

Whatever the causes—and it would be idle to pretend that there were only one or two—the fact is that agriculture did eventually produce surpluses large enough and consistent enough to sustain cities. Within those cities there grew a range of new industries, most with their own positive feedback loops which, among other things, produced more mouths to feed. All of the new industries affected agriculture indirectly and some affected it directly by providing it with new implements, new methods and new forms of transport. Cities changed forever the course and dynamics of nutrient flow between the farm and the consumer. In an agrarian settlement there was always the probability that some of the nutrient elements harvested in produce would return to the arable land by default. That was no longer possible from a city. Food wastes and human excrement were either returned deliberately or they were not returned at all.

CHAPTER 4

Town and Country
– Part 1 –

In Chapter 1 the comment was made that a diagram such as Fig. 1.1, which was intended to represent the functioning of an ecosystem, could also denote the functioning of human society. In some respects it can also depict the operations of a city. Indeed at least one city, Hong Kong, has been studied in detail specifically from an ecological standpoint.[1] Moreover, the evolution of a city involves a process of succession which, in many ways, is a formal homologue of succession in other types of ecosystem (see Chapter 2). In the early stages of civilisation a relatively small number of 'pioneer' industries are established. They consist predominantly of crafts—house construction, making tools and weapons, pottery, clothing etc. Progressively the number of industries increases and, as the city grows and the social environment evolves, more activities arise, some of which displace the pioneer industries as, for example, when iron metallurgy displaced bronze or motor mechanics displaced farriers.

Because these extra-agricultural industries did develop and cities did grow and multiply, the overall rate of food production was necessarily greater than that needed simply to sustain a growing agrarian population. Not only did an increasing number of farmers make their own direct impact on the natural environment, but so also did the burgeoning urban industries whose existence and sustenance depended on the farmers. These industries, in their turn, provided the farmers with better tools, tools that were crucial to the increase in yield/man-hour of labour that followed. And thereby hangs another positive-feedback loop.

Virtually every industry that was established in the town made its own demands and had its own impact on natural resources. Wood was used for construction, for fuel in cooking, heating dwellings (and in Rome, for heating baths), for pottery, and increasingly for the production of charcoal. Charcoal was indispensable in smelting from the beginning of the bronze age (or, in some regions, the 'chalcolithic' age) until the mining of coal in Europe early in the 13th century. Notwithstanding the discovery of coal, however, charcoal continued to be used for smelting in parts of Europe until late in the 18th century. It has been estimated that, in forty days, a single mediæval iron smelting furnace consumed the wood from 3 km² of European forest.[2]

The new industries that progressively came into existence evolved into vested interests, pressure groups, self-sustaining enterprises—and interacted with other such groups competitively, cooperatively, predatorily—consolidating the formal similarities to an ecosystem. There were other positive feedback loops, not all of them bad. Because of the self-stimulation of these networks, the pace of cultural change has been faster in cities than in any other human habitat. City dwellers became different from their country cousins, not innately, but because their environment was not only different, it was also changing faster. Over the course of time some city dwellers gained in understanding the abstract, but most lost the understanding of the land that is indispensable to a farmer or a hunter-gatherer. No product of human culture has matched the city in its capacity for intellectual stimulus; few have matched its capacity for squalour.

The development of new industries and the inevitable division of labour accelerated whatever division of wealth, status and living standards had been present at the birth of a city. Every industry needed energy, be it to haul a plough, fire pottery, sweep streets, manufacture arms, or transport goods; with the commercialisation of industry and trade, that energy was required to be as cheap as possible, whether it was paid for in money, in grain, or by barter. Slavery was an integral part of ancient civilisations. Slaves were the trophies sought in many wars and might well have been the first commodity bought and sold primarily for profit within ancient Rome. Ancient civilisations have been described collectively as 'slave civilisations'. Slaves served as farm labourers, as domestic servants, in the 'police forces' and armed services, as skilled craftsmen, as doctors, and as the sexual playthings of their owners. Some of that was also true of African slaves transported to the Americas to clear land, to establish and work plantations of sugar, coffee and tobacco, in the course of which they made their owners very rich.

Slaves and their children were the possessions, in every sense of that word, of their owners—to be treated however the owner chose. The ancient Roman town of Puteoli stipulated that one of the duties of the municipal funeral director was to torture slaves when requested by either private individuals or a magistrate. Torturing slaves, sometimes with extraordinary ferocity, was a relatively common practice in the Americas.[3] It is estimated that there were 60,000 slaves in Athens at the beginning of the 4th century BC and 2,000,000 in Italy at the end of the Republic, amounting to about 30-35% of the population in each case. This is a similar proportion to that (33%) in the southern American states shortly before the start of the American Civil War.[4]

The underlying economic motivation for slavery has persisted to the present. It showed itself not only in the African slave trade to the Americas and elsewhere,[5] but also, albeit it on a smaller scale, in the effective slavery of Australian aborigines on pearling luggers in the northwest of Australia at the end of the 19th century and on numerous Australian grazing properties until early in the 20th century. It caused the past exploitation of children to work in European factories and their present exploitation for the same purpose in much of the so-called Third World and, to some extent, even in Britain and the USA.[6] It has been reflected in the use of prisoners of war as a free labour force.[7] Frank or covert slavery has been closely associated with industrial development throughout history.

None of this could have happened in a society that was not supported by agriculture. Nor, presumably, would there have been the widespread sacrifice of animals, children, miscellaneous adults, prisoners of war, sometimes kings (and later, as kings came to recognise some of the disadvantages of the practice, royal understudies), all to placate or please those gods responsible, among other things, for sustaining soil fertility. The Aztecs raised human sacrifice to a major industry by conducting ferocious wars to capture prisoners for the express purpose of sacrificing them. About 20,000 prisoners annually are thought to have been thus despatched in the heyday of the practice.[8] The records are largely silent about the effectiveness of this method of sustaining soil fertility, but we can be confident that the results would have been better if the Aztecs had buried the bodies of their victims in arable ground.

Another dimension of the practice of sacrifice was to be found in ancient Rome. During the height of the gladiatorial season as many as 5000 animals, including large ones such as elephants and water buffalo, could be slaughtered in a single day as well, of course, as the hundreds of human victims of the arena. It is not suggested that this was done in the interests of soil fertility. Lanciani has described a 'carnarium', a collection of pits into which carcasses, human and otherwise, were thrown. When the pits were full, "the moat ... was filled with corpses ... until the level of the adjacent streets was reached",[9] a practice that might well have lessened some of the attractions of Rome as a place in which to live, or even to take a holiday.[10] The repercussions of these practices extended to recent times when Lanciani unearthed some of the carnaria. He described pits filled with a thick black viscous fluid of such overpowering stench that he and his workmen were obliged to stop their excavations at intervals—a testimony, if nothing else, to the need for oxygen in the composting process. If Lanciani had been able to postpone his excavations for a million years or so, he might have found oil instead of that putrid legacy of Roman entertainment.

These examples, although several millennia after the genesis of the earliest towns, and although straying a little from the central issue, illustrate aspects of a problem that must have beset human societies from the earliest days of civilisation—failure to recognise the potential value of most forms of biological material as a manure. Food wastes were certainly fed to pigs, and doubtless pig dung was widely used as a manure, intentionally or otherwise. But, although food wastes could become objectionable within a few days, they were not immediately so, and the incentive to dispose of them 'far away' was presumably less compelling than the incentive to dispose of sewage.

But what of human excrement, what of sewage? Human faecal matter is aesthetically objectionable to Man, even though communities, in particular impoverished city dwellers, have lived with and apparently become partly inured to appalling sanitation throughout much of history (Chapter 6). From the outset there must have been a strong incentive for sedentary communities to dispose of excreta in a more effective manner than their disposal of food wastes. The impact of dung on the growth of crops is faster and thus usually more obvious than that of undegraded food wastes. And so both the presumptions and the evidence are that, in the prehistoric period, and certainly throughout history, there have been attempts—serious attempts—to return human excreta from the town to the farm. It is impossible to quantify the proportion so returned, but we can be sure that, as cities became larger, the return of sewage to the farm became increasingly difficult.

So how was sewage disposed of in those early cities? Sewerage systems were in place by at least 4500 BP in the Indus Valley cities of Harappa and Mohenjo-Daro, which might even have been blessed with flushed water closets. Indeed, Mumford compares the sanitary arrangements of Athens very unfavourably with those of the Harappan civilisation two millennia earlier.

But it is Rome, probably more than any other civilisation of antiquity, that is associated with sophisticated hydraulic engineering and sewerage systems. Rome built the Great Sewer, the Cloaca Maxima, in the 6th century BC—the oldest monument of Roman engineering and still in use in the 20th century—before it had constructed any of the aqueducts or laid any of the pipes that were to bring water to the city. At that time, and indeed until the end of the 1st century AD, local water supplies from wells and rain-water cisterns were adequate. But that was not to last. Then, as now, the original water supplies were diminished by demand and overwhelmed by a growing population. And so aqueducts were built, bringing water for considerable distances, 132 km in the case of one supplying Carthage.[8]

The sewer carried away the excrement from public lavatories and from the ground floors of the houses of the well-to-do. The raw sewage was simply taken beyond the precincts of the city and, in this case, discharged into the Tiber. The not so well-to-do and, in particular, the impoverished residents of multi-storey tenements, had to make other arrangements. Their options were to use the public lavatories (for a fee) during the day, to carry the results of their efforts downstairs and deposit them in a cistern at the bottom of the stairwell, or empty a chamber pot from an upstairs window whence the contents drop, if not exactly as the gentle rain from heaven, nevertheless upon the place—or the passer-by—beneath. The one positive side in all this was that the cistern at the bottom of the stairwell was emptied by 'dung farmers', or those who would trade in the material, for application to farms.[11] It is not clear over what area of farmland this excrement was spread, but Mumford makes the reasonable suggestion that it was confined to farms close to the city and that those farms were eventually saturated.

A purpose of this bird's eye view of the genesis and early evolution of cities was to illustrate, however superficially, two principles that were enunciated earlier and will recur repeatedly. One was 'The First Law of Ecological Bloody-mindedness' which says, in effect, that every action on a complex dynamic system has some unintended consequences. The story of human cultural evolution has been just that. Every change in 'lifestyle', every technical innovation, every change to the law that Man has adopted has brought with it unintended and unexpected side effects, many of them undesirable. The usual response has then been to seek a remedy for the side effect(s)—to treat the symptoms. Then each remedy produces its own side effects which, in their turn, must be dealt with.[12] The result is a cascade, an explosion of technical innovation, social, legal, economic and environmental problems, and a social structure of progressively increasing complexity. Rarely has the remedy of abandoning the original change been implemented; if it had, the decision would probably have been lost from the historical record.

The second principle was mentioned in Chapter 3 when a comparison was made of the effects on soil fertility of withdrawing animal dung from a large area of pasture and applying it as manure to a small area of cropland. The transport of produce from farmland to be concentrated within a city is a similar process; the process has its most immediate and most obvious effects on the city. It is there that the first problems arise and the first pressure mounts to change the practice in some way to reduce the impact of stench and disease. And it is there that the next round of remedies for the symptoms rather than the disease are developed.

Some rough calculations—because of the number of assumptions involved, they can only be rough—using the ancient Sumerian city of Ur as a model, suggest that the area of land needed to supply such a city with its food was probably of the order of 950 km²—about 1000 times greater than the area of the city itself. The significance of this should never be underestimated. Its impact has shaped the course of development of agriculture and the management of cities throughout all of history.

In its simplest terms, the flow of nutrients from the farm to the city made an impact per unit area of the city 1000 times that of their removal from the soil. This discrepancy is 100 times greater than that of the hypothetical transfer of animal dung from grazing to arable land mentioned in Chapter 3. But it is not as simple as that, of course. The direct effects of taking nutrients from the soil not only occur at a rate 1/1000 the rate of deposition in the city, but they would not be apparent until the next season. But what if the farmland were more productive than that of the present example? If it were, say, ten times more productive, then the 'basic' area needed to support the city would be but 10% of that used in the previous calculation, that is about 95 km². The transport costs of bringing the food to the city would be less, but the soil would be depleted of nutrients ten times faster per unit area than the poorer soil of the first example. This would not necessarily be immediately apparent either. If the greater productivity resulted from a higher concentration of nutrient elements in the soil (as distinct from a more favourable moisture régime or absence of toxic factors), the impact on plant growth of this accelerated removal of nutrients would not be apparent until the proportion of a relevant element in the soil entered a range within which plant growth was a direct function of its concentration. This kind of relation is illustrated schematically in Fig. 4.1.

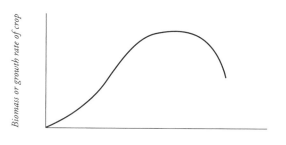

FIG. 4.1 *A theoretical representation of plant growth (rate or total biomass) as a function of the concentration of an essential nutrient in the soil.*

On the other hand, the effect on the city is immediate—literally overnight. Within 24 h of the consumption of food within the precincts of a city, the unused food trimmings—the 'food waste'—slaughter-house waste and, of course, human excrement must be disposed of in some way. This is not something that can be easily ignored, notwithstanding the appalling sanitary conditions that prevailed in all cities of the world until the end of the 19th century AD, and in the slums of many cities at the present time. It is a problem that became progressively worse as urban population densities increased, in particular with the construction of multi-storey buildings.[13] It is a problem that has had a profound influence on the course of history. It is a problem shared in principle with the feedlots of modern commercial 'agribusiness'. And perhaps it is worth restating that this is not a type of problem experienced by nomadic hunter-gatherer societies. (Somewhere in all this is the stuff of an aphorism, perhaps of the Confucian type: 'If people go to food—no problem. If food goes to people—big problem.')

All things considered, it is small wonder that so much attention has been focused on the symptoms of the disease, urban pollution, that is an immediate and direct consequence of feeding a city. It is characteristic of our collective thinking that urban sanitation on the one hand, and rural soil degradation on the other, have been seen and largely treated as two separate problems rather than as two components of a much larger one. The larger problem is that the flow of nutrients to a city and thence to its sewers and garbage dumps is essentially linear and ultimately unsustainable. Not that dealing with the larger problem was ever easy, although it was easier then with small cities than with large ones now. The example just discussed should illustrate one of the practical difficulties. The poorer the land supplying the city, the greater is the urgency to return manure from the city to the farms. But the poorer the land, the greater is the distance to travel both in bringing food to the city and in returning manure to the farms. In its heyday, ancient Rome reached a population of about a million which, of itself, was an extraordinary achievement. The logistics and costs of transport must always have been a factor in determining the size of any city; a city of one million inhabitants at that time might not have been possible without the Romans' propensity for constructing long straight roads and, perhaps more importantly, access to water transport. Nevertheless, no matter what other factors influenced the practices of the dung traders, it is scarcely surprising that, when sewage was taken from ancient Rome, it was deposited on nearby farms which duly became saturated.

But let us return briefly to the agricultural component of this process. I shall define land degradation as any process that diminishes, or has the potential to diminish, the capacity of land to support the total biomass, the average growth rate, or the range of species of plants under conditions of adequate water availability, light intensity, and within a range of temperatures appropriate to the plants that would normally grow 'there'.[14] As partly foreshadowed earlier, land can be degraded by the nett physical removal of soil, by the nett removal of nutrient elements free of soil, or by the nett accumulation of toxic substances.[15]

A principle already touched on in Chapter 2 now needs re-emphasis; it is that yields of crops or animal products are more than simply statements of mass of produce per area of ground. They have an indispensable time component; they are statements of rate, that is to say, mass of produce per area of ground per unit of time. Thus x tonnes of wheat per hectare, unless stated otherwise, signifies x t.ha^{-1}.y^{-1}; it can validly be expressed as an equivalent average rate of $x/365$ t.ha^{-1}.d^{-1}, or even per minute if some useful purpose were to be served by using those units. An extension of the same point is that the food requirements of a human population are also an expression of a rate, and that is true whether the population is growing, declining or is steady. Thus if a steady population of one million needs x kJ of energy and y tonnes of protein per day, a steady population of two million needs $2x$ kJ and $2y$ tonnes respectively per day, or $x/12$ kJ and $y/12$ t.h^{-1}—and these must be drawn from the soil at a rate which, because of wastage, is higher than that. So a steady population is not in itself a safeguard against exhausting the soil—it is merely less dangerous than a population that is increasing.

In Chapter 2, attention was also drawn to natural processes able to replenish nutrient elements removed from the soil in crops or animal products. The rate of replenishment clearly must and clearly did vary widely from one topographical situation to another; it varied according to the potential of a site to receive nutrients from smoke, from rain, from passing animals, from the growth of legumes—and so on. But if the rate at which elements are removed in produce (and by leaching and erosion) is less than the natural rate of replenishment, then the land could be farmed indefinitely without any form of manuring. To the extent that this might have happened, it clearly placed a premium on low yields, and hence on low population densities, under all but special circumstances such as regular flooding. There have been attempts to deduce what such a critical yield might be, where I use 'critical' to mean the maximum that can be sustained without manuring. For example, one estimate of the nitrogen requirements of wheat and likely natural sources of replenishment led to the opinion[16] that mediaeval English soil could probably sustain an annual yield of 1000 kg.ha^{-1}. This may well be too optimistic; mediæval yields, which in some situations are well documented, were generally about half that, even with conscientious manuring (see Table A4). Moreover some mediæval farms did indeed suffer from nutrient depletion and other forms of soil degradation.

Be that as it may, the argument is based on but one element, nitrogen. There are many other situations in which plant growth is limited by a different element.[17] If it is a macro nutrient such as phosphorus, there can be a reasonable expectation of some replenishment in bird droppings and, if there have been fires upwind of the crop, in smoke deposition and in rain (see Chapters 2 and 9). But there is no equivalent of biological nitrogen fixation for phosphorus. If the limiting nutrient is a trace element, such as cobalt, molybdenum or manganese for example, then the assumptions about a critical yield need to be different.[18]

There were also other contributors to soil degradation. The first cities were established in southern Mesopotamia between 6000 and 5000 BP in response, as Knapp puts it, "to the pressures of population growth, the social and organizational requirements of irrigated agriculture, and the demands of complex trading systems". Sedentary agriculture had spread south to the alluvial plain of the Euphrates and Tigris rivers by about 7000 BP. Although this was within an area of very low rainfall (less than 200 mm annually), the plain itself was subject to both seasonal and capricious flooding; it was very fertile as a result.[19] South of Baghdad, which is on the Tigris, the beds of both rivers are elevated above the surrounding plain, a condition that encourages the formation of swamps, the relatively frequent breaking of banks with its attendant extensive flooding and occasional chaotic changes of course. As always, the flooding deposited silt. It has been estimated that silt has been deposited to an average depth of about 10 m, at least over the northern region of the alluvial plain, during the past 5000 years,[20] that is to say, at an average rate of 2 mm per year.

Farming in such an environment required effective irrigation—which was indeed achieved. The Sumerians developed writing at least partly for the practical purpose of keeping accounts. Their records have provided historians with a valuable insight into grain yields over an extended period. These records, together with direct reports of salination and archaeological evidence of the types of grain cultivated, give a fairly comprehensive picture of the impact of progressive salination on Sumerian cereal cultivation.

In about 3500 BC, the proportions of wheat and barley harvested in southern Mesopotamia were approximately equal. A thousand years later wheat, which is less salt-tolerant than barley, accounted for about one sixth of the crop. Four centuries later again, that is by about 2100 BC, it had fallen to approximately 2% and, by around 1700 BC, wheat was no longer cultivated. As might be expected, yields were recorded by volume of grain, which I have converted to the following masses.[21] Around 2400 BC, that is by the time the proportion of wheat harvested had already fallen drastically, total annual grain yields averaged some 1955 kg.ha^{-1}. Three centuries later, they had fallen to around 1125 kg.ha^{-1} and by 1700 BC the corresponding average was down to 690.[22] These values give some insight into the extraordinary natural fertility of the region. The first yield, obtained when the effects of salination were already being felt is about twice that cited in the theoretical example given earlier for a self-sustaining yield of wheat. The average yield of wheat in modern Iraq is about 750 kg.ha^{-1}, but where traditional areas have been reclaimed, yields as high as 3400 kg.ha^{-1} have occasionally been obtained, although they more commonly lie within the range 1500-2500.[23] Australian average wheat yields for the decade 1980/81-1990/91,[24] fluctuated within the range 770-1700 kg.ha^{-1}. The Australian and modern Iraqi yields were dependent upon application of fertiliser but the Australian yields were subject to significant variations in rainfall.

This comparison underlines the remarkable productivity of the alluvial plain of the Tigris and Euphrates rivers and goes a long way to explaining why it was the site of the first unambiguous civilisation. But there was a fly in the ointment. The declining grain yields pointed the way to a declining civilisation. The salination responsible for the decline was an almost inevitable consequence of irrigation that was not supported by adequate drainage. With the notable exception of the Nile, salination eventually overtook virtually every ancient system of irrigation and has affected many modern systems. It is worth noting, nonetheless, that the impact of salination in ancient Mesopotamia was spread out over nearly two millennia. Extensive areas of Australia have succumbed to induced salination in less than two centuries, in many cases in less than one century, although some of the worst Australian examples stem from causes other than irrigation (see Chapter 9).

The intermittent flooding of the Tigris and Euphrates had two other major effects on Sumerian civilisation, effects of immediate relevance to the underlying theme of this whole discussion. The flooding and its attendant deposition of sediment caused seasonal swamps and severe leaching in some places but, overall, it sustained nutrient levels in the soil to the extent that the question of sewage disposal was almost irrelevant to soil fertility. It seems to have been predominantly salination, not impoverishment of soil, that wrecked Sumerian agriculture. The second effect was that of sedimentation in the irrigation systems themselves, a process that was evidently exacerbated by deforestation and excessive grazing in the catchment of the two rivers. By raising the ground level of irrigated fields, sedimentation made irrigation under gravity progressively less effective. And it blocked the irrigation canals themselves so that a population that had declined by as much as 80% in some areas through social factors and the impact of salination could no longer keep them cleared. The centres of Mesopotamian civilisation progressively moved northwards.[25]

As with the Tigris and Euphrates in southern Mesopotamia, the factor that, above all others, made settlement in Egypt possible was its river. The world's longest river at more than 6000 km, the Nile has perhaps also been the most generous. Herodotus, the so-called 'Father of History', is credited with the observation in the 5th century BC that, if the Nile were not to flood its delta, the Egyptians would suffer for all time.[26] The Nile valley and delta had been occupied by 7000 BP. It has been conventional to attribute the formation of the delta to regional climatic changes but a recent analysis suggests another explanation—a eustatic rise in sea level. During the Pleistocene, specifically 18,000-20,000 years ago, the sea was some 125 m below the present level. By 5500 BP the difference from the present was only -12 m. An effect of the rise was to reduce the gradient of the river near its mouth from 1 in 1450 to 1 in 5800, with the result that the flow rate of the river dropped and flood water had 'time' to spread over a wider area than previously and, in the process, to form the Nile Delta. At least by 8000 BP the delta had formed to the extent that it was capable of supporting grazing and cultivation without artificial irrigation.[27]

Like the Mesopotamian rivers, the Nile was subject to seasonal flooding, but with some important differences. In addition to their capricious floods, the Tigris and Euphrates were disposed to flood in spring and thus subject the flood plain—the farmland—to very high rates of evaporation in the summer. This, of course, concentrated the dissolved salts and was an important factor in the subsequent salination of the area. The Nile, in contrast, normally flooded in autumn. Moreover, having a more conventional attitude to life than either the Tigris or the Euphrates, the Nile kept its bed below the level of the flood plain. Accordingly, early irrigation methods were based predominantly on the construction of dykes to retain water after a flood rather than on long gravity-fed irrigation channels. The water table rose and fell with the river level and this, in addition to the topography, to the volume of floodwater, to natural drainage and to the other factors just mentioned, offered an effective safeguard against salination. The flooding deposited about 1 mm of sediment annually on the flood plain.[28] Although this was not greatly less than that left by the Mesopotamian rivers, it was evidently not enough to choke the irrigation canals, presumably because of their design and the velocity of the flood water.[29] But it was enough to compensate for the removal of nutrients in crops and livestock and the (presumed) failure of the settlers to balance that removal with manure.

These two examples, ancient Mesopotamia and Egypt, have been considered at moderate length because they offer an interesting comparison with one another, they have important elements in common and equally important differences. The regions were the sites of two of the world's earliest and, in several senses, most important civilisations. Both civilisations depended absolutely on their rivers and on agriculture irrigated from those rivers. One of them, Sumer, ultimately declined, apart from any cultural reasons, because of the failure of its agriculture. If the beginning of Egyptian civilisation is taken as starting with the first Dynasty (2900 BC), then it has so far outlived Sumer by about three millennia. There is no reason to suppose that the Egyptians were any better as farmers than the Sumerians. The Sumerians were not fools; their many remarkable cultural and technological achievements attest well enough to that. There is no evidence of which I am aware that either civilisation was any more conscientious than the other, or even conscientious at all, about completing the cycle and returning their cities' poo to the farms. It is simply that the Egyptians were lucky. But their luck began to run out in the 20th century with the construction of the two Aswan dams on the Nile and an explosion in their population.

It is impossible to quantify the extent to which early civilisations returned human excrement to the soil and difficult to know in some cases if any attempt was made to do so at all. What is certain, however, is that where agriculture has persisted over several millennia as, for example, in Egypt, Europe, China, India and South East Asia, it has done so partly by good management (including conscientious manuring), partly through luck, (the local climate and topography) and substantially because, over most of those several millennia, the human populations were small and their use of energy was very modest by to-day's standards. The dynamics of nutrient flow engendered by agriculture and sedentism were significant but not yet alarming.

CHAPTER 5

Town and Country
– Part 2 –

Two factors, probably above all others, determined the impact that agriculture has made on the biosphere, at least at the level of nutrient flow. They are the absolute size of the human population and its distribution between town and country. A large uniformly distributed agrarian population can farm 'for ever' (climate permitting) if it farms well. If it farms badly its impact is greater and its life expectancy is shorter than that of a small population adopting the same practices. Once the population is unevenly distributed, however, and farm produce has to be transported from the country to a city, new processes arise and demand attention if the community at large is also to have a reasonable life expectancy.

For reasons such as these, most attention from here on will be given to the industrialised world, its genesis and its evolution. It is the industrial world that acquired the highest proportion of city dwellers, that in due course produced and exported most food, developed a system of agriculture absolutely dependent on massive amounts of inorganic fertiliser and of energy. The 'less developed' agrarian world has its own sets of problems and they will not be ignored. But subsistence agriculture, although quite capable of causing severe soil degradation, is not in the business of taking nutrients from the soil and transporting them over great distances. With good management, subsistence agriculture can be sustainable; industrial agriculture, as it is currently conducted, cannot. 'Agribusiness' both suffers from and exacerbates the problem that nearly all who consume the food it produces live somewhere other than on the land that grows it.

We cannot say, as some have done,[1] simply that the growth of population forced certain changes upon farming practices and everything else followed from that, or simply that improved farming techniques produced more food and stimulated population growth and everything else followed from that. Like the argument about role of population pressure in the genesis of farming (Chapter 1), they are chicken-and-egg statements. The post-natal development of agriculture was part of a feedback system in which it was—and is—rarely possible to identify clearly a ubiquitous primary driving force, despite numerous claims that it can be done. The system drove and was driven by the invention of new devices such as the plough and the wheel, by the spread and interchange of crops and domestic animals, by the adoption of new farming practices such as crop rotation, by the use of legumes and dung as manure, by the construction of irrigation systems—and, of course, by the size, dynamics and distribution of the human population as well as its social and economic structure. It cannot be overemphasised that the greater the farm surplus, the greater the proportion of urban dwellers and hence the greater the need to invent and develop new industries if mass unemployment and all its social ramifications are to be avoided. One of the most powerful and dangerous positive feedback loops to have emerged over time has shown itself in the change of the primary objective of farming from growing food to making money.

In its simplest form, the plough or, more accurately, the ard, first appeared in South-West Asia about 5000 years ago. It took about 300 years to reach Britain, about 800 years to enter northern Europe (Denmark), nearly 3 millennia to get to the southernmost limit of South-East Asia and nearly 4 millennia to reach Japan, its easternmost limit in the 'Old World'. In this early phase it did not reach sub-Saharan Africa, Oceania or America.[2] The oceans and the Sahara Desert were effective barriers to its spread. The ard was a simple wooden device which, as Grigg[3] put it, "did little more than scratch the soil", but it was drawn by two oxen and it extended the area of arable land that could be worked by one individual. It thus increased the yield *per capita* and contributed to the surplus necessary to sustain towns and cities. By late in the first millennium BC, ards were fitted with iron shares in the Mediterranean and in China.

In the Christian era, most probably in the Middle Ages, ards were supplanted by heavy ploughs, implements with a coulter that cut through turf and a mouldboard to turn the ground.[4] These ploughs were drawn by at least four oxen and, in due course when a suitable harness was developed, increasingly by horses. Horses are faster than oxen and further increased the area that could be worked by one individual. Eventually ploughs were fitted with wheels.

The wheel is widely regarded as one of the major achievements of mankind; its reinvention is one of many accusations laid at the doors of those with bright ideas but who have not, in the eye of the beholder, 'done their homework'. Not only did the wheel revolutionise land transport, it opened the door to an enormous range of machines whose function depended on rotating parts. In due course it made a major contribution to the culling of wildlife and, for a while at least, it reduced the time needed to take victims of road accidents to hospital.

The wheel was in existence in Mesopotamia by about 5500 BP; it had spread throughout central Europe and North-West Asia by about 3800 BP and to India, North-East Asia and western Europe by about 2500 BP.[5] For our immediate purpose much of its significance lay in its ability to facilitate transport on the farm, between farm and town and within towns. Perhaps carts had a slight advantage in speed over pack animals but, more importantly, a single draught animal could move as much material as several pack animals and could probably sustain the effort longer. There would thus have been less need for a farmer (or a carrier) to maintain a relatively large herd of animals for transport. This, in turn, would have had no small effect on the farmer's time and on the area of land needed for pasture. Wheeled carts were probably used, for example, in the cereal harvest in Subir, a northern Mesopotamian civilisation in the third millennium BC[6] and for carrying dung in and around Mohenjo-Daro at about the same time.[7] But, notwithstanding the availability of carts, they were of limited use until there were roads good enough to carry them. Rome had its roads and its carts and its traffic noise but, in most of Europe, pack animals remained the principal means of transporting goods to cities until the 17th century AD,[8] apart from those cities that had ready access to water transport.

The problem of sustaining soil fertility in the face of the removal of nutrients in produce was evidently recognised in prehistoric times and some attempts were made to deal with it. It was certainly encountered and well documented early in the historic period and is described at length by the Romans, among others. As well as the deliberate transfer of dung to arable land, various systems of crop rotation were in use in the prehistoric period and subsequently. It is not my purpose to describe the details of this practice since that has been done often enough by others,[9] but there are some principles worthy of attention.

As the name implies, crop rotation is the practice of sequentially growing different crops on the same plot in successive seasons. In this it compares in principle, but differs greatly in time scale, with swiddening in which the plot rather than the crop is changed (Chapter 3). For crop rotation to be effective and still yield enough produce to meet the farmer's needs, it should make use of more than one plot or field in order to allow the 'primary' crop to be grown every year. Thus if there were two fields in use, crop A would alternate annually between fields 1 and 2; the other field would grow a different crop, pasture, or remain fallow. At its least, a simple system like this would diminish the likelihood of accumulating in the soil specific pathogens for Crop A.

What else it achieved would depend on what happened in the alternate years to the second field.

A more effective system involved three fields growing, for example, two crops with the third field fallow, under pasture or growing legumes destined to be ploughed in as 'green manure'. This practice can increase the number of crops grown on a farm and it also reduces the likelihood of building up specific pathogens. Its prospects of success in that regard are better than those of a simple two-field system if only because, with a symmetrical pattern of rotation, any one field grows a particular crop one year in three instead of one in two. That is a basic advantage that applies in principle no matter what happens in the other fields.[10]

If the third field grows a legume and the whole crop is ploughed in as a manure, it takes nothing from the soil except water (see below) but returns organic material whose nutrient value has been enhanced by the fixation of atmospheric nitrogen. If the crop is a pulse and the seeds are harvested but the stubble ploughed in, the same general argument applies but the numbers are different. In short, this is a very beneficial practice inasmuch as the soil is physically stabilised by the crops themselves which sustain or increase organic matter and raise the content of nitrogen in the soil.

If the third field is converted to pasture or if farm animals are brought in to graze on the stubble of a previous crop, the effect will depend on a number of factors of which time and timing are crucial. An arable plot obviously cannot make the change to pasture overnight; grass needs time to grow. So when the cattle or sheep are first moved onto the new pasture they will bring with them from their previous pasture a load of dung in the making. When that is deposited it will make a nett contribution to the nutrient status of this third field but, thereafter, the balance will remain unchanged if the total mass of the herd remains constant (an unlikely possibility) or nutrients will be lost if the mass of the herd increases (see Chapter 2). The dung will provide organic matter, important for soil structure and soil biota and, like urine, it will also supply nutrient elements in a form more rapidly available to plants than those contained in undegraded plant tissue (Chapter 2). It will also add water. The grazing will thus accelerate the conversion of plants or their remains to plant nutrients but, in all but exceptional circumstances, the grazers will extract some of those nutrients as a commission.

Then there is 'the fallow'. Broadly speaking, this takes one of two main forms. Either the plot is simply left to its own devices whereupon anything may grow—which is what happens in shifting agriculture—or the field is ploughed, kept free of weeds and the soil left bare. The benefits attributed to this practice are often lumped together under the description of 'resting the soil' but, with one exception, the specific mechanisms are vague. The benefits of the extended fallow or, more accurately, the period of regeneration in shifting agriculture are complex but, for the most part, easy to acknowledge (Chapter 2). The much shorter fallow used in a system of crop rotation is not so easily interpreted. To be sure, if the ground remains moist enough, residual organic matter will be at least partially degraded by the soil biota, yielding some of its nutrient elements in a soluble inorganic form. Although this process would occur anyway, the point here is that those elements are immediately available at a higher concentration than they would otherwise have been when the next crop is sown. And, if it had been a bare fallow, there will be fewer weeds to contend with when the soil is next sown. On the other hand, the loss of organic matter from the soil during a period of bare fallow has serious disadvantages (see Chapter 10).

It is often claimed that one of the benefits of a bare fallow is to help conserve water. The rationale is that living plants remove water from the soil by transpiration, the process of drawing water from the soil and releasing it to the atmosphere through the leaves; if there are no plants that cannot occur. Fallowing does indeed conserve water under some conditions, such as in soils at high latitudes with mild summers (as in Britain), but the effect varies with soil type, latitude, cloud cover, patterns of rainfall and, of course the type(s) of plants that would otherwise have grown in the field—in particular the depth of their roots. The value of a bare fallow is not always so obvious at lower latitudes where the intensity of solar radiation is high and where the surface of unshaded soils, especially those of dark colour, can become very hot and dry during the summer. Even though the deeper layers are insulated, evaporation from the surface will produce a gradient that causes diffusion of water to the surface. Moreover, bare soil is completely unprotected against wind or water erosion—the more so if it has been depleted of organic matter.[11]

Nevertheless the general practice of crop rotation is unquestionably beneficial in several ways. One level at which its benefits are perhaps not as widely acknowledged as they deserve lies in the effect of the practice on crop yields. In Chapter 4 attention was drawn to the need for recognition of time as a component in the expression of yield or, in other words, the recognition of yields as a rate. The point was made that high yields deplete the soil of nutrients faster than low yields and that, with some types of soil and climatic conditions, yields below a critical value can be sustained indefinitely without deliberate manuring. The significance of crop rotation in this context is that the practice reduces average yields. For example, a three-field system in which one field is always out of production for any of the reasons outlined will reduce the overall yield of the main crop(s) by 33%; a two-field rotation will reduce it by 50% relative to continuous cropping. In the absence of manuring this simple piece of arithmetic will extend the life of the soil nutrients by 33% or 50% over what could be expected from a system in which there was no fallow period.

At least since Roman times, crop rotation, fallowing, manuring with dung from farm animals and, albeit to a lesser extent, from their managers were practised fairly widely throughout most of the 'Old World'. Although the Romans seem to have been aware empirically of the value of legumes, these plants were not sown to any significant extent, at least in Britain, until the thirteenth century and then their use was unevenly distributed.[12] The agriculture of that time and those places was not of a generally high standard by many criteria, certainly not by economic criteria, but it did not obviously contain the seeds of its own destruction. It was able to keep pace with a growing population but, for the most part, it achieved this by bringing more land into production rather than by increasing the yield per area of land.

Yields of grain were often expressed as a relative measure, the ratio of seeds harvested to seeds sown. By that measure European grain yields supposedly did not change much from the time of the Romans to the Middle Ages, remaining around 3:1; modern yields in these units are 25-30:1.[13] There were rare occasions early in history when absolute yields were recorded (see Chapter 4) and by the Middle Ages there were numerous situations in which detailed records were kept. Mediæval European wheat yields were around 500 kg.ha^{-1}, which is about 12% of what can be achieved today in north-western Europe (see Table A4), but those low yields were not without their benefits. For example, Pretty[14] has analysed in some detail the records of 14 manors belonging to the Bishop of Winchester for the period 1283-1349. Mean gross yields (515, 530 and 755 kg.ha^{-1}) and yield/seed quotients (4.0, 2.3 and 3.5) respectively for wheat, oats and barley were low.[15] Inasmuch as the theme of Pretty's review was the sustainability of mediæval agriculture—the manorial system persisted for many centuries—these low yields are significant (see also Chapter 4). The productivity of livestock was also low; for example cattle provided only 550-685 litres of milk annually.[16]

Low yields can extend the life of a farm, but other factors are also needed and were evidently in place in these English manors. They produced a wide range of foods and other materials and they supplemented their produce from wild resources such as deer, boars, birds, fish, hares and rabbits and a variety of plant products of which acorns and berries were conspicuously important. There was thus reasonable insurance, at least at the level of sustenance, against failure of part of the system. They also used natural products such as the dung of wild animals and, in suitable locations, seaweed as manure.

The farming was very labour intensive, and manorial estates were sizeable communities which consumed most (about 98%—see below) of their produce.[17] If only 2% of the produce were removed for sale then the remaining 98% was redistributed in some way within the estate. Moreover, the nutrients removed for sale were replaced, at least to some extent, by the use of wild resources (see above). There was deliberate use of dung, both human and farm animal, as manure. This was commonly left in heaps for some time before redistribution—and was thus subject to some loss of nitrogen and other nutrients, mainly by leaching—but at least the leachate had to go somewhere on the farm.[18]

In other words, the manorial estate, although different in its social structure,[19] shared with European agrarian villages a relatively diverse agriculture and a high degree of self-sufficiency. But these generalisations should not be applied uniformly. In the manorial system, as in all others, soil quality varied from one location to another, as did the standard of the husbandry. Some farms were successful, others were not and had to be abandoned because of soil degradation. Soil impoverishment was a common experience in Roman Italy, especially on large estates owned by absentee landlords and worked by slaves,[20] and it almost certainly contributed to the decline in the population of the Roman Empire that began in the second century AD.

Increasing the total yield of farm produce, be it by extending the area of farm land or by increasing the yield per area, will support a bigger overall population, but it will not necessarily support a proportionately larger urban population. For that to happen there must be an increase in the proverbial surplus, and that requires an increase in the yield *per capita*. This, in its turn, means that the farm labourer must work harder and/or be provided with better tools (to increase yield per man-hour), or he/she must work longer hours. Working harder or using better tools will not automatically improve the yield per area of land (the 'specific' yield; it is more likely to increase the area that is worked. Increasing the specific yield requires a refinement of method or a change in the characteristics of the crop. The surplus can also be increased, of course, by attacking the other term in the equation, that is to say by reducing the amount of food consumed by the farm population. At different times and in different places, all three methods have been employed.[21]

Output *per capita* did vary widely from one place to another over the first millennium AD. For example, it has been estimated that in southern Scandinavia at the end of the first millennium BC, one man could plough a field of some 600 m² (0.06 ha) in about a day, amounting to 30-40 man-days to plough the 2-3 hectares needed to support a family.[22] On the other hand,

Langdon has based some calculations on an assumed "optimistic" ploughing rate of an acre (0.4 ha) a day in mediæval England.[23]

Slicher van Bath summarises technical developments in agriculture from the time of the Roman Empire to the 13th century AD thus:

"In the Roman era there were hamlets with square fields. On these fields a two-course rotation was followed; they were ploughed with a 'sliding' plough with no mouldboard, drawn by oxen. In about the sixth century AD, when the climate had already become much rainier, the people of western Europe went over from the prehistoric walled fields (a type of square field) to strip cultivation, by which much better drainage was secured. Still better results were achieved with the help of the wheel plough with fixed mouldboard, with which it was possible to plough the fields into a markedly ridged form. Now, for the first time, the boggy districts, such as river valleys, could be tilled. In the eighth century the peasants in the densely populated areas changed over to a free three-course rotation, which resulted in raising the production of food for human consumption. During the same period, the increase in population drove others to individual farming ventures in the form of square fields with scattered dwellings (enclosures).

The tenth century saw the introduction of improved harness, which enabled horses to be used for ploughing. Thanks to the three-course rotation, there was enough fodder to keep more horses. In all likelihood it was through this augmentation of the sources of energy that the ensuing rise in population from the eleventh to the thirteenth century was made possible."[24]

There are no reliable population records until the 17th century, but a number of reasonable guesses and interpolations point to a decline in European numbers from the second century AD, when the Roman Empire had about 45 million. Two centuries later the population was about 36 million and, by 600 AD, it had fallen by another 20% to 29 million.[25] (These numbers are slightly greater than those represented in Fig. 5.1 (below), but the trends are similar.) There were many causes of this drop, all of them interrelated—of course. There were social and political causes. Disease and a decline in agricultural productivity made their respective marks—as did attacks of increasing frequency and severity by equestrian nomadic tribes. Overall, the distribution of population between town and country within a wide area embracing Europe, much of South West Asia, Western Russia and parts of Northern Africa[25] point to a general farm surplus, early in the 8th century, of 2% or less. By late in the 15th century it had increased to something like 3-6% (see Appendix).

Reasons for this increase included various technical innovations, the use of horses as draught animals,[26] and changes in the management of farms.

Society was still overwhelmingly agrarian but farm surpluses had increased about threefold in eight centuries. Such a rate of change might not have set the world on fire. An annual surplus of 6% still left 94% of the produce more or less on the spot, its nutrients available for future crops and pasture—and some of the 6% surplus was deliberately returned to some farms as manure. But something was going on, something of considerable significance.

As I argued earlier, merely bringing more land into production (by clearing forest or by reducing the area under fallow) would not of itself increase the surplus. That achievement demanded increased output per worker. But, if claims that nett specific yields did not increase from the time of the Roman Empire to the Middle Ages are valid, then the increase in total production was achieved by extending the area farmed. The added surplus, such as it was, sprang more from increased productivity of the farm workers than of the soil.[27]

It is tempting to wonder what was the primary driving force (or the major limitation) in determining the pace of change in agriculture—technology, population pressure or attitudes of mind. But to ask the question in that way would be to confront yet another chicken-and-egg dilemma. The change from hunting and gathering to farming spawned a new type of ecosystem. Like any other, this one embarked on a process of succession in the course of which it became more complex. New 'species' appeared and exerted their influence on the dynamics of the system, but the overall feedback mechanism remained positive and everything accelerated, slowly at first and with intermittent setbacks, but inexorably. Some of the more obvious components of this system were—and are—the total size of the human population and its spatial distribution, human birth and death rates, general standards of health, the total area of agricultural land and its distribution between pasture and various types of crop, the yield of farm produce per area of land and per man-hour of labour, the opportunities for employment off the farm, the status of technology, the energy supply, the speed and carrying capacity of transport systems, the money supply and its distribution, the frequency and severity of war and, of course, attitudes of mind and social conventions.

One would be brave or foolish to nominate any of these factors as the primary cause of a change in population dynamics or of an increase in farm surplus. This is true even in apparently simple cases such as a sharp decline in population (the apparently simple events are often negative). For example, the drop in population, especially in Europe, in the 14th century (see Fig. 5.1) can be attributed directly to bubonic plague. But the disease reached pandemic proportions because of a number of factors—the size and distribution of the population, standards of nutrition and levels of immunity, the state of sanitation etc, which in turn can be attributed to the farm surplus, the availability of employment, the extent of understanding of the disease which in its turn depended on the passage of time, the range of human activities, the level of technological sophistication—and so on.

Moreover, systems like this do not develop smoothly. They cannot. They evolve in spurts and dribbles. There are time lags in the output of one component's reaching and becoming the input of another. Equally importantly, each process within the system has, in the language of reaction kinetics, its own characteristic rate constant. The growth of cities has more often than not exceeded their capacity to dispose of sewage. Opportunities for employment will fluctuate, increasing as new industries are born, decreasing as industries become mechanised and 'efficient'. The fluctuations can be dampened by (government) controls as, for example by limiting the growth of a city to a pace appropriate to its services—to what these days is called 'infrastructure'. But throughout most of history the basic process has been essentially *laissez faire* with some alarming results and very disturbing portents.[28]

The practical upshot of all this is that farm output and surpluses did increase, especially in Europe, the population grew despite some setbacks, the absolute size and proportional population of cities increased. With the passage of time England became the pace-setter of European agriculture with progress being affected by technological development, the pace of which was modest until the late 18th century—until the 'Industrial Revolution' in other words. By about that time the 'three-course rotation' had largely given way to a four-course embodying the use of clover and turnips.[29]

The changes in English agriculture from at least the beginning of the 18th century until the middle of the 19th were affected by social, political and economic factors among which the 'enclosures' loomed very large. This practice amounted to the progressive exclusion by fences or hedges of peasant farmers from land that had previously been open common. The reasons for the enclosures were partly managerial—for example four-course rotation was not feasible on common land. There were, needless to say, also one or two other motives.[30]

Whitlock has pointed out that some form of enclosure had been going on 'almost from time immemorial', but it was in the 18th century that the process received the blessing of Acts of Parliament.

"Between 1702 and 1762 246 private Enclosure Acts, involving some 400,000 acres, were passed by Parliament. With the return of the Tories to power, the flow became a flood. In the next forty years, from 1761 to 1801, no fewer than 2,000 Acts, involving more than 3 million acres, found their way onto the statute book. And from 1802 to 1844 nearly another 2,000 Acts legitimised the enclosure of a further 2,500,000 acres. The torrent culminated in a General Enclosure Act in 1845."[29]

The process dispossessed a large number of 'small' peasants who were then obliged to work as farm labourers on a wage that was not enough to feed them and their families (see Note 21). At the same time the farms on which they were employed became more commercial and more economically 'efficient'. The accelerating mechanisation of industry and of agriculture in the late 18th century saw both farm surpluses and the rate of migration to the cities increase.[31] Towns and cities held most of the English population by mid-19th century although, in absolute terms, there were more farmers and farm employees than at any previous time.[32] The type of calculation of surplus used in the Appendix is now more difficult because of the increased importance of trade. For example, by early in the second half of the 18th century, Britain had become a nett importer of wheat.[33] Nevertheless we can reasonably conclude that the surplus amounted to more than 50% of total production. Somewhere in the world farms were producing more wheat *per capita* than they had in the past. But all of this was still a long way short of the enormous surpluses of late twentieth century agribusiness.

Some characteristics of the English farming of that time are reflected in figures for land ownership for two Kentish villages, Dunkirk for the period 1827-1828 and Hernhill in 1840. In Dunkirk 84% of the population owned land of 4 ha or less, amounting in all to 4.6% of the farm land associated with the village. In Hernhill 44% of the population owned land of 4 ha or less, amounting to 3.6% of the village farm land. In 1841, 88% of Dunkirk's male population were either farmers (9% of the total) or employed by farmers. By 1851 the proportion had fallen to 73%, (12% farmers). The corresponding proportion for Hernhill in 1841 was 84% (15% farmers) and in 1851, 66% (15% farmers).[34] If, as is likely, the surrounding farms supplied most of the food consumed by the non-farming inhabitants of the two villages, the average surplus needed to supply the villages would have been of the order of 10-15% in 1841 and 30-35% in 1851. It is a big change in a short time, especially as it almost certainly underestimates the total increase in output, an output dominated more and more by the demands and financial inducements of feeding growing cities.

The growing farm surplus, industrialisation and urbanisation were influenced, not only by technological developments, but also by social factors and attitudes to money. The preachings of Martin Luther and John Calvin affected protestant views of the morality of making and hoarding money[35] and, in due course, Adam Smith discoursed on how to get it and what to do with it when you had it. Moreover European society was now becoming aware of science and its potential for application to technology—and hence to making money—in a way that was completely new. These factors, together with all the others that I have touched on, and doubtless many that I have not, combined to produce the Industrial Revolution, to increase farm surpluses and urbanisation—and to affect population dynamics (Fig. 5.1).

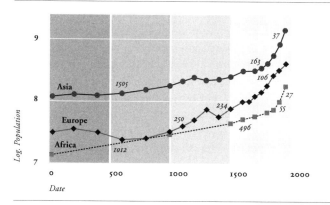

FIG. 5.1 *Semi-logarithmic plots of the human populations of Asia (circles), Europe (diamonds) and Africa (squares) for the period 0-1990 AD. Numbers above or adjoining the curves denote the doubling time (years) of the population for that part of the curve. Plotted from estimated populations as tabulated by Livi-Bacci (1992).*

The decline in the European population from 200 to 600 AD ensured that there was no overall increase of any significance in Europe for the whole of the first millennium. Asia, on the other hand, increased over that period with an average doubling time (t_d) of about 1500 years, or an average annual growth rate of about 0.05%. There is not enough information about Africa to allow comment about possible fluctuations in population during this period but, on the figures that have been proposed, the African population increased over that millennium at an average annual rate of about 0.07% (t_d = c 1010 years).

Both Asia and Europe were hit by a major pandemic of bubonic plague that began in Asia in the 13th century and entered Europe early in the 14th (see Chapter 6). The effect on the Asian population was less severe than on the European and, again, there is nothing in the available estimates to suggest that there was any significant effect on the dynamics of the African population. Both Asia and Europe experienced a second pandemic of bubonic plague early in the 17th century but this was much less devastating than the first, inasmuch as it suppressed but did not reverse growth of the two populations. The suppression lasted about 50 years after which both regions resumed growing at about the same rate as prevailed before the outbreak. In Asia this was equivalent to an annual increase of about 0.43% (t_d = 163 years) at which it remained constant for 200 years—until the beginning of the 20th century when it entered its explosive phase. Europe, on the other hand, recovered from the plague with an annual growth rate of about 0.28% (t_d = 250 years) which lasted for a century or less—and then came the sharp increase that coincided with the Industrial Revolution. For the next two centuries Europe grew at a remarkably constant rate of 0.65% annually (t_d = 106 years)—about 50% faster than Asia. Again, there was no discernible effect of this second outbreak of plague on the dynamics of the African population which continued to grow from 1500 to early in the 19th century at a steady rate of 0.14% a year (t_d = 496 years), after which it entered its explosive phase.

Growth of the European population was accompanied by urbanisation—slowly at first, rapidly from late in the 18th century. As Slicher van Bath has commented, the middle of the 19th century marked the end of a predominantly agricultural period in western European history. Thereafter the tune was called more and more by secondary industry, farm surpluses increased rapidly, migration to the cities and the colonies accelerated—and the problems of sewage disposal entered their explosive phase. Indeed, as the Fourth Law of Ecological Bloodymindedness says, "For every increment in the agricultural surplus there is a corresponding increment in the volume of urban sewage." The next chapter begins the process of illustrating some of the expressions of that law.

CHAPTER 6

The Sanitary Imperative

A major theme of this book—indeed it might be stated as an axiom—is that sedentary communities are necessarily confronted by a sanitary challenge merely by virtue of their sedentism. If that challenge is not met it will undermine health. The manner in which it is met will, in turn, determine very broadly the way in which the supporting agriculture is managed. The size of a community will profoundly affect the details of that determination, but it will not change the principle. (Communities largely dependent on aquatic —especially marine—sustenance can sometimes escape fortuitously from either or both of the consequences.) In this and the following chapter I have tried to illustrate some important features of the process with the major emphasis on the 'West'—Europe and some of its colonial derivatives—up to the end of the 19th century. It was in those regions that the phenomenon forced the final metamorphosis from what might be called traditional farming to modern 'agribusiness'.

The unit in the process can be one individual; a graphic example is a prisoner in solitary confinement. If he is fed, however poorly, within a day he will be obliged to defaecate[1] and urinate. If the cell has no toilet facilities, as has happened, it will quickly become foul; if the cell measures 3 x 2 metres and it takes, say, a quarter of a hectare of land to sustain the prisoner, the concentration factor (see Chapter 4) will be about 400. Since it is his own excrement that he is in the midst of the prisoner might not contract a new disease but he will quickly become demoralised. His jailers might not care about that but, depending on climate, location and season, the mess will probably attract rats and flies, and the prison warders might have some concern for their own health and the state of their working environment. The hazards caused by leaving the prisoner to his own devices under such conditions can be avoided only by specific action on somebody's part—either by installing a lavatory of some kind and washing facilities and/or by manually removing the excrement. The incentive for taking this action will almost certainly spring from a wish to avoid the sanitary and aesthetic problems of not doing so rather than from a perceived need to return the excrement as manure to the farm. And so it has been with all settlements—with individual farms, with villages and with cities, although from all of these there have been examples of returning some human excrement to farms.

Man, in common with all other animals—vertebrate and invertebrate—has been exposed to infectious diseases throughout the entire course of his evolution. These diseases have been a determinant of both human evolution and history, but it is probably fair to say that their impact on cultural evolution over the past 10,000 years has been far more dramatic than their contribution to biological evolution over the preceding several million years. We do not know the infectious diseases to which prehistoric Man was exposed but we can make some reasonable guesses. In principle, of course, hunter-gatherers would have encountered the same types of pathogens and parasites as their agrarian and urban descendants, when 'type' is used broadly to designate bacterial, viral, rickettsial, fungal, protozoal and metazoal agents. But the frequency of exposure, the dynamics of transfer and proliferation, and the predominant diseases were necessarily different from the patterns of infection and infestation in sedentary communities. Moreover then, as now, those patterns changed with latitude, climate, season and, of course, the genera of animals with which the human population came into contact. These infectious agents made their mark on human evolution by progressively eliminating hosts that were innately susceptible or which lacked the capacity to develop an effective immune response. It is no great advantage to a pathogen to kill its host—especially if it has difficulty in moving on from the corpse to the next victim.[2] Partly for that reason the selective process was not confined to the human hosts—it included the pathogens. What was evolving was not simply a species, but rather a subset of an ecosystem.

With agriculture and sedentism the goal posts were moved and a new set of players came onto the field. The archaeological evidence about the type and incidence of infectious disease in prehistoric communities, both nomadic and settled, is limited of course. But such evidence as there is, together with our present understanding of the nature of infections and infestations, leads firmly to the conclusion that the impact of infectious diseases became very much greater with the advent of sedentism. Densities of population and the time spent in one place were very important. This latter—quite literally the residence time—affected the likelihood of contact with sources of infection and, of course, the probability of establishing and augmenting pools of infectious agents. All of this can be incorporated into an ecological principle: when an environment is changed by any means, not only does the balance of plants and animals change, but so also does the balance of pathogens, both generally and for any one host species. In other words, if the nature of a habitat changes, any species—Man, kangaroo, salmon, tomato—can expect to encounter a different range and different balance of potential pathogens. (See the Appendix for a brief classification of major types of transmissable disease and infestation.)

People from all cultural periods, all life styles and all types of subsistence would have encountered pathogens and parasites, but the probability of contracting a disease varied widely—very widely—from one set of circumstances to another. As Cohen[3] has observed, the commonest group affecting nomadic bands may well have been the zoonoses. Within this group the specific pathogen would have been different according to habitat, but certainly there would have been a greater range to chose from in the wet tropics than, say, in dry temperate regions.

A nomadic band of hunter-gatherers, like any other species of wild animal that moves about in small groups, could be expected to suffer from mild chronic infections and infestations that made little serious impression on their general state of health. But from time to time, if the bands moved to a new area, if other human bands or mobs of other species came to theirs, or if a potential pathogen mutated to a virulent form, the nomads would encounter a new disease that would be seriously debilitating, would kill some of their number and leave the survivors with partial or complete immunity to that disease. The status quo would have been restored and the system would have evolved to a new level of complexity with an additional pathogen and a modified host.[4]

Probably the greatest threat to nomadic peoples came from the adventitious pathogens (Group 4, Appendix). Physical injuries were ever present and the chance of infection always high, especially from wounds suffered in fighting. It is about as certain as anything can be when speculating about the distant past that, for much of the time, infected wounds were a significant cause of the death of nomadic hunter-gatherers.

The comments made above about adaptation and immunity apply particularly to the specific human pathogens (Group 3, Appendix). In general, these agents could be expected to cause mild chronic infections interspersed from time to time with virulent outbursts. The means of transmitting a pathogen, the severity of the illness it caused, and the size of the host group would all be critical determinants of the epidemiology of these diseases and their impact on the human population. For example, airborne respiratory infections should have been much more significant for a group that spent part of the year in an enclosed space, such as a cave or huts, than for a small nomadic band in a mild or hot arid climate. In contrast, the incidence of sexually transmitted diseases would be largely indifferent to the physical environment but sensitive to the sexual mores and social customs of the community.

Gastrointestinal diseases should not have been a major problem for nomadic bands but could well have become troublesome with even brief interludes of sedentism. The likely incidence of these diseases would then be determined by the standards of sanitation and hygiene adopted by the community (see the quotation from Dawson in Chapter 3). But with farming and sedentism the dynamics of infections changed as did the relative importance of the various groups of infectious agents.

With the adoption of sedentism, Man established a new set of conditions that challenged his health in several ways. His nutrition was for the most part poorer than it had been in his hunter-gatherer days (see Chapter 1), he spent a substantial part of his time in buildings at densities conducive to the spread of airborne infections, and he had to live, in one way or another, with two important products of his daily activities—his excrement and his food scraps. It should be obvious by now that what he did with these commodities left more than a little to be desired. The accumulation of faeces in streets and back yards had the immediate effect of raising gastro-intestinal diseases from the minor status they had held in nomadic societies to a major threat to health and to life. Accumulated garbage attracted scavenging animals, particularly rodents;[5] their presence, together with the common practice of farmers and villagers of living in close contact with their domestic animals, introduced human communities to a wide range of new diseases (especially those of Groups 1-3, Appendix). There is no reason to suppose that the absolute importance of earlier infections was diminished—people still suffered injuries at work and, although a smaller proportion of them might have been mauled by wild animals, they now fought wars rather than short skirmishes. Moreover, the outcome of these wars was decided by pathogens perhaps as often as it was decided by force of arms. All in all, infectious disease had become a much greater threat to well-being and to survival than it could possibly have been for nomadic hunter-gatherers.[6] In more general terms, when Man became sedentary he created an environment very different from the one in which he had evolved. It was an environment with a completely new set of problems —and he did not understand them. Those problems were ecological as well as social and economic. Agriculture and sedentism created a social and economic hierarchy of a scale and complexity that was impossible in hunter-gatherer bands. The hierarchy included, of course, an impoverished class which experienced—and continues to experience—most of the destructive impact of sedentism on human society.

— Europe —

European life was moulded by disease from the time of the first settlement to late in the 19th century. In many ways it still is. To our knowledge, no disease has affected a human population as much as bubonic plague, in particular the European pandemic of the 14th century. It is unlikely that any disease has had such a profound effect on human behaviour and the human psyche as did the Black Death on Europeans. There had been several major outbreaks in the European arena before the 14th century. In the 6th century the Roman Empire suffered a pandemic of proportions comparable to that of the Black Death; according to some sources 100 million people died from the plague over a period of 50 years. But that was 800 years before the mediæval pandemic, nearly two centuries longer than the interval between the Black Death and the present.[7]

The 14th century pandemic apparently began in Asia and was brought to Europe first by traders along the Silk Road. The traders carried, among other things, marmot furs infested with the oriental rat flea, *Xenopsylla cheopis*, which carried the plague bacterium. After their initial overland journey the furs were transported by sea in rat-infested vessels from the Black Sea port of Kaffa to virtually all the ports in the Mediterranean. The disease spread overland from there through Europe to the British Isles and Scandinavia, and maintained pandemic proportions from 1347 to 1352. Both urban and rural populations were affected.[8]

The disease is an extreme example of a zoonosis (Group 1, Appendix). The causative agent is the bacterium, *Yersinia* (formerly *Pasteurella*) *pestis*. The normal host is a rodent but the disease can exist in the wild as a pathogen of any of several genera. Although the source of the 14th century pandemic was probably the marmot of central Asia, the pool of infection for its spread through Europe was the black rat, an animal that found itself very much at home in sedentary human communities and on ships. *Yersinia pestis* is normally spread throughout a rat population by its vector, the flea which, when it bites an infected rat, ingests some of the bacteria. These then multiply in the flea's digestive tract to the extent that they form an obstructive plug of cells. The unfortunate insect can no longer feed properly when this happens and eventually, more in desperation than disgust, it vomits the bacterial plug into the next host's bloodstream.

The disease is lethal for rats; when the fleas run out of their preferred hosts they look for somewhere else to go and something else to bite; under the circumstances of which I write, that usually meant *H. sapiens*. The bacteria spread through the lymphatic system but, in some cases, the bloodstream is directly infected. This produces a form of the disease known as septicaemic plague in which death occurs sooner than in the commoner form. Both forms of the disease are transmitted from person to person by the human flea or by direct contact into a skin lesion of some kind. A third manifestation of the disease, pneumonic plague, occurs when the bacterium invades the lungs from where it is transmitted through airborne droplets, sputum etc. Pneumonic plague has the highest mortality rate and the fastest course of development. Overall, the incubation time from infection to the onset of symptoms is 1-6 days.

The mediæval epidemic infected 25-30 million people out of a European population of 80 million. The mortality rate was 70-80%; in all, 20 million died in Europe over the six-year period.[9] The causes of bubonic plague, or for that matter other infectious diseases, were not understood until late in the 19th century. According to McEvedy,[8]

"During the period of the Black Death people were inclined to attribute the disease to unfavorable astrological combinations or malignant atmospheres ('miasmas'), neither of which could be translated into a public health program of any kind. More paranoid elaborations blamed the disease on deliberate contamination by witches, Moslems (an idea proposed by Christians), Christians (proposed by Moslems) or Jews (proposed by both groups)."

The theory that became dominant and remained so until late in the 19th century—at least in Europe and its (former) colonies—was the Miasma Theory which originated in classical times and enjoyed the support of Hippocrates in the 5th century BC, of Galen in the 16th century AD, and of William Harvey in the 17th century, to single out some of its better known adherents. In essence the theory attributed the spread of infection to some unspecified volatile substance(s), in short, to 'bad air'. The persistence of the name 'malaria' for a parasitic disease is a relic of that belief. Despite its assignment of responsibility to bad air, the theory allowed for person to person transmission and was responsible in due course for the adoption of quarantining.[10]

Asia had its feet more firmly on empirical ground with regard to the aetiology of bubonic plague. For example, an old Sanskrit text advised householders in Hindustan to clear out as soon as rats fell from the roof and died. A late 18th century Chinese poem, written during a bubonic epidemic by a poet who succumbed to the disease a few days later, contains the lines,

Few days following the death of rats
Men pass away like falling walls.[11]

There is merit in gaining some feeling for the sanitary impact of sedentary life, a feeling over and above a dispassionate acknowledgment of a problem, its dimensions and its underlying causes. Cipolla[12] has described the creation, between 1348 and 1700, by the states of northern and central Italy of "the most advanced system of public health and hygiene in Europe". This was achieved through a series of public health magistracies in the major cities. The term 'magistracy' is used because these bodies had their own jurisdiction, judicial system, police force and prisons. They were inspired, if that is the right word, by the Black Death. Their paramount riding instruction was to prevent further outbreaks of bubonic plague and, when that was not possible, to contain them. They were guided by the tenets of the Miasma Theory and thus their attention was focused largely on sources of bad smells, a preoccupation that has left some graphic accounts of sanitation in 17th century Italy. Yet there were detailed accounts from the early 17th century of streets effectively blocked by household waste and human excrement, of 'privies' discharging "into certain horrible backyards and open courtyards which look and smell so disgusting that this alone would be enough to bring on the plague when it is very hot."[13] Obviously not all of northern Italy was as bad as this suggests or these towns would not have occasioned such comment—but this was in one of the most hygienically advanced areas of 17th century Europe.

Italian agriculture suffered from a shortage of manure. Peasants from farms near the towns bought cartloads of human excrement from the people employed to empty the cesspits. In essence this type of activity was fairly common throughout Europe until the mid 19th century or later, but it was not particularly popular, at least in early 17th century northern Italy. The peasants wanted the 'solid stuff', not the 'watery stuff'. They spent 'a great deal of time' collecting and storing human excrement which was, as Cipolla puts it, "a raw material of primary importance". A pertinent comment on all of this comes from Cipolla himself. "The most pathetically tragic aspect of this business, however, was that of the people, whose poverty was so abject that they collected the manure they found in the streets where they kept it until they had accumulated a sufficient quantity to sell."[12]

Such descriptions should not be confined to Italy. Samuel Pepys commented adversely in his diary about the unpleasant consequences of having his cellar flooded with the contents of his neighbour's 'house of office'. But his wife, Elizabeth, also kept a diary in which she gives a more graphic account of hazards of city life in 17th century England. In an entry in February, 1661, she described going to a public bath house—and then her experience while walking home afterwards.

"Some discourse there is on how long the benefit of the bath should last and no certainty do I have of it. But once I did know the precise time that it did have effect. For on my way home did I get messed with a pot of turds which did descend with no warning, and not quick enough was I to keep from being covered with the filth. And no means for cleaning my clothes, or returning for another cleansing, though well did my state call for it. That night Samuel and I did talk of the shitty hazards of the town, and not just the turds that do fly, and the ordure that does pile high on each corner. But so much excrement each day does get produced and no ordered place to put it—which does offend Samuel.
... 'Tis said that some there are of high birth who do make over the special room for the taking of the turds, and too, that 'tis possible to flush all down with water and to make drains to carry all hence.* (* See note 14.) But too fantastic does this seem to me. ... And the turds and what to do with them do continue to try us and our ingenuity."[14]

During the three years of 1740-1742 European mortality rates were abnormally high. The reasons were complex and the immediate causes of death numerous, but the underlying influences were exceptionally cold winters, especially that of 1739/40, summer droughts, poor harvests, increased food prices, and migration of farm workers to the towns in search of work and food. Because of the severe and protracted winters, people spent more time indoors and thereby increased the transmission of airborne and louse-borne diseases. Post[15] has examined this episode in detail and has provided some revealing first hand descriptions. Until indicated otherwise, all the ensuing quotations are from citations by Post.

Ireland, in particular, suffered a severe 'demographic crisis'. A letter written in May 1741 to the Lord Primate over the *nom de plume* 'Publicola' stated that "Multitudes have perished and are daily perishing under hedges and ditches, some of fever, some of fluxes[16] and some from downright cruel want in the utmost agonies of despair." He went on to describe a "... helpless orphan exposed on the dunghill, and none to take him in for fear of infection."[17] A pamphlet published in Dublin in 1741 described "... roads spread with dead and dying bodies; mankind of the colour of the docks and nettles which they fed on." Another writer using the *nom de plume* of Triptolemus declared that "... whole thousands have perished, some of hunger, and others of disorders occasioned by unnatural, unwholesome and putrid diet."

On Post's assessment the majority of Irish deaths were caused by dysentery and typhus; the latter was the predominant louse-borne disease of the time and it affected all social classes. Death rates varied with locality, of course, but in many areas they matched or exceeded those in Europe during the Black Death. For example, the pastor of Cullen, in Limerick, reported that 38% of his parishioners had died.

The dungheap near the back door was a common feature of rural living throughout much of 18th century Europe. It was a splendid reservoir of enteric organisms and possibly viruses which contaminated the soil and drinking water, especially when this was drawn from a nearby well. It did not have to be so but it was—through ignorance and because the impact of a family on a small area around the cottage was too great for the natural recovery processes in soil to have any chance of dealing with it. The problem was exacerbated by the cold winters when microbial activity in the soil was brought virtually to a halt, but the inhabitants of the cottage continued to eat and to defaecate. The soil became saturated with excrement and the substances that were leached from it. This material provided a growth medium for enteric bacteria which, when the temperature was high enough to allow significant microbial activity, suppressed the growth of aerobic soil saprophytes by denying them oxygen. And of course, there was a continuous addition of new enteric organisms. It was indeed a vicious circle and it is small wonder that dysentery was endemic in much of 18th century rural Europe. This was a symptom solely of sedentism, it was not necessary to invoke urban living.

But what of the cities? It is probably fair to say that, at least until the mid-nineteenth century, all Western cities suffered from several components of a major sanitation problem. The drainage systems could not properly handle storm water since they were usually blocked by refuse. But heavy rain washed the filth from the streets into the most accessible water course which was quite often a source of drinking water. Household garbage accumulated in the streets; human excrement accumulated in cesspits and all too often in back yards, on the streets and in dumps. The water supplies were inadequate both in volume and in quality. All of these factors interacted in ways that made cities singularly unhealthy places, especially for the poor.

It was a common practice in Western cities of the nineteenth century to remove both garbage and 'nightsoil' at intervals that varied with the wealth and social status of the district. In London, for example, nightsoil was removed at daily intervals in the best districts, less often and less regularly in the poorer ones.[18] The nightsoil men and the garbage 'rakers' sold some of their stock for use as manure, some was burnt, some was dumped on land and some into the sea. There is no shortage of accounts of life and death in 19th century European cities. Nor was there any shortage of contemporary comment—bitter or satirical—in the form of cartoons—especially in the London weeklies—in paintings and, of course, in literature.[19]

On the 13th of February 1849, John Stephens, the editor and proprietor of the *South Australian Register* and the *Adelaide Observer*, delivered a public lecture in Adelaide on hygiene and sanitation. His purpose was to bring to public attention the unsatisfactory state of sanitation in Adelaide, then a mere thirteen years old, and warn the inhabitants what they would be in for as the town grew, if they did not mend their collective ways. The lecture attracted so much interest that he published it as a booklet[20] which he made available free to subscribers to the two papers. Much of what he had to say referred to conditions in Britain, since that was the road down which he saw Adelaide travelling if it did not take heed. Stephens described in some detail the state of London's streets, its overflowing privies, the blocked drains and sewers—and the stench.

The stench had more than an aesthetic impact. Within the context of the Miasma Theory it was the source of infectious disease. And, indirectly, so it was. By ensuring that the windows of small overcrowded houses were kept closed, it greatly increased the spread of airborne infections. And whether the windows were open or closed, crowded dwellings were a godsend to lice and fleas and the diseases they carried with them. Stephens was well aware of the horrors of overcrowding. He cited some population densities (Table 6.1) and went on to say:

> "In Liverpool, in houses not exceeding twelve feet square, with one bedroom and a low attic, there are often found from twenty to thirty persons huddled together. But those who will not listen to the living will not refuse the testimony of a voice speaking as from the grave. Mr John Johns, who died from fever caught in visiting the sick in Liverpool, says, 'In one small cellar, without a window, I saw eighteen persons lying on wet, dirty straw. In one house I counted eighty-one, in another sixty-one—in every stage of fever—on filthy straw in the corners.'"

and, reporting the observations of the Rev. J. H. Woodward on a 'sanitary visit' to his parishioners in Bristol,

> "Another house was tenanted by five families with eleven children; two married couples slept in one of the rooms; five of the children were ill of the small-pox, and in the midst of these a quantity of fruit was prepared for sale in the public streets."

These are some of the consequences of system of commercial agriculture.

A dispassionate summary of an effect of occupancy levels in 19th century Britain is shown in Fig. 6.1.

TABLE 6.1 *Some 19th century very high urban densities*

	Density (individuals.km⁻²)	References
England around 1840		
Bethnal Green	23,000	(a)
London, 'Part' of	94,000	
Liverpool, 'Part' of	128,000	
Another part	254,000	
Various, 1894		
New York (10th Ward)	130,900	(b)
Prague (Josefstadt)	121,350	
Paris (Bourse)	108,550	
Bethnal Green	91,325	
A modern city state		
Singapore	4,587	(c)

Sources: (a) Stephens (1849); (b) Lampard (1973); (c) World Resources Institute (1994). Compare these densities with those in Table 1.1.

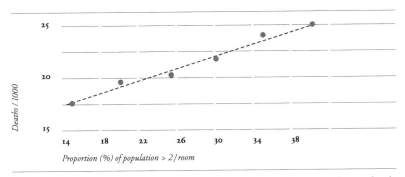

FIG. 6.1 *London's Death rates from all causes as a function of housing density for the period 1855-1892. The numbers on the horizontal axis represent the proportion (%) of the population with more than two persons per room living in tenements of less than five rooms. They are plotted in groups of 5% denoted by the top of each percentage range. The point corresponding to 15% represent 15% and below; the point corresponding to 40% represents all rates above 35%. Plotted from statistics listed by Wohl (1973).*

Stephens turned to the matter of excrement and what was done about its disposal. The "parochial medical officer" declares that he has not seen and does not know of a single water closet in the whole of Bethnal Green. The parish of St Mary's in Plymouth has an open shed used "as a privy by both sexes, and all ages, ... and there is neither drain nor sewer". "Dwellings of the lower classes in Nottingham" have no private lavatories at all. "In Manchester there is only one privy, on an average, to 250 persons."

Nineteenth century Britain had not improved much, if at all, over 17th century Italy, a point that has been made by Cipolla, yet in the late 19th century Britain had the best system of public health and one of the lowest death rates in Europe. Stephens reminded his audience that throughout Adelaide there were localities where "outbuildings" were shared by several houses, and that in some parts "there are numbers of houses without privies, or any place of deposit for their refuse". The latter were mostly rented houses.[21] Stephens acknowledged the wisdom of "the Rev. R. Parsons of Ebley, Stroud, (who) said, not long ago, that the scavenger who swept the street and cleared the drain did more for the public health than the most skilful physicians."

TABLE 6.2a *Mid 19th century death statstics from selected regions of Great Britain and some 'Western' countries*

Region and social class	'Average age at death'
London:	
'gentry' and 'professional persons'	44
'sons and daughters of toil'	22
'Artisans':	
Parts of London;	16
Parts of Liverpool and Manchester	<16
Liverpool, 1847	
'gentry'	43
'shopkeepers'	19
'artisans'	15

	Life expectancy at birth	
	(1800)	(1850)
England	37.3	40.0
France	33.9	39.8
Sweden	36.5	43.3
USA (white population)	-	41.7

Compiled from Stephens (1849) and Livi-Bacci (1992).

English births and deaths had been officially registered since about 1839; I have summarised in Table 6.2 some of the Registrar-General's mortality statistics as given by Stephens. More generally, the average English annual death rate from 1600-1800 hovered around 28/1000 with intermittent peaks up to about 48/1000. By the mid 19th century it was down to about 23/1000.[22]

TABLE 6.2b *Child mortality*

Proportion (%) of children dying before the age of 5 years (average for a 16-year period)	
Birmingham	48
Glasgow	43
Hull	'nearly' 50
Leeds	48
Liverpool	53
Manchester	51
London	41
England	26

Compiled from statistics cited by Stephens (1849). Stephens himself died at the age of forty four.

— The Antipodes —

For an exercise such as mine, Australian cities and their supporting agriculture have the advantage of a short but well documented history. The processes that I am seeking to identify were compressed into a small time scale and proceeded very rapidly indeed by the standards of the 'Old World'. By 1850, the year after Stephens's lecture, the settlement of Sydney was 62 years old and had a population of about 40,000. At that time the European population of the colony was increasing with a doubling time of about 7.6 years.[23] During the first half of the 19th century, Sydney's population increased from about 2,500 to 44,000; by 1881 it had reached 223,000 (including the suburbs). The average exponential growth rate for the period 1800 to 1881 was 5.5%, corresponding to a mean doubling time of 12.6 years. The corresponding rates for London and Manchester were 1.8% (t_d = 38.5 y) and 2.44% (t_d = 28.4 y) respectively (see Fig. 6.2). Sydney's growth rate over that period was so high that, regardless of the level of understanding of sanitation or the standards of sanitary engineering, it was impossible for the municipal provision of such amenities as clean water and the removal of wastes (excrement and household garbage) to keep pace with the population. It is scarcely surprising that Sydney and other Australian cities—especially Melbourne[24]—had their share of enteric diseases that, in due course, imposed their own demands on the cities' administrators and on the populace at large.

Sydney's attitudes to sanitation were inevitably influenced by events in Britain. For example, the concept of 'public nuisance', under the pervasive influence of the Miasma Theory, was defined in Britain in 1846 to include any "premises, pool, ditch, gutter, water course, privy, urinal, cesspool, drain, ashpit, animal accumulation or deposit so left as to be a nuisance or injurious to health"[25] and was duly transported to New South Wales. An outbreak of cholera in Britain in 1831 prompted the New South Wales Quarantine Act of 1832. The convict ship Hashemy reached Port Jackson in mid 1849, having lost sixteen convicts to cholera en route. Its arrival not only united the NSW community against the practice of transportation, but it changed the perception, if only of the middle classes, of excreta and garbage from that of a nuisance to a serious threat to health.[26]

According to the census of 1851, Sydney had a population of 44,240; there were 8583 houses in the city of which 89.9% were occupied with an average number of 5.7 individuals per dwelling. That number, however, says nothing about the size of the house. Many working class homes reportedly had but two rooms that were sometimes occupied by a family and lodgers. By 1856, with the city's population at 53,258 (+20%), there were 9605 houses (+11.9%) of which 92% were occupied by an average number of 6.0 individuals per house.

By 1850, about 27% of Sydney's houses were served by piped water; the other 73% obtained their water from public street pumps, from wells or from private vendors. There were some water closets in 1851 but, as Coward puts it, they were 'very rare' and to be found only in the mansions of the wealthy, and some of the best hotels and private boarding houses. Not all the houses with piped water had water closets. Houses without piped water, three quarters of the total, also had no drains. The water closets, many cesspools and streets drained into the Tank Stream which had been the colony's first source of drinking water. By mid century it was the principal sewer.

Until about 1880, the commonest type of lavatory was the cesspool.[27] This was an open hole that might or might not have been lined with brick or stone. A seat was mounted over it and covered by a shelter of some kind—a rude shelter would not be an inappropriate description. There was commonly one such amenity to a group of houses. They overflowed from time to time, especially after rain. They were infrequently and irregularly emptied—or partly emptied—by professional 'nightsoil' men. An official enquiry of 1852 estimated there to be about 6000 cesspits in Sydney at the time and assumed that, on average, they were emptied about once every three years.

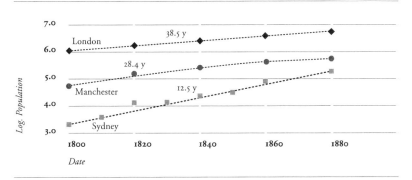

FIG. 6.2 *Semi-logarithmic plots of population growth of London (diamonds), Manchester (circles) and Sydney (squares) for the period 1800-1880. The numbers above each plot are the doubling times of the relevant population. In the case of Manchester the number refers to the period 1800-1840. Plotted from statistics tabulated by Coward (1988).*

Sydney's health, or at least that of the white population, was generally better in mid-19th century than in contemporary European cities, and much better than European health during the previous century. Its advantages stemmed more from the good luck of a relatively low population density and a mild climate than from good management.[28] The NSW Registrar-General began compiling vital statistics in 1856; for the period 1857-1880 the average death rate in the City of Sydney was 25.15 ± 1.08 per 1000 of population. The corresponding rate for the suburbs was slightly less and, for rural NSW over the same period it was 13.98 ± 0.70. For all NSW over that same period, children under the age of five accounted for 44% of all deaths.[29] (The 1992 NSW mortality rate was 7.49 per 1000 and, for all Australia, 7.04 per 1000.[30]) The health of the early European settlements in Australia has been discussed at some length[31] and I shall not attempt even to summarise what has been said on that topic. My immediate concern is the light that mortality statistics shed on standards of sanitation at the time.

An urban death rate that is high by modern standards and, moreover, nearly twice that of the country suggests strongly that infectious diseases arising directly from crowding, poor ventilation and poor sanitation were having a major impact. The high infant mortality carries the added implication that personal hygiene was bad. Into the bargain, deaths from typhoid fever and diarrhoea were high. For example, for the 5-year period to 1885, 'typhoid' and 'diarrhoea' accounted for about 11% of all metropolitan deaths for the same period.[32] All of this was duly recognised at the time, albeit largely within the framework of the Miasma Theory. The theory was wrong, but it justified action that at its heart was a belated commonsense response to a difficult problem of Man's own making—and by the criteria of mortality statistics and the aesthetics of urban living, the response worked. Campaigns to get rid of cesspits began in the 1870s and, from 1885-1915, death rates declined both in the City of Sydney and in rural NSW. By 1915 metropolitan and rural death rates were almost equal, being 10.90 and 10.31 per 1000 for city and country respectively (Fig. 6.3). Infant mortality (under 5 years of age) declined over the same period at a rate corresponding to 0.5% (of total deaths)/y.[33]

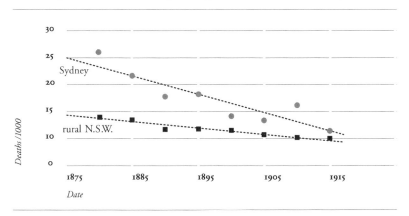

FIG 6.3 *Crude mortality rates for Sydney (excluding suburbs) and rural NSW for the period 1880-1915. The elimination of cesspits and the installation of a comprehensive sewerage system in Sydney began seriously in about 1880. Circles, city; squares, country. Plotted from statistics tabulated by Coward (1988).*

The pressure to improve sanitation came mainly from the middle classes who saw their own health threatened by the living conditions of the working classes. So what was done? The response was to dump the stuff—and that meant moving the problem—'somewhere else'. The offending materials of course were mainly human excrement and household garbage, but they were by no means all. By the time Sydney became serious about improving its sanitation there were various industries in and around the city whose wastes were offensive, even if they were not necessarily dangerous. They included abattoirs, small butchers, and various industries engaged in converting animal carcasses into marketable commodities such as leather, tallow and fertiliser. There were others that were not necessarily offensive but, as it turned out, were dangerous. These included small dairies that were the source of many an outbreak of typhoid fever—especially when, with entrepreneurial zeal, they used contaminated water to dilute the milk they sold.

One of the imperatives in (eventually) establishing an extensive functional sewerage system was a copious and reliable supply of running water. That in itself was a major engineering undertaking but, among its more obvious social and environmental impacts, must be included the effect of an immediately accessible water supply on personal cleanliness[34] and on public expectations. Those expectations, together with growing urban populations, progressively locked cities and their water supplies into another positive feedback loop. But, before proper sewerage systems (Chapter 8) were installed, another facility was established. This was the ubiquitous removable can, the dunny-can, the can that was taken from the back of an outhouse through a trap-door—sometimes when the said outhouse was in use, much to the alarm of the user. This was the dunny that has been the inspiration of many a joke, many a cartoon and many a nest of redback spiders. It was a big improvement on the cesspool but it was not without its problems, including the basic question of where to put the stuff when it was removed. There were designated areas, of course, but 'night cart' operators were not always the most fastidious or most scrupulous of businessmen, and were not above dumping it on convenient vacant lots if that saved time. This was apt to incur the displeasure of nearby residents. Some was bought by Chinese market gardeners who, true to their traditions, saw the value of the commodity as fertiliser. Eventually they were denied that option by government regulations. Some was taken to sea for dumping, but the residents (usually well off) of those districts that had a suitable wharf or jetty objected to such traffic near their homes.[35] Superimposed on all of this, the geographical growth of the city increased the practical difficulties of finding somewhere to dump the nightsoil within the time available (see Chapter 7). But the urban sanitary problems of Australasia, Europe and North America were largely overcome, at least for the time being, through various measures that were dependent on government legislation. Where there had been good will under the earlier *laissez faire* conditions, it had not been either extensive or powerful enough to overcome the prevailing financial interests and social inertia.

— The 'East' —

In the late 1930s a survey of agricultural and sanitary practices in Northern China was undertaken by a team of Americans. The survey was subsequently reported in some detail by Scott[36] from whom the following outline is derived.

The annual death rate for the whole of China at the time was about 30/1000. Life expectancy at birth was 34 years compared with 60 in contemporary Britain and 61 in the USA. Less than one third of the population reached the age of 60 compared with two thirds in Britain and the USA. About 42% of all deaths occurred in children under the age of 10. Scott comments that these statistics refer to years when there was no major catastrophe, that, in normal times, the health and life expectancy of the peasants was worse than the numbers suggest and much worse when there was a crop failure or shortage. In Tinghsien (about 200 km southwest of Beijing) gastro-intestinal diseases accounted for 18.5% of deaths. In addition chronic debilitating faecal-borne infections and, in particular helminth infestations, were widespread.

The economic significance of human excrement was well recognised by Scott and by earlier authors whom he cites. Indeed, Scott's account is replete with information and comment that is not only sobering but of enormous potential for reviving a flagging dinner-time conversation. To wit:

"Faust reckoned the faecal output per person per year to be worth about half a crown to nine shillings. The output of the entire population would then be worth between 50 and 80 million pounds sterling annually at 1924 market prices. Reisner reckoned that the manurial output of 400 million Chinese per year would be: nitrogen 1,800,000,000 lb; phosphoric acid 570,000,000 lb; and potash 685,000,000 lb. The cost of these, if bought as commercial fertilizers, would be over 100 million pounds sterling."

The area studied by Scott and his colleagues was the district of Lungshan, about 37 km east of Tsinan. Within 8 km of Lungshan there were 139 villages with a total population of 95,000 giving an average density for the area of about 470.km^{-2}. In contrast to the situation in 19th century NSW or Europe, the health of Chinese peasants in the 1930s was worse than that of townspeople. The incidence of gastrointestinal diseases was higher on the farms than in the towns, in part because the farms both recycled their own excrement and imported more (often as 'faeces cakes') from the towns. The average size of a farm in North China at that time was about 2.3 ha; it supported a family averaging 6.5 members. There was thus a fairly heavy concentration of people whose output was applied over many generations to farmland, and it was augmented by the urban faeces. The logistical difficulties of disposing of this material in a way that was both hygienic and able to meet the needs of agriculture can be appreciated from some of Scott's quantitative observations. They are in the same vein as the previous quotation and have similar conversational potential.

The average daily size of a Northern Chinese stool was found to be about 250 g wet (about 60 g dry). The farm population of the region was 124 million—good for a daily output of 31,000 tonnes faeces. The corresponding yield from the 26 million urban residents was 6,500 tonnes.[37] The region as a whole was thus faced with the problem of disposing of 37,500 tonnes of faeces daily. Given the commitment to using it as fertiliser, it is small wonder that gastro-intestinal diseases were endemic throughout the region.[38] Logistical difficulties also predictably affected the distribution of manure; farms close to towns received more than did those further away. The effect of this was readily apparent from the relative sizes and general health of crops; it was another manifestation of the 'sheep camp phenomenon' (Fig. 1.2).

Although the use of human excrement as fertiliser underpinned the high incidence of gastro-intestinal and related diseases in North China at that time, those diseases could have been largely avoided with a better understanding of their nature and modes of transmission. The lack of such understanding is graphically illustrated in the following passage from Scott.

"The young children in many, and possibly most, cases defaecate wherever they happen to be playing, in the streets, on the threshing floors or in the courtyards. Their trousers are usually made with an open seat, so they can do this without trouble to themselves or anyone else. ... In the heavily infested families[38] 87 per cent reported that the children habitually defaecated in the courtyards, and so also did 68 per cent of the lightly infested families. Some even reported that children defaecated on the room floors. While such stools may have been swept up and put in the pit-latrine, this did not happen in all cases and, even where an attempt to clean them up is made, there is usually a small portion left behind with its quota of eggs. Very often the children's stools are eaten by dogs or fowls ... (who) not only track out the faecal material on their feet, but also pass, in their own excreta, ascaris eggs which they have eaten."

During the first half of the 20th century, cholera, bubonic plague, typhoid and paratyphoid fevers, dysentery and hookworm were among the major infectious diseases of India. Hookworm was found in 40-70% of the population in various regions. India had some modern sewage treatment facilities, but they could deal with only a small part of the total Indian output. In 'French Indo-China' there was extensive use of nightsoil as fertiliser. Epidemics of cholera were frequent. An examination of 1250 individuals found 71% to be infested with *Ascaris*, 77% with *Trichuris* (whipworm) and 50% with hookworm. Korea also made extensive use of nightsoil in its agriculture. During the Japanese occupation (until 1945) cholera was kept in check but dysentery and enteric fevers were common. In Korea and Japan, as well as China, fluke diseases were common because of the practice of eating raw fish grown in ponds fertilised with human excrement.

I have probably said enough to make the point that in China, as well as some of its neighbours, public health in the early—mid 20th century was worse than Sydney's in mid 19th century and comparable with that of Europe in the preceding centuries. The unfortunate aspect of the Chinese situation is that it arose from a very old empirical recognition that, if farming were to persist, the nutrients removed from the soil in food must be returned to it. In that they had been largely, but not wholly successful, but they did not understand the complexity of the ecosystem with which they were dealing and were themselves a part. They did not understand that pathogens and parasites are a normal part of any ecosystem; if conditions are changed to favour the parasites, then their hosts can expect trouble. Moreover, the dimensions and characteristics of the whole system were changed profoundly by the inexorable increase in population densities.

The approach adopted by the American team in response to their survey began with an acknowledgement of the importance of human faeces in Chinese agriculture and the economy. They explored methods of treating the excrement that would diminish its threat to public health while retaining its value as fertiliser. They concentrated on composting techniques and emphasised that the element most at risk in Chinese farming was nitrogen—since that was easily lost either as ammonia or through denitrification. Composting methods were modified to take that into account while simultaneously producing temperatures high enough to destroy ascaris eggs and some pathogenic bacteria. The project was interrupted by the Japanese invasion of China. It was a common sense approach that, had it been widely implemented, would have gradually reduced the incidence of faecal-borne diseases, and might have preserved a system of agriculture that had demonstrated its staying power (see also Chapter 7). It contrasts with that of mid-19th century Sydney—and Europe—which turned to sewerage in a long overdue response to the sanitary problems of sedentism. By the criterion of public health that approach worked well (*e.g.* Fig. 6.3)—but there were unintended consequences. There was a sting in the tail.

The Road to the Sea

— The Old World —

For much of its history, western European agriculture has been sustained in part by the application of manure to arable land. As I emphasised in Chapter 2, when that manure comprised the dung of farm animals—as most of it did—its application amounted to the transfer of nutrients from one farm or one part of a farm to another.[1] The produce that was sold and transported to a city, be it cereals, vegetables, fruit, eggs, milk, wool[2] or livestock, removed nutrient elements from the farm. If the farm were to survive in the long term, those nutrients had to be replaced. Until about the middle of the 19th century, replacement was assisted by the recovery of urban waste—human excrement, the dung of city horses and sometimes city garbage.

Probably nowhere in Europe was this done more diligently than in Flanders where the practice was well established by the Middle Ages and where the sale of such commodities was itself an industry of some significance. By the 16th century about 60% of a Flemish farmer's expenses went to the purchase and application of manure. The high rents paid in kind for 16th century farms close to Antwerp have suggested that the productivity of those farms was sustained by their relatively easy access to Antwerp's dung. There is a reference in a lease contract of 1574 to the use of nightsoil on a farm. In early 17th century Holland, magistrates in Groningen directed that farmers in nearby peat-marsh settlements must manure ploughed peat-marsh with city refuse. By the 18th century Flanders had an extensive trade in manure that by now included city refuse, pigeon dung, oil cakes, ash, and the mud dredged from canals. By the 19th century there were "great stores of manure on the Schelde between St Amand and Baasrode … whence the excrement from Dutch towns was transported by barge". Slicher van Bath comments on the surprise of foreign visitors "at the care with which the Flemish collected manure", an activity that, among other things, benefitted urban sanitation.[3]

Europe was using well tried methods but problems were arising. By mid 19th century western Europe had about twice as many people living in towns and cities as on farms.[4] With increases in both the absolute and relative size of cities, the practical difficulties of collecting nightsoil and returning it to farmland increased. I am not aware of any reliable statistics about this problem, but it must have been the case that, as the cities grew, the proportion of urban excrement that was recycled diminished. This was not simply a function of the amount of waste. When the waste is human excrement that has to be collected from the streets, from tubs or from cesspits, and then transported through the city by horse and cart, some of the logistical aspects of the problem become more apparent. The bigger the city the further the waste must be carted to a farm or whatever dumping ground was used, and the longer that took. The bigger the city the smaller is its perimeter relative to its area (for roughly constant shape) and to its population. The dumping area is unlikely to increase in proportion to urban population. It was inevitable that, as cities grew, the use of nightsoil as manure became progressively restricted to nearby farms.[5] The sanitary corollary (see Chapter 6) had an important economic component. Until at least the middle of the 19th century and the introduction of effective sewerage systems, most of Europe—and especially England—was an uncompromising capitalist society. Getting rid of human dung was a commercial operation rather than a public responsibility. The cesspits of the rich were emptied much more often than those of the poor.[6]

When there are large losses of an element during storage, handling and transport, as is conspicuously the case with nitrogen, the proportion of urban waste recycled has an effect on soil impoverishment that is small in comparison with that of the size, both relative and absolute, of cities (see also Chapter 14). For example, annual losses of soil nitrogen, attributable simply to the consumption of wheat in European cities early in the 8th century, were probably of the order of 3 kg.ha^{-1}, whether the amount of urban nightsoil collected and returned to cropland amounted to 10% or 50% of the total performance. By the middle of the 19th century with its much bigger population and higher proportion of urban dwellers, the amount of nitrogen lost was likely to have been around 16 kg.ha^{-1}—again for 10-50% recovery. To put it another way: with the same set of assumptions, about 60% of the nitrogen contained in wheat grain was lost in the 8th century through urban consumption and processes associated with it. By mid 19th century that proportion had increased to about 75%, and it did not matter much whether 10% or 50% of the cities' wastes were recycled.

On the other hand, much less phosphorus is likely to be lost in passing, so to speak; the proportions not returned to the soil would have been only around 7% in the 8th and 30-40% in 19th century. But the increase in the proportion of phosphorus lost as time passed and the cities grew was much greater (20-30-fold) than the corresponding increase in the loss of nitrogen (about 6-fold). (See Table A5 and Fig. 14.1.) Even with the uncertainties of the calculations, the specific rates of loss of both nitrogen and phosphorus caused by eating wheat in the 8th century were evidently so low that they could have been balanced indefinitely by 'natural' replacement.[7] The situation in the 19th century was different and would have been affected by, among other things, the proximity of farms to industries and the transfer through the atmosphere of nitrogen oxides produced in combustion (see Chapter 8). The situation in mid 19th century Europe was by no means desperate. Crop rotation was still practised—usually 3- or 4-course rotation—and there was scope for growing legumes. But there was writing on the wall, and it was ominous.

By then the European population was growing with a doubling time of about a century—faster than at any previous time in history (Fig. 5.1). Cities were growing much faster than that. We have seen that over the space of forty years from 1790 the ratio of urban:rural workers in Britain changed from 1:2 to 2:1 (Chapter 5). For most of the 19th century European cities were growing with doubling times of the order of 30-40 years and, in the 'New World', a good deal faster (*e.g.* Fig. 6.2). In the face of such dynamics, the nutrient deficits just discussed would inevitably have 'blown out'—to borrow a phrase from present-day economists—had they been left to their own devices. But in addition to the problems inherent in the population dynamics, the 19th century saw technical developments that complicated the situation even further. Worsening urban sanitation (Chapter 6), already appalling, demanded that something effective be done about getting rid of the stuff. That something was the construction of sewers flushed by a reticulated water supply. Like all new technologies, this one, even though some of its essentials had been used in Rome a couple of thousand years earlier, had a range of consequences, some of which were entirely unintended and unforeseen. The sewers at first simply transferred raw sewage somewhere else—nearly always a river or the sea. Urban water consumption increased sharply, sanitation improved, death rates fell (*e.g.* Fig. 6.3), and water pollution took on a new set of characteristics that concealed some of its problems for another century.

Like any new industry, hydraulic engineering demanded skills, employed people, stimulated the economy, attracted more people to the city—people who would themselves need food, water—and sewerage. The process added to the complexity of urban society and generated yet another positive feedback loop. But the sewage was no longer available to any significant extent as manure, not for reasons of theory or principle, but rather for reasons of convenience and cost. Indeed, from the time the first modern sewers were built there have been some 'sewage farms' and well-intentioned attempts to use sewage as fertiliser. Most of these attempts were beaten by costs and by the NIMBY principle. Today they are also beaten to a large extent by the chemical toxicity of metropolitan sewage.[8] Sewers consolidated Man's growing ecological role as an agent for transferring nutrient elements from the soil to the sea.

The 19th and early 20th centuries brought a potential crisis for European agriculture—indeed for Western agriculture generally. But farming did not collapse through soil impoverishment because other sources were found for essential elements lost in sewage. The specific need for phosphate fertilisers had been recognised empirically by late in the 18th century when bones were crushed and spread as a manure,[9] but this was simply an extension of the principle of using the dung of farm animals. It was taking something that had been removed from the farm and returning it—and it was a drop in the ocean of what was needed. The response of the Western world was to find other sources of phosphorous—to fertilise farms with phosphorus obtained from the ubiquitous 'somewhere else'. Guano was used as such a source early in the 19th century, as was 'basic slag', a by-product of the steel industry.[10] Mineral phosphate deposits were mined systematically from mid-century. At the end of the century the German chemist, Fritz Haber, devised a method of synthesising ammonia from nitrogen and hydrogen for which achievement he was awarded the Nobel Prize for Chemistry in 1908. It is almost impossible to overestimate the importance of these technical developments. Without them neither the world's human population nor those of its farm animals could ever have reached their present sizes, nor could the distribution of population between town and country have produced the enormous disparity in 'favour' of cities that now prevails throughout the industrial world. Soil impoverishment would have seen the agricultures of industrial western countries reduced to shadows of their former selves. Whether or not the avoidance of that disaster is seen as a cause for rejoicing, for a sigh of relief or for some less favourable response will depend in part on one's perception of time scales. But whatever one's perception, it is difficult to deny that the treatment of the symptoms brought with it a daunting range of unintended consequences.

China has a greater area of mountainous and highland terrain and proportionately less level land suitable for agriculture than any other geographically large nation. It has roughly the same total area as the United States but only about 11% of Chinese land is arable compared with about 30% in the USA.[11] Until the early historic period northern China was moist and densely wooded; what is now the Gobi desert was steppe or woodland. Clearing began early and continued until most of northern China was deforested and semi-arid.

At the end of the 19th century China had a population of about 420 million distributed over an area of some 9.3 million km², a national average population density of 45.km⁻² or, taking the proportion given above, slightly more than 4.ha⁻¹ arable land.[12] It was an agrarian nation and, although there were many cities of which some were very large by 19th century standards, there are some conflicting reports about the actual size of the largest Chinese cities and the extent of Chinese urbanisation.[13]

By the end of the century the average size of a Chinese farm was about 1 hectare.[14] Citing simple averages in this type of situation is always an oversimplification (see Chapter 13) but, in anticipation of the next section, a few individual examples should be sufficient to emphasise the enormous contrast between China and Australia at the time.

"A £10 licence might secure the lease of 1,000 acres for a year, or, in the case of a specific grant, might bring the Australian Agricultural Company one million acres in 1824"[15]: "... in mid-1844 Boyd was in possession of 14 Monaro runs totalling approximately a quarter of a million acres; for these 14 runs he was paying only four licence fees—a total of £40 per annum."[16] Sheer size was itself sufficient to prevent systematic manuring of such farms.

Traditional Chinese farming has probably supported more people at a higher density for longer than any other system of agriculture. But it was not 'sustainable'. In some regions and in some situations it made its contribution to land degradation. In late imperial times, for example, upland farming—unlike lowland paddy farming was "commonly quite reckless about mistreatment of the soils".[17] But, overall, it inevitably encountered the many effects and ramifications of population pressure, or more fundamentally, the positive feedback interaction between food supply and population (Chapter 1). In China's case the collapse of traditional agriculture took a course slightly different from and slightly slower than Europe's, but the principles were the same and the pressures were similar. Those pressures showed themselves in conflicting demands for commodities that would formerly have been used in manures, in soil erosion following deforestation and increased intensity of cultivation and, as in Europe, in the logistical difficulties of using as manure the waste of growing cities. By the middle of the 20th century the fate of traditional farming was largely sealed by the industrial aspirations, the stupidities and, in the case of the Great Leap Forward, by the madness of Mao and his governments.[18] But in the late 19th century the traditional system was still largely intact, although it was under strain.

By the end of the 19th century there was widespread leaching of nutrients from soils in the south and parts of central China. Erosion from hillside farms had helped sustain those in the valleys as had flooding on the plains. But, despite that, "hundreds of square miles" of previously cultivated land had been abandoned because of erosion.[19] Nevertheless, traditional farming was still productive in most of the formerly arable land. The essential characteristics that made this possible were, briefly, a clear recognition of the value, both agricultural and economic, of everything that could conceivably be used as manure, a careful relation of tilling and cultivation to topography and to water supply, intercropping and multiple cropping—and hard monotonous physical work for long hours.

China's system of canals had its genesis more than two millennia ago. They were lengthened and joined into networks over centuries; they provided a transport system that could not be matched for speed, convenience and carrying capacity until the 20th century. Van Slyke[11] has commented that as a feat of engineering, the Grand Canal surpassed the Great Wall.

The canals supplied water for irrigation; canal mud was used directly as manure, as indeed it was in 18th century Flanders (see above). The Chinese applied it at a rate of around 160 tonnes.ha^{-1} every second year.[20] It raised the level of the ground and brought with it nutrients and, of course, water. Mud from a canal that had flowed through a village and contained its wastes was valued more highly than mud from canals that had flowed only through farmland. Where there was no canal, it was a common practice to carry both soil and subsoil into a village and mix it with the village wastes to make compost that was subsequently carried back to the fields. Mud was not only spread uniformly over fields before planting, it was also spread between rows of, for example, beans—to prepare the ground for a second crop, such as cotton, planted between the rows before the beans were picked.

Intercropping and multiple cropping were normal East and South East Asian practices; failure to take proper account of them can seriously distort attempts to quantify the productivity of the agriculture of those regions. Intercropping, by planting a second crop before the first was finished, extended the growing season and lessened the danger of a complete crop failure through disease or a bad season. Intercropping at the density used in China was possible only with human labour. At the end of the 19th century Western agriculture was becoming much more mechanised, but there were no machines that could sow seeds on a significant scale between close rows of established crops. It was inevitable that the West's progressive mechanisation would demand not only bigger fields but, crop rotation aside, monocultures within those fields.

Chinese use of urban nightsoil for manure is not only legendary, it was a commercial enterprise; the contractor paid for a licence to collect the stuff and subsequently sell it to farmers.[6] Nightsoil was taken daily from the cities to the countryside, much of it by boats on the rivers and canals. Inevitably the farms closest to cities benefitted most from the trade—ultimately at the expense of the soil in the more distant farms.

King cited analyses of excreta and the amounts of nutrient elements discharged per million of population[21] and proceeded to make the following comments. They are instructive for revealing something of the attitude of an American soil scientist at the beginning of the 20th century.

"Man is the most extravagant accelerator of waste the world has ever endured. His withering blight has fallen upon every living thing within his reach, himself not excepted; and his besom of destruction in the uncontrolled hands of a generation has swept into the sea soil fertility which only centuries of life could accumulate, and yet this fertility is the substratum of all that is living. ... The rivers of North America are estimated to carry to the sea more than 500 tons of phosphorus with each cubic mile of water. To such loss modern civilization is adding that of hydraulic sewage disposal through which the waste of five hundred millions of people might be more than 194,300 tons of phosphorus annually, which could not be replaced by 1,295,000 tons of rock phosphate, 75 percent pure."

Chinese roofing thatch that had passed its expiry date (3-5 years in the Canton area) was used as fertiliser. The thatch was applied to soil directly as a mulch, it was composted or it was burnt according to need[22]; in the last case the ashes were used. A second item was earth from the floors of compost sheds, the walls of disused cesspits, the floors of dwellings and sometimes also their walls. In this case the main aim was to use the nitrates that arose partly from oxidation of nitrogen compounds in the soil. The nitrogen content of floors was also boosted by the practice of infants' relieving themselves *in situ* (see Chapter 6). This soil was sometimes applied directly to fields, sometimes incorporated into compost and sometimes bought by merchants who wanted to extract the nitrates for the manufacture of saltpetre. The Chinese had a long history, not only of intensive farming, but also of making explosives.

A remarkable system of integrated agriculture, still in use in the 1980s after about two thousand years of development, is the so-called 'dyke-pond' system as practised in the delta of the Zhujiang River in southern China. The delta covers some 12,000 km^2; in the 1980s it supported 1,200,000 people at a density of about 17.ha^{-1} of cultivated land. The sustainability of the dyke-pond farms depended on recycling produce within the system and complete compensation by urban organic wastes of the relatively small proportion of produce that was sold to towns. An interesting and detailed account of the farms and their functioning is available.[23]

By the end of the 19th century, the three main islands of Japan had a total population of about 45 million and about 865 km² of arable land. Although there were similarities between Chinese and Japanese farming practices, there were also significant differences. Japan was much more urbanised than China. The Tokyo-Yokohama complex, for example, had a population of 1.39 million, a size that placed it in 5th place in Lampard's Top 30 for 1890.[13] Indeed Edo (the earlier name for Tokyo) was 'approaching' 1 million by the beginning of the 18th century[24] and may well have beaten London across that particular line. From 1603 to 1868, the period of the Tokugawa Shogunate, Japanese farming had been essentially feudal with a heavy emphasis on rice, which was not only the staple food but was also the medium used by the aristocracy to measure their incomes. Early in the Meiji era (1868-1912) Japanese contacts with the West increased and Japanese farming practices began to change. The changes were reflected in implements and techniques, in the establishment of agricultural schools and experiment stations, and in a Bureau of Agriculture that, among other things, kept statistical records. For example, a new plough design led to deeper tillage of rice fields and better drainage that, in turn, facilitated the application of commercial inorganic fertilisers. Some farmers were also experimenting with new varieties of rice. By about 1889 rice yields for some individual farms reached 10,200, 11,400 and 12,300 kg.ha⁻¹, although the averages for the districts in which these farms were situated were much less than that.[25] There was writing on this wall too but, as yet, only in lower case. By 1895 the average rice yield for all Japan was about 2100 kg.ha⁻¹.

King[14] reported that about 56% of the cultivated land in Japan was allocated to wet rice but, where topography and season allowed, the fields were drained after harvest and planted with other crops. At the turn of the century the Japanese were still in the same league as the Chinese in their commitment to using human dung as manure. King describes loads of nightsoil taken from Tokyo and Yokohama "carried on the shoulders of men and on the backs of animals, but most commonly on strong carts drawn by men, bearing six to ten tightly covered wooden containers holding forty, sixty or more pounds each". He also noted screens set up along country roads for the immediate benefit of the traveller who might have been caught short, but really for the ultimate benefit of the farmer who used the product of the said traveller's discomfort to manure his fields.

Like the Chinese, the Japanese traditionally used a wide range of organic materials as manures. They have a very high ratio of coast line to total area and, at the time of which I write, they made a lot of use of marine products including, perhaps especially, seaweed. According to the Japanese Bureau of

Agriculture, about 10 tonnes of organic manures were applied annually per cultivated hectare. That figure did not include commercial inorganic fertilisers which by then were coming into use, but it did include human excrement at rates of up to about 4 t.ha^{-1}.

There have been two very different responses to the sanitary challenges discussed in the previous chapter. The traditional one was to take the nightsoil out of the town and distribute it over farmland. Europe followed that course to varying degrees for much of its history but, by the end of the 19th century, whatever resolve it might have had was beaten by the logistics of maintaining the practice in huge cities. The poorer inhabitants of those cities paid a terrible price in health and life expectancy, especially during the period in which the old system of disposal was breaking down. After various transition stages the industrial nations replaced the traditional system with sewerage. This worked wonders for urban public health but it necessitated profound changes in the management of Western agriculture.

The Eastern approach, especially that of China, was similar to early Europe's—only more so. Because of its commitment to recycling human excrement, China seems not to have experienced sanitary conditions as bad as those of Europe's urban slums and, partly because of that same commitment, it did not have as high a proportion of urban dwellers. In 1949, 90% of the population was still classified as 'rural'. It did pay a price in the health of its peasants, but a smaller price than that paid by the European urban poor, at least until the second half of the 19th century. But by the second half of the 20th century the Chinese were also challenged by the dynamics of an exploding population and the logistics of applying their traditional remedies to the sanitary problems of expanding cities.

Finally I should mention one other example of a farming system distinguished by its longevity—the indigenous agriculture of New Guinea. Agriculture arose there independently some 9,000 years ago and has persisted in what we must assume to be its essentials until the 19th century in the lowlands and the 20th century in the highlands, when white colonial administrations both introduced and catalysed change.

The main island is home to more than 1000 different languages, about a quarter of which are of the Austronesian[26] group. The other 75% are restricted to Papua New Guinea and, of those, about 700 are spoken only by small communities, mostly in the highlands, many of them unintelligible to communities in adjacent valleys. That is sufficient in itself to imply that the people in those areas were essentially sedentary and that their day to day movements did not take them far from home. Given the general nature of the terrain, that is not surprising.

The type of farming practised in the highlands to the middle of the 20th century was shifting agriculture, swiddening or 'slash-and-burn' agriculture—all synonyms. It was probably the first type of farming practised wherever agriculture arose independently and in many regions where it was introduced early. In most of the world it was superseded by fixed-field cultivation, but in New Guinea it persisted for a good 9000 years.

Although the details vary with location, altitude and population density, the essentials of New Guinea's traditional shifting agriculture are more or less as follow. Undergrowth is slashed and trees are felled over the area of an intended garden—perhaps a half to one hectare. Some of the wood is removed to be used for the construction of houses or fences; the rest is stacked in the centre of the cleared area and, in due course, burnt. The garden plot, which is quite likely to be on a steep slope, is not tilled in any conventional sense. Crops are planted in holes dug with a digging stick (a 'dibble') or, in some villages, with a wooden spade. As many as forty different crops might be sown in one garden—in some regions systematically but often in a manner that can seem to be random. It is intercropping with a vengeance. Gardens are usually fenced to exclude pigs.

Manuring is not practised to any significant extent and, partly for that reason, the productive life of a garden is short. An abandoned garden enters its fallow period during which the forest regenerates and plant nutrients are replaced naturally through processes such as those referred to in Chapter 2 and in Note 7 (above). Fallow periods can be as long as 40 years before a plot is re-used. In contrast to gardens, orchards last for many years and are inherited. They are commonly located in gullies where the soil is moist and the trees are relatively sheltered from wind. Animal protein is obtained principally through hunting small animals or feral pigs (New Guinea has no large native mammals), fishing and, at least on the part of children, catching insects. There are occasional ceremonial feasts for which large numbers of domestic pigs are slaughtered.

In 1964/65 an American geographer, W. C. Clarke, spent almost a year living with a clan called the Bomagai-Angoiang and other Maring-speaking people in the New Guinea highlands. He subsequently described the farming and social structure of the Bomagai-Angoiang[27] from which account the following summary is drawn.

The settlement was based in the Ndwimba basin, at an altitude of about 1,100 metres on the north east side of the Bismarck Range. In 1964 the population was 154. The people were of small stature, the average height of males being about 156 cm and females 146 cm; the average weight of males was 49 kg and of females, 41 kg. Adult males consumed on average about 11,000 kJ daily (*cf* Chapter 1) of which about 3% was supplied by meat. Clarke estimated their population density to be equivalent to 0.3-0.35 persons per hectare of land then "subject to cultivation". He further estimated "that on the average each fully active man has 1.2 acres of productive garden". In some other parts of the highlands densities were about three times greater than this and land was in short supply. Analyses of various soil samples indicated phosphorus (total?) to lie within the range of a 'trace'-225 kg.ha^{-1}, nitrate nitrogen 3-9 kg.ha^{-1} and ammonium nitrogen, approximately 25-465 kg.ha^{-1} for the top 18 cm of soil (see Chapter 11 for some comparative analyses of other soils). Clarke identified some 17 *major* crops in the gardens, of which the root crops, sweet potato, taro and cassava accounted for about 60% of the total. The monoculture yields of some of these crops can be very high, for example 7-14 t.ha^{-1} for sweet potatoes and up to 18 t.ha^{-1} for taro.[28] Not all crops were always planted in the same garden.

A map of Ndembikumpf contains the only reference made by Clarke to any aspect of sanitation; paths to men's and women's "feces fields" are marked. Houses were dispersed through the surrounding forest and were not necessarily near the owner's several gardens. Indeed, it was normal practice for most gardens to be situated 30-50 minutes' walk from the house. Moreover, a house was abandoned every few years as it succumbed to rot and a new one built somewhere else, again without any obvious relation to the position of the owner's gardens. In the face of these arrangements and the 'feces fields', systematic manuring with human excrement was clearly not very likely. Nevertheless enteric diseases do not seem to have been a serious problem, apart from a dysentery epidemic in the 1940s, apparently introduced by traders.[29]

Two questions suggest themselves. In the context of the 'Third Law', how did New Guinea's sanitary practices affect farming? And why did shifting agriculture not evolve in New Guinea into sedentary—or fixed-field—agriculture as it did in most other parts of the world?

Any answer to the first question must recognise that, because human dung was not taken to the gardens, the farmers had no option but to use a form of swiddening. No other type of farming could have lasted during those 9-10 millennia. The ecological dynamics of shifting agriculture have some qualitative features similar to those of nomadic hunter-gatherers, but the two systems differ widely, of course, in the area of land required for sustenance. In turn, the low population density, the daily movement of individuals over a considerable distance to and from gardens, the porous soil, the warm humid climate that condemned houses to a short life and promoted rebuilding in different sites, and an abundant supply of clean water, all meant that the population, although essentially sedentary, was rarely seriously threatened by enteric diseases.

The short answer to the second question may well be that the population remained below the level necessary to force a change to fixed-field agriculture. The significance of population size or, more specifically, density is that it affects the duration of the fallow period. Given a 1.5-2 year life of a garden, there must always be some gardens in a state of preparation. The larger the population, the more gardens needed to sustain it. Because there were definite limits to a clan's territory, an increase in the number of its gardens past a certain point would shorten the fallow period—the time available for recovery. Shortening the fallow period would, in turn, diminish the yield from a garden re-established in that area.[30]

And that leads to another question. Why did the population of New Guinea not grow to a level that would have forced it to change to sedentary agriculture? Here I can only guess. Tribal warfare in New Guinea seems to have been something of a national sport (until it was banned by a somewhat straight-laced Australian colonial administration) and it undoubtedly made its contribution to keeping the population in check. Given the effective isolation of many of the highland clans, as reflected in the extraordinary number of indigenous languages, it is possible that no clan ever reached a 'critical mass', if such a state is necessary, for the technical developments needed for the step from shifting to fixed-field agriculture. If that were indeed so, then a barrier would have been reached and the positive feedback relation between population and food production that exists elsewhere with sedentary agriculture would not apply. In other words, we could reasonably assume that deaths by war and by natural causes were complemented by behavioural responses to a limited food supply (Chapter 1). Those responses might well have included the use of contraceptive plants and other measures that limited birth rate and perhaps infant survival. But since contact with the 'outside world', things have changed.[31]

— The New World —

The husbandry of European and Chinese farmers up to late 19th century stands in stark contrast with 19th century colonists of Australia. Geologically Australia is among the oldest continents on earth. Its age and its weariness made it vulnerable to the assaults of European colonists and their descendants. Among other things, white settlement in Australia has caused more soil degradation in two centuries than most of Europe achieved in five or more millennia. Why?

There are many reasons. One that enjoyed favour several decades ago, and still does to some extent, was that the early settlers were from the British Isles and tried to farm in a British way—an approach that was not suited to Australian conditions. That has some limited truth, but it is a very small part of the story. Australia shared with North America the distinction of invasion and colonisation by the British, but there were some very important differences in the two situations. The first British migration to North America was voluntary—although once the colony was established, the English recognised a good jail when they saw it. The despatch of British convicts to America ended with the War of Independence but, because at the time it was quite easy to achieve convict status, England was not short of such individuals. It was therefore decided that the Great South Land might accommodate a few felons and thus relieve hulks on the Thames from the possible opprobrium of overcrowding. Accordingly the First Fleet arrived in Port Jackson in January 1788.

In both North America and Australia there was at first plenty of land, access to which needed little more than doing something about the natives and the trees. But in both colonies the natives were at first curious and friendly to the new arrivals. Indeed in America it is doubtful if the Pilgrim Fathers would have survived their first winter without aboriginal help.[32] At Port Jackson the Australian Aborigines bartered goods with the settlers and were generally disposed to be helpful. In turn, the colony's administration disallowed stealing small possessions from the Aborigines, but stealing land was quite acceptable. Apart from climate and the stark contrast of most of the Australian natural environment with anything in the British Isles, a major difference between the experiences of the English settlers in America and Australia was that aboriginal Americans had a well developed agriculture whereas aboriginal Australians had never needed to farm in any conventional sense of that word. Accordingly, at least in Massachusetts, the Pilgrim Fathers were able to learn from examples of Indian cultivation. Apart from that, the early white approach to farming in America was essentially European but compressed, as Gras[33] has noted, into a short space of time. At first, particularly in New England, there was a tendency to build small, self-sufficient 'nucleated' villages, similar to those that existed in Europe up to the time of the Industrial Revolution. The underlying philosophy and the methods of farming changed with westward migration.

The first settlers in New South Wales had no local example on which to draw. They arrived in summer, and although they brought food with them, by the time they arrived that food was somewhat past its prime. There were, however, fish aplenty and, as we have noted, the Aborigines were helpful. In fact, had they been even slightly better understood, they probably would have remained so—but that is another story. Nevertheless, Port Jackson might not have survived the first few years as a coherent settlement without additional supplies from Britain. There were questions in Britain about the value of continuing to support the colony, but by 1800 food production had grown sufficiently to allay such doubts.

In 1788 Port Jackson had 7 horses, 7 cows, 29 sheep, 19 goats and 74 pigs. Maize—the second significant American influence on Australia—had been brought out on the First Fleet; by the end of that year, 1788, there were two and a half hectares of the crop. By 1792, the year in which Governor Phillip left, there were 478 hectares of maize, 84 hectares of wheat and 10 hectares of barley, grown for the most part on about 68 farms of 10-12 hectares, mainly to the west and north west of Sydney. By 1794 the colony was self-sufficient in grain—for the time being. This self-sufficiency was achieved despite, in Bolton's words, "the inexperience of many farmers, the inadequacy of their technology and a tendency to rapid soil exhaustion in the absence of adequate manuring".[34]

Maize was better suited to the coastal climate of Port Jackson than either wheat or barley and, because it was planted at a different time of the year from the other two cereals, it provided the means of a continuous two-course rotation with wheat. Consequently, in the absence of conscientious manuring—a theme that will occur repeatedly—the soil was quickly impoverished. But maize was neither as popular nor as satisfactory as a staple food as wheat. By 1797 wheat had replaced it as the colony's principal crop. By 1805 there were 5140 hectares under cereal cultivation, 517 horses, 4,325 head of cattle, 5,123 goats, 20,617 sheep and 23,050 pigs. The increases in livestock came partly from Britain and partly from local breeding. Although the emphasis to this point had been on self-sufficiency, some secondary industries, including shipbuilding were by then well established.[35] Someone somewhere was producing a significant surplus of food. This had been achieved in 17 years with a workforce consisting predominantly of convicts.[36]

Soil impoverishment and a growing population led to a spread of the colony to the Hawkesbury River so that by 1795 some 500 settlers were farming the river's flood plain and immediate surroundings. But it was not until the Blue Mountains were crossed, and the grassy plains to the west encountered, that Australian pastoralism really began to develop its distinctive flavour. A road to the new town of Bathurst was in place by 1815 but, for the next three years, the surrounding area was grazed mainly by excess stock from the coast. Then another drought and the impact of overstocking made themselves felt and the migration began in earnest. At first it was to the Western Plains, soon south west to the Limestone Plains and the Monaro,[37] and before long, over most of the southern and eastern parts of the continent.[38] Westward migration took farmers progressively into regions of decreasing and—no less importantly—more erratic rainfall. This in itself was to have profound implications for the fate of Australian soils (see Chapters 8-10).

Voluntary migrants were attracted to Australia in increasing numbers from early in the 19th century, mostly in response to advertisements about a healthy climate and cheap land. Indeed, land was initially so cheap that it was often more important for an impoverished settler to acquire livestock than to own land. Having got the first, with good management he could be reasonably confident of making enough money to lease or buy the second before too long.[39] The cheapness of land encouraged indifferent exploitation just as it did in North America. Money was the yardstick. Then, as now, there were conflicts between good husbandry on the one hand and, on the other, economists and those whose judgment of any enterprise is based first and foremost on its economic 'efficiency'. Referring to the two-course rotation mentioned above, Jones and Raby[35] comment,

"This system was wasteful of land because it greatly taxed the soil while few resources were allocated to restoring fertility. Instead land that was exhausted was allowed to go out of production and cultivation was shifted to fresh land. This system was usually practised within the boundaries of a single holding.

As a result, farming at this time is often presented as elementary and inefficient. ... There is, however, a danger that technical (in-)efficiency may be confused with economic (in-)efficiency. What may have been best practice in a technical sense may not have been so in an economic sense. Critics of early New South Welsh farming, as of colonial American farming, have tended to judge it by the standards of farming in Britain or northern Europe, where land was relatively scarce and the other factors of production relatively abundant. In New South Wales it was just the opposite."

It was the spread of population outwards from Sydney—and from the other major Australian ports—that brought the Europeans into real conflict with the Aborigines and the trees. All the segments of this story are complex and it is not my purpose to try to discuss any of them in detail.[40] They are processes that have occurred to some extent in almost every other part of the world, but in Australia they are writ large.

Clearing forests was one of the first tasks facing aspiring settlers and, as government policies sorted themselves out, the process of clearing was aided by subsidies of one kind or another. Clearing forests was approached with a zeal that may reasonably be called diabolical. It still is. In the first 100 years of white settlement more than 25% of the forest of New South Wales was cleared,[41] and the other states were also doing their bit. The cheapest way to kill trees was to ringbark them. They died on their feet and remained standing until they were burnt, pushed or blown over, or succumbed to termites. Clearing forests was the first step to erosion, especially if the forest litter and understorey were burnt, as was often the case. Then, as now, the timber that was felled was undervalued. Much of it was burnt simply to get rid of it, a significant proportion of the nutrient elements that it contained being lost in the smoke; some was used in extraordinary ways.[42] In due course, deforestation led to dryland salination of large areas of inland Australia (see Chapter 9).

Although the original convict settlement in Sydney had to give its first priority to subsistence, and thus to mixed farming, it was not village farming of the traditional European type. There was evidently no systematic attempt to return the dung of farm animals—let alone the human dung—to the soil (see Chapter 6). The resulting impoverishment of the soil accelerated exploration and, as more land was settled, the emphasis was consistently on specialised commercial farming with the aim of sending produce to centres of habitation and selling it. Moreover the acres were broad, especially in rangeland, and even if the will had been there, the logistics of manuring with dung were daunting. Of course, many early settlers were out of range of regular dependable transport and were obliged to grow and hunt food for their own consumption. Others who were within access of a settlement and depot were sometimes supported by the government until they were on their financial feet.[43] But there were times when some, especially women and children when the men were away, went both cold and hungry.[44] Australia has never practised traditional 'village agriculture', to any significant extent, a tradition that included consuming most of the produce of the farm on the farm.

The focus of attention of new settlers from early in the 19th century was on pasture and on wheat. Both preferences were influenced by the climate and types of soil but, to a large extent, the preference for pasture was encouraged by its relative ease of initiation and management. It was much easier to let animals run on grassland than to plough and sow and harvest and thresh. It was not long before the colony was exporting wool to England. For a time horticulture, in particular vegetable gardening, became the province of immigrant Chinese who mostly lived close to towns. True to their tradition, they did conscientiously manure their plots. Migration northwards along the coast widened the range of crops.

Clearing forests, displacing and killing Aborigines encouraged the multiplication of kangaroos and wallabies; these marsupials saw the new grasslands as a good thing and, in consequence, were shot in huge numbers by pastoralists. Indeed, in the second half of the 19th century, some state governments enacted laws requiring landholders to kill the marsupials. But the marsupials were not a primary cause of land degradation—that accolade rested with placental mammals.

W. K. Hancock[45] quotes the reminiscences of one, William Crisp, the seventh of twelve children sired by Amos Crisp who came to the Monaro District in 1835. Crisp had this to say:

"Jimenbuan in the early days was very different from what it was after the passing of the Sir John Robertson's Land Act, which gave selectors the privilege of taking the land selected before survey. Some of them would put on more stock than the area they selected would carry. ...

Before the passing of the Land Act ... the Matong Creek for about five miles above and below its junction with the Jimenbuan Creek was a succession of deep waterholes, there being no high banks, and grass grew to the water's edge. Hundreds of wild ducks could be seen along these waterholes, and platypus and divers were plentiful. Five years after the passing of the Act the whole length, instead of being a line of deep waterholes, became a bed of sand, owing to soil erosion caused by sheep."

Hancock identified some of the prejudices here and elsewhere in Crisp's reminiscences and he, Hancock, attributed immediate blame for the erosion to cattle rather than to sheep. The cattle, he said, ate out the native kangaroo grass and made the pasture more hospitable to sheep. But he had no quarrel with the essence of the account. Then, as now, overstocking was a problem. Then as now, there were those who recognised the problem and spoke against it. It was a problem exacerbated by the variability of rainfall in Australia. Land that could support a relatively high stocking rate in a good season could support none in a drought. It is perhaps at this level that the comment that English settlers did not understand the Australian environment has most substance. Another is that they thought it would be quite a nice idea to import some rabbits.

The contribution of rabbits to soil erosion in Australia was so far reaching and so profound that one could be forgiven for thinking that their importation was a deliberate act of sabotage. It was not, but it was an act that had its origins, if not in outright stupidity, certainly in a disastrous lack of understanding of either the Australian environment or of the way in which natural systems worked. It was the kind of mistake that was repeated in later years with other imports.[46] Indeed in the 19th century, 'Acclimatization Societies' were established in many countries, including Australia, for the specific purpose of importing exotic flora and fauna. But back to rabbits. In a brief account such as this I can do no better than quote Ashton and Blackmore[41] on their introduction.

"On Christmas Day, 1859, 4 hares and 24 wild rabbits arrived in Hobsons Bay, Victoria, aboard the Lightning. They were destined for Barwon Park where its owner, the eminent squatter, Thomas Austin, was to breed them for blood sport. During the 1860s Austin and his irate neighbours were to spend tens of thousands of pounds attempting to stop the plague which had emanated from Barwon Park."

Their spread across the country was not without help. They were good to eat and provided free or cheap meals for poor rural inhabitants as well as an income for professional trappers. Needless to say, the trappers took care not to exterminate all the rabbits in their operating area, although it is unlikely they could have done so even if they had tried. Trappers contributed to the migration of rabbits from time to time by offering them a lift to new pastures. Rabbits ringbarked saplings, dug extensive warrens that, together with the removal of grass, made soil erosion inevitable. Rabbits outbred, outgrew and out ate every other herbivore with which they found themselves in competition[47] and, in yet another demonstration of abysmal thoughtlessness, pastoralists systematically killed wedgetailed eagles and dingoes, the native predators that might have done most to keep rabbit populations in check.

Even after their introduction to Australia, rabbits would not have become nearly so devastating had they not been continually aided and abetted by the European settlers. It is worth speculating for a few moments on the likely sequence of events had a couple of dozen rabbits been dumped on favourable land, say the Western Plains, and then left entirely to their own devices in the complete absence of European settlers. It is as well not to worry too much about how that might have been achieved in practice.

For a start, they might not have survived long enough to breed. There is a very good chance that the Aborigines, the dingoes and the eagles would have seen to that—and if they missed any, snakes would no doubt have been interested. But if the rabbits did manage to establish themselves, what then? Their breeding rate would probably still have outstripped almost everything else and there would probably would have been a local explosion of their numbers. In the course of that explosion they might have eaten bare the areas immediately adjacent to their warrens. But here some important differences emerge between this hypothetical situation and what actually happened. Rabbit dung remained in their grazing territory, as indeed it did in the real situation but so, in effect, did their bodies. They might have been eaten by Aborigines, dingoes or eagles but, as emphasised in Chapter 2, the dung of those predators was spread more or less at random throughout the area. Little if any of the original soil nutrients would have been removed. The rabbits that in reality were eaten by human beings led to the deposition of human dung a long way from the original grazing area and, as well, millions of rabbit hides were exported. A lot of nutrient was removed from the soil by the business of rabbiting.

But that is only one difference. In the real situation the assault on the pasture was mounted by sheep or cattle and sometimes also by macropods, as well as by the rabbits. The grass could not grow fast enough to keep up with the whole demand, especially when a lot of that demand progressively depleted soil fertility by exporting nutrients. But, in the hypothetical situation, the grass would not have been able to keep up with an exploding rabbit population either. So then what?

Leaving aside an Aboriginal involvement, which is more difficult to guess, the rabbit population would have exploded, the dingo and eagle populations would have increased in rough proportion (together by about 10% of the total increase in rabbit biomass), the combined effects of predation and the depletion of pasture would have produced a collapse in the rabbit population[48] followed by a roughly proportional decline in the predator populations (again by about 10% of the decline in rabbit biomass). Then there would have followed a series of damped oscillations in all three populations until some kind of dynamic balance was achieved—a rough approximation to a steady state (see Chapter 2). In the course of those changes there would have been changes in the species composition of both plants and other animals in the area in question. It is possible that some animal species in direct competition with the rabbits would have been driven to extinction. But in the end, a new balance would have been achieved, there would have been no significant nett removal of nutrients from the land, and it is unlikely that there would have been any significant long-term soil degradation—UNLESS there had been heavy rain or strong winds during the initial period of soil denudation produced by the first explosion of the rabbit population. It cannot be overemphasised that the damage done by rabbits would not have reached such enormous proportions had dingoes, eagles—and Aborigines—not been driven away or exterminated, had the rabbits not been harvested and had their grazing not been supplementary to—or supplemented by—farm animals that were themselves harvested.

There were contemporary critics of 19th century farming practices in Australia, in particular of overstocking; many are cited in recent Australian environmental histories.[40] Few, however, are as interesting as Joseph Jenkins—at least in relation to the claim that it was the application of British farming methods that caused so much environmental damage in Australia. Jenkins was Welsh. He was born in 1818 and emigrated to Australia in December 1868, arriving in Melbourne in March 1869. He had been a successful and respected farmer in Wales but he migrated to Australia on his own—apparently to get away from his wife. While in Australia he worked as a 'swaggy', an itinerant farm labourer, in Victoria. He returned to Wales in 1894 and died there four years later at the age of 80. He kept a diary[49] in which, as well as factual observations, he recorded comments on farming practices that he encountered. Some of those comments are complimentary, most are critical—especially of the manner in which Aborigines,[50] horses, farm labourers and the land were treated. He repeatedly despaired at the failure of the colonial farmers to manure the land. His comments are significant because they are not those of an 'expert' or a government official or an historian. They were made by an experienced farmer working as a labourer within the system. A few quotations should suffice to make the point.

"The country's soil and its climate cannot be surpassed by any country, but the law and the lawless cannot be so classed. So wanting are high principles, humanity and morals, that people from all classes of society hold that no honest man should set foot in Australia. ... Consequently the land is neglected and exhausted. Presently one man is allowed to hold a million acres of land with good surface-soil without obligation to employ a single labourer, while the same land is neither rated nor taxed. On the other hand, the small farmer has to pay a tax of 1s in the pound to support public roads, although there are no roads serving the squatters" (February, 1873).

"There are three characteristics peculiar to farmers in this Colony—exhausting the land, abusing horses and exploiting the labourer." (July, 1877).

"It is no use talking to the farmers about preparing farmyard manure to fertilize the land." (September, 1877)

"I was surprised to be asked to sow oats before the wheat. The farming practices here are quite different to what they are at home. Thus, they neglect to cultivate the land, they speak against farmyard manure as a means of nourishing the land, and against preparing fodder for the winter and dry summers. ... Farming activities are wholly neglected during ten months of the year, so that workers are unwanted during the whole of that period" (May, 1879).

A pervasive attitude to forests is stated succinctly:

"The folk here are very religious on Sundays, and believe that God is only present in chapel, while the devil is in the bush" (June, 1879).[51]

"I am saddened when I look at the exhausted land all around me. Should the population of the world continue to increase, a bread famine will inevitably emerge. This will not take place during my lifetime, as I have not many more years to run, but it will in years to come unless the land is properly cultivated and remunerated for the crops it produces. Ploughing, sowing, and harvesting, are simple processes in farming, but no one can farm well without judgment and plenty of good labourers. This farm consists of fertile land, and the tenant has been here four years, but during this time the manure has been allowed to lie in heaps around the yard, some nine of them average sixty tons each, while the paddocks close by are exhausted for want of manure" (April, 1881).

In addition to mismanagement, to the introduction of exotic flora and fauna with all that implied, to the apparently inexhaustible supply of land and the resultant disregard with which it was treated, and in addition to the vagaries of the climate, there was another critical factor that underpinned the degradation of Australian soils. I have already referred to it—but it deserves emphasis. Whereas 'Old World' agriculture had existed for many millennia predominantly as a subsistence enterprise in which most of the produce remained on the farm, in the 'New World', especially in Australia, the emphasis almost from the outset was on commercial farming in which farms specialised in a single type of product—predominantly either cereals or livestock. Nearly all the produce was taken away from the farm and sold. Australia never had the equivalent of, for example, 8th century Europe where the farm surplus was about 2%, nor did it have a period of six centuries during which the surplus probably did not exceed 5 or 6% (Chapter 5 and Appendix). In two centuries the Australian farm surplus shot from zero to about 99% (see Chapter 13).

Wheat yields can provide a useful indicator of how soil is standing up to cultivation. Nineteenth century Australian yields need to be treated with some caution because they varied very widely with region, with season and even between neighbouring farms.[52] In New South Wales they were affected by drought, by rust and by the progressive movement of wheat cultivation inland from the coast. For example, the coastal region accounted for almost 50% of the state's total area under wheat in 1861, but by 1878 the proportion had dropped to less than 3%. Over approximately the same period the entire area of wheat cultivation in New South Wales increased from 43,700 to 94,300 hectares while the population went from 367,000 to 734,000, a doubling in

18 years.[53] But estimates of average Australian wheat yields (as distinct from those for NSW) from 1815 to the end of the century indicate a drop from 1280 to 500 kg.ha^{-1}.[54] The rate of decline in the yield was not uniform, but it was inexorable. It was a clear indication of soil degradation through impoverishment, erosion and, in some places, through salination. This is an extension to the whole Australian scene of the soil impoverishment that rapidly showed itself in and around Port Jackson and added a note of urgency to subsequent exploration. From the beginning of the 20th century yields began to increase again.

A similar thing happened in America, for broadly similar reasons, but the effect was quantified earlier. Gras[55] cites reports of wheat yields "in an inland county of New York" in the range of 1340-2680 kg.ha^{-1} in 1775 dropping to about 500 kg.ha^{-1} by 1845. He describes falling yields in Kentucky, Virginia, Ohio, Michigan, North Carolina and Minnesota. In all of those states the rainfall was more predictable and the virgin soil was, in all probability, better than that in the Australian wheat areas. But the problems of mismanagement were much the same. Gras had this to say about it:

"The use of fertilizers is significant. At first there was no thought of putting anything on to the soil. Barnyard manure was dumped into rivers and cotton seed thrown away. Ashes were heaped up rather than spread over the land. And leaves were left to blow into corners where they were of no service."

PART II

CHAPTER 8

Water
– Part 1 –

Water must lie at the heart of any assessment of human ecology, especially in the broad context of agriculture and sedentary communities. Much has been and much will doubtless continue to be written about it. My objective in this and the following chapter is to identify some of the principles underpinning human consumption of water and give some feeling for their dimensions and implications.

Water is extraordinary in a number of ways, nearly all of which stem from the polarity of its molecules and their capacity to form hydrogen bonds with one another and with polar groups on other types of molecule. It is this polarity that underpins, for example, the abnormally high freezing and boiling points of water. By way of comparison, the hydrocarbon butane, which has a molecular mass of 58—more than three times that of water—has, at atmospheric pressure, freezing and boiling points of -138.4°C and -0.5°C respectively. A molecule structurally much closer to water is hydrogen sulphide (H_2S); at atmospheric pressure its freezing and boiling points are -85.5°C and -60.7°C respectively. Moreover, water not only has an abnormally high boiling point but it also has a very high specific heat —the amount of heat required to evaporate (a specified mass of) it. Evaporation of water is thus of paramount importance in preventing overheating of terrestrial organisms —both plants and animals—and, of course, of the planet itself.

Water is much more than just a solvent that accounts for 60-65% of the mass of a mammalian body or about two thirds of the mass of a bacterial cell or transports solutes through soil to the roots of plants. Without its strong polarity there would be, for example, no such thing as a lipoprotein membrane to hold a living cell and its organelles intact, and to select what stays inside, what stays outside and what passes between those two zones. Without water, proteins would not adopt the conformation that enables them to act as enzymes—assuming they could exist in the first place. Water supplies the oxygen released by the photosynthesis of green plants and water is produced by

the respiration of animals and other aerobic heterotrophic organisms. These
are just a few of its more obvious biological roles. It is a molecule that is indispensable to life in any form that we understand that term; there is no other
single inorganic compound that can even approach water in this respect. It is
not a substance to be treated lightly.

Earth is the only planet in the solar system known to have water in the liquid state although, in the distant past, there was evidently sufficient liquid
water on Mars to form rivers. Mars still has water in its polar ice caps. The total amount of water on Earth, if all in liquid form, would amount to some 1.4
$\times 10^9$ km^3. More than 97% is in the oceans and, of that, about 81% is in the
southern hemisphere. Ice accounts for about 2% of the total water, groundwater 0.6-0.7%,[1] lakes and rivers less than 0.02%. About 46% of the total volume of lake water, however, is saline; consequently rivers and freshwater lakes
hold only around 0.009% of the planet's water. About 600 km^3 of water is estimated to be contained within the tissues of terrestrial plants and animals.
Finally the atmosphere, which is the medium that above all others ensures
that the pigeonholes are not sealed and that there is exchange between them,
contains on average about 0.0009% of the total global water.[2]

Water has been and continues to be a critically important influence on the
earth's topography. It erodes mountains and produces valleys; it transports
the eroded material to the sea at a rate that, 'discounted' for human contributions, is estimated very conservatively to be of the order of 9×10^9 tonnes annually.[3] In times of flood the amount of sediment carried by a river is greatly
increased but much of that sediment is deposited on land, producing and fertilising flood plains. Egypt, probably more than any other ancient civilisation, depended on this process for its existence and continued to do so until
the construction of the Aswan High Dam. Water is also the most important
transport agent in the soil erosion induced by human activities.

Water is not only an integral part of food production but, because of the
nature of civilisation, it has also become virtually indispensable in the disposal of the most obvious waste products of food. Because the seas and oceans
were seen to be vast, and because rivers flow, throughout all history bodies of
water have been used as convenient receptacles for waste. Sometimes this was
by default when urban wastes were (and are) washed by rain off streets into
waterways. It was deliberate, however, when sewers in ancient Rome were
constructed to discharge into rivers or the sea, and it is deliberate today when,
in some places, the practice is extended to include even lakes. So it has been
with industrial wastes such as the effluents from fruit canneries, abattoirs, as
well as from secondary industries and some types of mine. A dominant reason
for these practices has always been convenience, which translates into costs.

The rationale—or the pious hope—has been that the body of water was large enough to dilute the waste beyond the point of nuisance and/or it was flowing fast enough to carry the stuff off somewhere else—out of sight and hopefully out of smelling distance.

Consumption does not significantly change the amount of water on the planet but, either through causing pollution or physical inaccessibility, it does affect the availability of water for many purposes. Natural aqueous environments are ecosystems in their own right. The functioning of these systems is affected by processes such as outlined in Chapter 2 but, in addition, aquatic systems have distinctive characteristics caused especially by effects of temperature on the solubility of gases (see Appendix).

— Sewers and Sewage —

The primary aim of sewers, at least in the 19th century, was to provide a more effective method than any that was available previously, of getting filth off the streets of large cities and squirting it somewhere else. That somewhere else was mostly a body of water. The dominant guiding principles of the process were that the sewage would be diluted; the more critical requirement lay in those words 'somewhere else'. It was not that urban and industrial wastes did not already find their way into water. They did so with a vengeance and, among other things, produced in mid-19th century England a river, the Calder, whose water could serve as a functional alternative to ink—if the writer could stand the smell.[4] They were responsible for inflammable streams and the 'Big Stink' of the summer of 1858-59, an occasion on which the Thames was so foul that the Houses of Parliament were obliged to close down. As Rosen[5] put it " ... the stench from the Thames assumed the dimensions of a national catastrophe".

Such sewers as existed anywhere early in the 19th century were little more than drains that might or might not be covered with bricks. Their primary function was to drain surface water and its suspended muck to a nearby river or the sea. It was that suspended muck that caused the trouble. A flushing toilet had been designed at the end of the 16th century. An improved model with two valves was patented in 1778 and came progressively into service in wealthier households early in the 19th century, but these devices were still obliged to discharge into cesspits, partly because the normal flow rates of water through the sewers were too small to cope with material that gained access by intent rather than by accident. Hamburg was possibly the first city to install (in about 1843) a complete sewerage system flushed regularly by river water.[6]

Thereafter similar systems followed progressively with various modifications throughout the world.

This type of sewer greatly improved urban sanitation (Chapter 6) and lessened water pollution in the immediate vicinities of cities. But it did so by collecting the waste and discharging it, with no treatment other than a coarse screening, at a single point 'somewhere else'. Writing in *The Lancet* in 1856 in relation to an outbreak of typhoid fever in a London school, William Budd, a physician, commented that "The sewer may be looked upon, in fact, as a *direct continuation of the diseased* intestine".[7] In large measure that is true whether or not the intestine is diseased, but there are some critically important physical and chemical differences in the two environments. Let us begin with the bowels.

The human alimentary canal provides a superb environment and culture medium for some types of bacteria and some types of proto- and metazoa. Its contents are moist, highly nutritious, at least from a bacterium's point of view, and are incubated at 37°C. There is not normally much microbial activity in or above the stomach. During the preliminary digestion of food, the stomach contents are mixed with hydrochloric acid at a concentration that can reach 0.01 M—not the kind of environment that many microorganisms enjoy. Once the partially digested food passes into the intestines, however, the acid is neutralised and the contents become, if anything, slightly alkaline. This is where the bacteria really get down to business.

Air is ingested with food and, in the upper part of the bowel, there is some microbial respiration. There might be a few strict aerobes at that point but, if there are, they don't last long. The oxygen is rapidly exhausted and the environment becomes completely anoxic. So much so that 95-99% of intestinal microflora are strict anaerobes and, of those, some are so sensitive that they are killed by even brief exposure to atmospheric oxygen.[8] These organisms can find life quite challenging once faeces has been voided.

The important organisms are facultative—those that respire when oxygen is available and resort to fermentation when it is not. The causative organisms of cholera, typhoid fever and the various dysenteries are all facultative. If they were not they could not meet the dual requirements of multiplying in the intestinal tract and, at the very least, continuing to metabolise in the aerobic environments of water, milk and some foods. The alternative mechanism of transmission is some form of resistant dormancy such as sporulation in bacteria or encystment in protozoa. The strictly anaerobic spore-forming bacteria, however, are rarely responsible for infectious enteric diseases, although some of them are very dangerous in other respects (see Appendix). Viruses might be present but they will not multiply in an extracellular environment. Many will remain infectious for significant periods, however, and some (*e.g.* hepatitis viruses) can remain viable in and be transmitted via aquatic hosts such as oysters. But the faeces of a healthy individual is not a threat to public health—with the important qualification that some people can be asymptomatic carriers, especially of *Salmonella typhi*. Communal sewage, on the other hand, must always be considered dangerous.

Viewed chemically rather than biologically, faeces consists of a range of macromolecules—proteins, polysaccharides, polynucleotides, lipids—that are present as cellular components of the intestinal biota. There are metabolites of small molecular-weight, also within the cells of the biota, and extracellular metabolic end-products that have not been assimilated by the host. There is likely to be partly or undigested cellulose and perhaps small amounts of oligosaccharides and peptides, as well as some inorganic salts. There are gases including carbon dioxide, hydrogen and methane.

Throughout the entire process of digestion within the alimentary tract and the subsequent impact of voided faeces on water, the availability of oxygen and the patterns of microbial energy metabolism dominate events. The term 'respiration' has both a physiological and a biochemical connotation. The physiological meaning, which is the common one, generally implies inhaling oxygen and exhaling carbon dioxide. The biochemical meaning, on the other hand, implies the transport of electrons from a metabolite along an electron transport chain[9] to an electron acceptor. Needless to say, the commonest and ecologically most significant electron acceptor is oxygen which, in the process, is reduced to water. As we have already seen, under anaerobic conditions some bacteria use inorganic ions, notably nitrate (NO_3^-) and nitrite (NO_2^-) which are reduced to elementary nitrogen (N_2) and sometimes to ammonia (NH_3), and sulphate (SO_4^{2-}) which is reduced to sulphide (S_2^-). Some specialised bacteria, the methanogens, reduce CO_2 to CH_4 (methane). But in the absence of oxygen the commonest type of energy metabolism is fermentation. This also involves the oxidation of one substance at the expense of another but the mechanisms are different. To summarise, hydrogen atoms are transferred enzymically from one organic metabolite to another until substances that cannot be reduced any further accumulate. This is the essence of fermentation; it is a type of process that is widely used industrially (*e.g.* to produce acetone, butanol, ethanol, lactic acid, etc). The products of fermentation are commonly relatively small molecules. If they contain sulphur or nitrogen they are usually smelly—which lies at the heart of one of the characteristics of sewage. But these fermentation products are manna from heaven for aerobic and facultative bacteria—provided the said bacteria can lay their hands on some oxygen.

When voided, especially into water, the strict anaerobes will be either killed or inhibited by oxygen, and they will then fall prey to the facultative bacteria that were formerly their bedfellows. The cells of these anaerobes provide a high proportion of the nutrient substances that sustain the deoxygenation and other ecological changes in water. As well as that, of course, there is a significant contribution from urine to the total nutrient balance sheet, principally from urea and creatinine as sources of nitrogen, and from inorganic phosphate.

In industrial countries, domestic sewage also contains other substances, notably detergents, which provide about half the total phosphorus. Urban sewage often contains industrial wastes, some of which are metabolisable and contribute to eutrophication, and some of which are refractory and toxic. Among other effects, toxic substances can inhibit the treatment of sewage.

Obviously the composition of sewage must vary with circumstances. The biological waste will reflect diet but, that said, it is reasonable to expect that, over time, the average chemical composition of the excremental component of urban sewage will remain relatively constant for a particular city or, perhaps more importantly, a particular culture. In addition to the various synthetic substances that are emptied into sinks and drains, the overall concentration of all constituents is also affected, of course, by the volume of water that the sewers carry.

In a number of Western countries the flow rates of urban domestic sewage are equivalent to *per capita* water discharges ranging from about 160 to over 400 litres per day.[10] Total phosphorus concentrations from various sources that I have encountered range from about 4-23 mg.l^{-1} (median 10.5 mg.l^{-1}). A value of around 11 mg P.l^{-1} would thus seem to be a reasonable point of reference; it is a good 500 times greater than total phosphorus concentrations in oligotrophic streams.[11] Total nitrogen concentrations in raw sewage can be around 40 mg N.l^{-1}.

Apart from statistical variability, the total concentration of an element, that is to say the sum of the concentrations of the element in all of its prevailing chemical states, does not tell the whole story as far as its contribution to eutrophication is concerned. For example, the biological significance of nitrogen can be affected by how much of the element is in the forms of nitrate (NO_3^-) or nitrite (NO_2^-) and by the outcome of any competition between denitrifying bacteria on the one hand and, on the other, organisms that might wish to assimilate the nitrogen from those ions. Not all phosphorus compounds are immediately assimilable by microorganisms but, given time, most would be hydrolysed—usually by microbial action—to a form that is. The commonest assimilable state of phosphorus is orthophosphate ($H_2PO_4^-$,

HPO_4^{2-} or PO_4^{3-}—depending on pH) but some salts of orthophosphate are effectively insoluble in water (calcium phosphate for example) and thus prone to sedimentation. For these and other reasons, there is competition between chemical and physical processes and assimilation of phosphorus by the biota. Obviously the physical state of the waterway plays a major role in determining the ecological outcomes of all of this.

It might be said that the aim of sewage treatment is to ensure that the major ecological changes associated with the discharge of biological wastes into water are confined to a restricted zone—the treatment works—and to accelerate the breakdown of those wastes with copious supplies of air.

The principles and methodology of sewage treatment are outlined in the Appendix. Suffice it to say here that a technological development, sewerage, brought about monumental improvements in urban sanitation (Chapter 6) but generated new problems of its own. Because most of the world's sewage is discharged into water, water pollution is inevitable, but its nature and severity are affected in the first place by how effectively the sewage is treated. It should be emphasised that most treated sewage effluent contains phosphorus at concentrations similar to those of some microbiological growth media—sufficient to support dense algal populations. Whether or not that happens will depend, among other things, on the fate of the phosphorus—whether it is discharged into the sea or into fresh water, and the extent to which the effluent is diluted after discharge. Bear in mind that dilution of the order of 350-fold would be required to bring the phosphorus concentration of conventional tertiary effluent down to the level of an oligotrophic river. Conventionally treated sewage is still a water pollutant because of its capacity to cause eutrophication. It is not as severe a pollutant as raw sewage because it has much less organic matter and, at worst, a lower count of pathogens. At best there should be no viable pathogens.

— Some Examples —

In this section I want to outline examples of water pollution in which the dominant tendency is towards eutrophication. In all cases the process is tied—sometimes directly, sometimes indirectly—to agriculture and the movement of nutrients. I shall not discuss possible remedies at this stage where my primary purpose is to identify problems and to convey some idea of their magnitude and extent.

I should also issue a word of warning. The remainder of this chapter contains a good deal of quantitative information about the magnitude of water resources, rates of consumption and so forth. It should need only a moment's reflection to appreciate that such data are, at best, reasonable approximations. While flow rates of rivers can be assessed fairly accurately, estimating volumes of groundwater must, in most cases, involve a significant amount of guess-work. The widespread installation of meters enables rates of urban water con-sumption in developed countries to be measured but, in developing countries where such meters might not be available, guesswork again takes priority. Similarly, part of the rural population of developed countries, although pro-portionately small, depends on rainwater tanks for its water supply. Such tanks might or might not be included in assessments of water availability.

Accurate assessment of withdrawals for irrigation depends on two fac-tors—the availability of meters and the honesty/objectivity of the irrigators. The statistics do not indicate the frequency with which both requirements occur simultaneously. And finally, one can encounter simple arithmetical er-rors. For example a table listing Australian water consumption for all pur-poses in 1987 (see below) shows a total 645 Gl (4.6%) greater than the actual sum of the tabulated numbers—a discrepancy which, through some inconsis-tent 'rounding up', possibly reflects an acknowledgment of uncertainties in the estimates but is largely attributable to a number of errors in addition.[12]

One of Australia's two longest inland rivers, the Darling, provides what is perhaps an extreme example of how not to treat a rural river—as distinct from those rivers that have been effectively destroyed by urban and industrial wastes. But first, a background comment. Excluding Antarctica and Green-land from the comparison, Australia is the world's driest continent. The aver-age annual rainfall for the entire land mass is around 465 mm; the next lowest continental region, Asia, averages about 600 mm. The highest rainfall is in South America with some 1630 mm per year. To make matters worse, how-ever, evaporation and rates of infiltration into dry ground in Australia are so high that it also has the lowest proportional runoff into its rivers (12% of the rainfall). The next lowest is Africa with 38%; the highest is again South Amer-ica with 57% runoff.[13] But as well as that, rainfall, especially in inland Austra-lia, is extremely variable.[14]

A 1987 report assessed Australia's annual total water consumption as 14,600 Gl of which irrigation accounted for 70%.[15] In that year Oceania had 1,957,000 ha of irrigated land compared, for example, with 27,605,000 ha in North and Central America. By 1997 the irrigated area had increased to 2,988,000 ha for Oceania, 2,679,000 ha of which were located in Australia. By then total Australian water consumption had risen to 24,060 Gl annually, with irrigation accounting for 75%.[16] The following paragraphs illustrate some of the ramifications of such statistics.

The Murray-Darling Basin—or drainage division—covers about 80% of New South Wales, about 60% of Victoria, a significant area of southern Queensland and of the south eastern corner of South Australia; its total area exceeds a million square kilometres. The average runoff into the basin is around 6% of the rainfall—half the national average. Nevertheless the area contains three quarters of Australia's irrigated farmland, a proportion that partly explains some of the troubles it is experiencing. The Basin grows half of Australia's farm produce. The Murray flows in a generally westerly direction and obtains most of its water from the western slopes of the southern segment of the Great Dividing Range. The Darling begins nominally near Bourke in north-western New South Wales but it is fed by numerous streams in northern NSW and southern Queensland. For most of its length it flows roughly south west and then turns south before joining the Murray at Wentworth on the Victorian border. The Murray discharges into the sea near the town of Goolwa, south east of Adelaide in South Australia.[17] In an 'average' year, the Darling contributes about 12% of the total flow of the Murray.

The streams that feed the Murray and the Darling support numerous towns and, for the most part, receive their (treated) sewage. Only three towns discharged treated sewage into the Murray itself by 1992 and they were investigating methods of disposal on land.[18] Thirteen major towns are situated on the Murray downstream from its confluence with the Darling and depend on the Murray for their water. In addition there are five large pipelines providing water to other South Australian cities including Adelaide. As well as Bourke, the Darling proper supports eight smaller towns and provides irrigation for large areas of farmland before it joins the Murray. Most of the water of the Murray-Darling system is used for irrigation.[19] The Darling flows through the driest region of NSW—predominantly through country with average annual rainfalls of less than 250 mm. Cultivation would not be possible without irrigation.

The state of the Darling is affected, of course, by the management of the rivers that flow into it. One that has received a good deal of publicity is the Gwydir, a river that supports several towns and, more significantly, large areas of cotton cultivation. To these ends the Copeton Dam was completed in 1976. The dam wall is 113 m high and its gross capacity is given as 1364 Gl.[20] The average annual flow into the dam is some 445 Gl but, by 1994, irrigation licences amounting to 525 Gl had been issued. The same problem can be expressed in different units; the NSW Department of Water Resources initially thought that the dam could irrigate 33,000 hectares of farmland but, by the time the issue of new licences ceased, 86,000 hectares had been licensed for irrigation.[21] One might wonder if the people most immediately involved were surprised to find that the dam was full rather less often than they had originally expected. In early 1995, after a drought, the dam held 2% of its capacity.

But not only do the irrigators draw water from the Copeton Dam, after heavy rain they can also take 'unregulated' water from the river below the dam. In the larger plantations—which account for most of the area under cotton—the water is not used immediately but instead is stored in very large private reservoirs. Many of these plantations have been levelled with the aid of laser technology, a degree of precision considered necessary for proper control of the flow of irrigating water. The plantations are massive monocultures sustained, not only through irrigation, but also through heavy application of pesticides and fertilisers; when it does rain properly, substantial quantities are leached into the river.

While it might not be possible to produce reliable statistics of the effect of withdrawals on the flow of the Murray below the Darling,[19] there is not much doubt about the effects of these processes on the flow of the Darling itself and on rivers, such as the Gwydir, that join it. One disturbing effect is loss of wetlands, specialised ecosystems that, among other things, play an important role in purifying rivers in flat country. Since the explosion of irrigation in the region, the Gwydir wetlands have contracted from about 70,000 to less than 5,000 hectares and the remaining segments are considered to be under stress. Recent estimates indicate that about half of the planet's wetlands have 'disappeared'.[22]

In 1991 the NSW Department of Water Resources had 31 monitoring sites on rivers in the Darling catchment. Of those sites, one gave a median value for total phosphorus in the 'low range' (less than 0.02 mg.l^{-1}), five were in the 'medium range' (0.02-0.05 mg.l^{-1}) and twenty five were in the 'high range' (above 0.05 mg.l^{-1}).[23] The flow rate of the Darling was measured at Burtundy, the last town before the confluence with the Murray, for all of the period 1st January 1971–1st January 1991, except for about 9 months in 1985. For roughly 29% of the measured period—some 5½ years—the flow rate cannot be distinguished from zero on a small scale graph.[24] On one occasion in 1985 the Darling was reported to have 'flowed backwards' after rain because irrigators upstream had 'pumped a hole in the river'. During drought in 1994 the river stopped flowing again.

Over the period August 1976–January 1991, 492 samples were collected at Burtundy for estimations of total phosphorus concentration. The median value of those determinations was 0.29 mg.l^{-1}, the median of the first decile (the lowest 10% of determinations) was 0.11 mg.l^{-1}, and the top decile (the highest 10%) was 0.53 mg.l^{-1}. Even the lowest group of estimations gave a median value more than twice that used to classify rivers as 'high' in total phosphorus. By way of comparison, the median of daily total phosphorus determinations in the effluent of the Australian Capital Territory's principal sewage treatment works (see Appendix), for the month of August 1995, was 0.14 mg.l⁻¹.

At the other extreme, the median concentration of the top decile of samples of the Darling is about 10% of the total phosphorus concentration in the low range of raw urban sewage. It is small wonder that this river has experienced some of the most savage cyanobacterial blooms and serious general eutrophication of any river used for human consumption and for watering stock. Some lakes and ponds can match it but it is rare for a major rural river to do so.[25]

By agricultural standards Australian inland soils in their natural state are deficient in phosphorus. The pollution of the river with this nutrient element can be attributed entirely to human interference—to excessive withdrawals so that flow rates and the capacity to dilute pollutants are dangerously low, to the replacement of some of the abstracted water with sewage effluent and to the leaching of excess phosphate fertiliser from riparian farmland.[26] It is a litany of collective mismanagement compounded by powerful commercial pressures, by conflicting interests and past rivalry between the states of NSW and Queensland—and by a lack of plain commonsense. It is yet another 'tragedy of the commons'.

None of this discussion has addressed other characteristics of the Darling—its turbidity, its oxygen, its nitrogen content, its salinity, its general ecology—but probably enough has been said to identify the essentials of a serious water problem closely tied to the management of agriculture. It is characteristic of this type of situation that attention is focused overwhelmingly on the apparent end of a process rather than the earlier stages. I say 'apparent' because the entire sequence of events is too complex to have a clearly defined beginning and end. In the present example the 'end' is the state of a river, something that is easily recognised and an entity in which the consequences of widely dispersed activities are concentrated. This is not to say that those activities pass generally unnoticed, but it is to say that it was the state of the river that, above all else, drew public and official attention to them. In this case the dominant direct effects on the river were produced by farming practices; sewerage systems made a supplementary contribution. In other situations, however, it is the disposal of sewage and, as in previous centuries, urban run-off that are the immediate and dominant causes of water pollution.

By the late 1980s about 60% of the total population of the OECD countries was connected to sewage treatment. Of those countries the range of the populations so favoured was 1% for Turkey to 98% for Denmark.[27] In other words, the sewage of 40% of the population, some 333 million people (1990 figures), was not treated before it was discharged somewhere. At a guessed daily discharge of 300 l per person, the total *daily* flow of *untreated domestic* sewage thus amounted to about 100 Gl—slightly more than half the *annual* throughput of Melbourne's large Werribbee sewage farm.

Enclosed bodies of water surrounded by large populations at high densities are obviously going to suffer the most. More than 130 million people live permanently on the Mediterranean coast and their numbers are swelled by another 100 million tourists in the summer. Their sewage is discharged into the Mediterranean. Seventy to eighty percent of that discharge was still untreated in the mid 1990s.[28] Sewage is not the only pollutant to enter the Mediterranean but it, together with runoff from coastal towns, is the primary cause of direct toxic pollution and of eutrophication. The Mediterranean suffers from widespread chronic eutrophication and zones of acute pollution. One of many consequences of both these conditions is the accumulation of algal toxins to dangerous levels in the tissues of edible shellfish. Other consequences are much as already outlined and include a sharp decline in the biodiversity of the sea. One animal that was common in the Mediterranean is the striped dolphin. For the two years up to 1992, several thousand of them died, apparently from an infection with a morbillivirus.[29]

One of the worst affected areas lies at the top of the Adriatic Sea which suffers badly from a high population density settled around a relatively small volume of more or less stagnant water. To rub salt into the wound, the River Po discharges into the northern Adriatic. As Pearce put it,

"The northern Adriatic's main source of freshwater is the River Po, which drains a large heavily populated and intensively farmed region of northern Italy. The river is heavily polluted with nutrients. Each year some 5000 tonnes of phosphorus and 100 000 tonnes of nitrate and ammonia reach the sea from the Po—about ten times as much as 50 years ago."[30]

These general types of effect are common in enclosed waters and in open coastal waters throughout the densely populated regions of the world.[31] About 60% of the world's population lives in coastal areas; changes in coastal marine ecology are certainly not confined to the densely populated regions. For example, coral in the Great Barrier Reef off the Queensland coast has been suffering in a number of ways that have become increasingly obvious over the past several decades. This is an informative example because, despite a number of significant towns and cities, the region overall is sparsely populated by current standards. The annual rainfall is everywhere above 800 mm and, in most of the immediate coastal belt, above 1200 mm. The discharge of sediments and nutrients (N and P) into the sea along the eastern coastline of Queensland has been the subject of a comprehensive study,[32] part of which I have attempted to summarise in Table A9.

Grazing land is clearly the greatest contributor to the total discharge because there is much more of it than of the other categories. But cultivated land is by far the worst in releasing materials from a specified area, although it is only slightly worse (30%) than towns in discharging phosphorus. As a first approximation, it would seem that if all the land in the survey area (450,515 km²) were still pristine, the discharge of sediment, nitrogen and phosphorus into the sea would occur at about half their present rates. But that is too simple. Most of the land classified as pristine is hilly or mountainous; most of the grazing and cropping land is relatively flat. An estimate that discharge rates are now about four times those that prevailed before European settlement is probably not unreasonable.[33] Worldwide, there is a close correlation between population density in a river catchment and the amount of nitrogen and phosphorus carried to the sea by the river system.[34]

On a still wider scale, during the last two decades of the 20th century, all oceans experienced increases in the incidence of the so-called 'red tides', a comprehensive and somewhat misleading term for blooms of phytoplankton. These blooms commonly include dinoflagellates and cyanobacteria that produce toxins capable of killing fish and mammals both directly through ingestion of the plankton and indirectly through the consumption of fish or shellfish that have assimilated the toxins.[35]

Like the cyanobacterial blooms in rivers and lakes, 'red tides' are an example of eutrophication caused, in most cases, by pollution from the land. It is difficult to estimate accurately the total quantities of nutrients discharged into the oceans but attempts have been made—as summarised in Table 8.1 for the elements, nitrogen and phosphorus.

Human activities are thought to be responsible for about half the total bound nitrogen that enters the oceans.[36] The sources of this so-called 'anthropogenic' nitrogen include the obvious ones of sewage, agricultural and urban runoff, industrial effluents—all of which directly enter a waterway. Moreover, inasmuch as deposition from the atmosphere in rain and snow is an important contributor, the sources of nitrogen include combustion.

In 1988 India had 212 cities with populations over 100,000. Seventy one of them were sewered. There were about 240 cities with populations in the range 50,000-100,000. Twelve were sewered. India's twelve largest cities collectively generated 6.5 Gl of sewage daily of which only 1.5 Gl was collected. New Delhi discharged daily some 200 Ml of untreated sewage into the Yamuna river, subsequently to join the Ganges. Overall about 70% of India's water supply is considered to be seriously polluted.[37] The slums of Rio and their attendant problems of water supply and sanitation have received a good deal of publicity since the so-called 'Earth Summit' in that city.[38]

TABLE 8.1 *Estimates of global annual rates of transfer of nitrogen and phosphorus into the oceans*

Source	Mass of element (10^6 tonnes)		Ref.
	'Natural'	*'Human'*	
Nitrogen			
N-fixation	14-42		
Rivers	14-35	7-35	
Atmosphere (rain, snow)	14-35	14-35	
Organic N		28-70	
Totals	42-112	49-140	(a)
Rivers	15	7-35	(b)
Phosphorus			
Rivers	1	0.6-3.75	(b)

Sources: (a) Cornell *et al.* (1995); (b) World Resources Institute (1992).

All of this reflects on public health for essentially the same reasons as the appalling sanitation of Western cities up to late in the 19th century (Chapter 6), but the problems are greatly exacerbated by the size of present day populations, population densities and, of course, economics. A consequence is an estimate by the World Health Organization that, in the developing countries, 75% of all illnesses and 80% of child deaths are associated with inadequate sanitation and contaminated water supplies.[39]

There are also other consequences, needless to say. The loss of wildlife from Europe's rivers scarcely needs mention even though the state of many of those rivers is better than it was a century ago. Africa is not short of problems with its waterways and supplies, but I shall mention just one. Lake Victoria is one of the world's large freshwater lakes. Among other things it has been an important source of food and employment for Kenya, Tanzania and Uganda. It is now becoming eutrophic and otherwise polluted from sewage discharge, farm runoff and industrial wastes. There has been a series of incidents in which much of the fish population died from asphyxia when large sections of the lake became anoxic.[40]

So far this discussion has focused predominantly on the consequences of polluting water with nutrient substances. The arguments that I want to advance to this point took shape in Chapter 6. They are first that sedentism, *ipso facto*, threatens sanitation and will usually, of itself, pollute water. Sewerage is a very effective way of improving urban sanitation and, in some circumstances, also the quality of a town's water supply. But, in the majority of cases, the sewer achieves this by transferring the problem of water pollution somewhere else. Just how serious the new problem is depends on the overall dynamics of the system—in particular the size and density of the human population and the dilution factor of the effluent. This generalisation is qualitatively true whether or not sewage is treated.

The broad ecological principles are much the same whether sewage is discharged into a river, a lake, or the sea—but the numbers are different, and the sea is not normally a direct source of drinking water.[41] A river or a lake with a substantial human population in its catchment or on its banks can very quickly become polluted beyond the point at which its waters are fit for consumption. This is a chronic problem for European rivers where some, for example the Rhine, are drunk and excreted several times over before they reach the sea. Stringent purification is needed before such waters are consumed.

But, over and above the question of potability, pollution will reduce the biodiversity of a body of water, whether that pollution is caused by heavy metals, toxic synthetic compounds or by nutrient substances with their attendant eutrophication. The species lost will include those that are used as human foods. Moreover, otherwise edible species that survive may well be toxic from accumulated microbial toxins, synthetic compounds, heavy metals or even radioisotopes. A long term consequence of discharging sewage into the sea is to place further stress on the world's fish populations, exacerbating the impact—already serious—of overfishing. A comment made earlier in this chapter deserves repetition at this point. The thrust of the recent discussion, and of public concern, has been on water pollution—an obvious and potentially dangerous phenomenon at the 'end' of a chain of human activities. That chain of activities also has a 'beginning' that is not as immediately obvious and is rarely the first reason for tackling water pollution, although in specific types of situation (*e.g.* Chapter 9), it is widely acknowledged. I refer to soil degradation, a topic discussed in Chapter 10.

In the next chapter we will consider some other interactions between farming and water.

Water
– Part 2 –

Even apart from the contemporary range of curious additional meanings that have been bestowed on 'consumption' and 'consumer', the more literal meaning of consumption can have a range of physical implications that depend on just what is being consumed. For example, the consumption of food by animals gives rise to heat, more animals, and a range of substances that, for the most part, are chemically different from those that were consumed. In the same vein, the consumption of various forms of energy by human communities produces heat, light, movement, new structures, deforestation, soil degradation etc, and new chemical compounds, some of which are significant pollutants. Consumption of water, on the other hand, changes its chemistry only under special conditions and has but a minute effect on the amount of water on the planet.[1] It does, however, influence its distribution and the dynamics of that distribution and, in doing so, it might render some sources of water unavailable. Perhaps more obviously, as we saw in Chapter 8, a significant indicator of water consumption by human communities is that the water becomes polluted.

Water consumption is commonly measured by the rate at which it is withdrawn from the relevant source or the rate at which it is discharged (*e.g.* into a sewerage system). Although human activities profoundly affect the amount of water reaching the sea from a river catchment, the physical structure *per se* of cities does not necessarily do so to any great extent. There are some qualifications, of course. Where there was once soil into which rainwater could infiltrate, there are now sealed roads, roofs and stormwater drains. In most cases the water that falls on these surfaces runs off, ultimately into the sea or a lake, and carries with it a range of pollutants. A city thus can increase the runoff more or less by the volume of water that would otherwise have infiltrated the same area of soil. On the other hand, many cities have parks and gardens that are watered from time to time; the water used for that purpose should not run off to any significant extent. The balance of these two processes will be determined by the relative areas of covered surface and of irrigated gardens, and by rainfall patterns.

The situation is very different for agricultural irrigation. Agriculture in general contributes significantly to river pollution through runoff and leaching but, on balance, irrigation is a nett remover of water. Irrigation water is 'lost' by infiltration followed by evaporation from the soil and transpiration through the leaves of plants, two processes that together have the misfortune to be known as 'evapotranspiration'. Depending on the geology of the land and the irrigation techniques used, irrigating water is also likely to find its way into groundwater (see below) and some might flow back into a river. But, on balance, extensive irrigation reduces the volume of water directly reaching the sea via a river.[2]

As well as the Darling River (Chapter 8), two other examples of the loss of water through irrigation deserve mention. The first is the Colorado River. About 64% of the runoff from its catchment is lost to irrigation and another 32% is lost by evaporation from reservoirs.[3] Very little of this river reaches the sea. The second example is provided by the Aral Sea, a very large lake or inland sea in the south of the former Soviet Union. The two rivers, the Amu and the Syr, draining into the lake from the mountains to the south east were diverted in the late 1950s to irrigate cotton plantations because the Soviet Union had embarked on a programme of 'cotton independence'. Before the diversion of the rivers, the Aral Sea lay at the heart of a complex and productive ecosystem. For many centuries the water level in the lake had remained at 50-53 m above sea level. The shoreline provided a rich habitat for birds, there was some 250,000 ha of wooodland and, in the delta of the Amu River, productive perennial pasture. The Sea itself was slightly saline (about 10 g.l^{-1} total salts) and oligotrophic with a large diverse fish population (see Appendix) that, among other things, yielded annual commercial catches of some 45,000 tonnes. The rivers provided water for drinking and for irrigation, the latter being restricted to the flood plains.

The expansion began in 1956 with a goal of 1.5 million tonnes of cotton annually. By 1960, diversion of water for cotton irrigation had reduced the total flow into the Sea to a rate below the losses to evaporation and seepage. In 1960 the lake surface was 53 m above sea level, its surface area was 68,000 km^2 and its volume 1090 km^3. By 1989 the corresponding measurements were 39 m, 37,000 km^2 and 340 km^3. By 1992 the two rivers had effectively stopped discharging; recent estimates put the yearly loss of water at about 27 km^3. More than 30,000 km^2 of lake bed is now exposed, and what once were ports are up to 80 km inland. Because the water was originally slightly saline, the dust that had been the lake bed contains a high proportion of salt. Wind erosion of the bed is put at about 10^8 tonnes annually. Some of the dust has been detected as far as the Himalayas and, closer to hand, salination merely from dust deposition is likely to put about 5 million hectares of cropland, principally the cotton that started it all, out of production. The salinity of the Sea had increased nearly threefold (to 28 g.l^{-1} by 1989)—almost in proportion to the reduction in volume. Most higher forms of aquatic animals have disappeared and the fishing industry has been destroyed.[4] The accumulation of pesticides drained from the irrigated lands into the local rivers—still used as a source of drinking water—has seriously undermined the health of the local people.[5] In Uzbekestan new land is still being sown with cotton as old land becomes too saline.[6] All things considered—including its role in the trans-Atlantic slave trade—one might be forgiven for concluding that cotton is one of the most destructive crops ever to have been cultivated.

Surface water cannot be exploited to anything like the extent required by the present human population without the construction of dams. But dams cannot increase the amount of available water above the theoretical maximum—the 'exploitable yield'—which is the total runoff from the 'dammable' part of the catchment.[7] If the dependent population grows to the point at which its water consumption equals the exploitable yield, then that catchment has nothing more to offer. Building more dams within the same catchment will not help. The options available to a community in this situation are essentially:

1 Reduce the size of the population.
2 Reduce the *per capita* rate of water consumption.
3 Supplement the supply by damming a river in another catchment.
4 Tap into groundwater if it is available.
5 Purify and reuse waste water.

Of these options only the first offers the possibility of long-term stability but, in most countries at the time of writing, is the least likely to be adopted in the short term. Up to a point, the second is relatively easy to implement in the industrial world, if only through a suitable charging system. The third is simply an invitation for a city to grow even further; it is another example of a vicious circle. This type of relation between the size of a city and its water supply is to be found in most of the large cities of the world. One of the best—and most hair-raising—examples is Los Angeles.[8] The fourth option is a variant of the third with similar general implications plus others of its own. Consideration of the fifth option is currently being forced on many cities and, in effect, has been practised in parts of Europe for some time. All options, except the first, amount merely to buying time if the human population continues to grow.

To put these generalisations into perspective we need to have some idea of how much water is used and for what purposes. As in Chapter 8, statistics on water consumption should be treated with reserve—at best as useful approximations. Agriculture is the most voracious consumer of water, a relation that links water consumption to population size more tightly through food production than through any other pathway (see below and Table A6). Only in Europe and North/Central America might agriculture not account for more than half the total water consumption; in those regions, the potential for industrial water pollution is high. The real nature of irrigation, however, is clouded by these simple statistics. Some additional insight can be gained be including another criterion, the annual rate of withdrawal of irrigating water *per agricultural worker.* This type of material is limited by the availability of the relevant demographic information, but there is enough for the period around 1987 to allow some useful comparisons.

Suitable data are available for ten sub-Saharan and four north African countries, and for Madagascar.[9] The lowest sub-Saharan rate (kl/agricultural worker/year, to the nearest 5 kl) was that of Burundi at 25. The highest was South Africa (4745) and the median of the ten was 57.5. Madagascar, which devotes 99% of its water consumption to agriculture, used 4700 kl. As might be expected, the north African countries were thirstier, ranging from 1760 for Algeria to 7290 for Egypt (median 3330 kl). But the comparisons with Australia and the USA are stark; the corresponding consumption rates in those two nations were 24,615 and 63,700 kl respectively. Irrigation in both nations falls largely within the domain of 'agribusiness'.

Irrigation rates per area of farmland can also be instructive but they too should be regarded with some caution. Table 9.1 summarises such rates for several selected regions.

TABLE 9.1 *Approximate irrigation rates of cropland in selected regions*

Region	*Rate of application ($Ml.ha^{-1}.y^{-1}$)*
World	9
Africa	11
USA	11
Australia, pasture + cropland	5-6

Areas of 'cropland' for the first three entries were obtained from World Resource Institute (1992), (Table 18.2); the volumes of water applied were derived form Table A6. The areas of Australian farmland were obtained from Australian Bureau of Statistics (1992a and b); volumes of irrigation water are from Table A6 and Australian Bureau of Statistics (1992a). Because of uncertainties in definitions of land use, in particular of 'pasture', and of proportions of groundwater applied, I considered it unwise to attempt any greater precision than shown in this table. Moreover it is likely that the Australian application rates listed above are too low (see text).

In this comparison Australia emerges as the most parsimonious of the regions listed, but there are some uncertainties in the numbers and in the definition of 'cropland'. Australia's national average, however, is lower than the corresponding rates for each of the three states irrigating with water from the Murray River (in Ml.ha^{-1}.y^{-1}: NSW, 8.4; Victoria, 6.7; South Australia, 13.7).[10] Irrigation of much farmland in Australia makes use of both surface and groundwater and the proportions are not easy to assess nationally. Some feeling for the use of groundwater, however, can be gained from statistics for the Namoi Valley in northern NSW (see Chapter 8). Estimates of groundwater withdrawals increased from 29 Gl in 1978-79 to 200 Gl in 1982-83. Nearly all of it was applied to cotton. Official allocations of water for irrigation for the same area in 1983 ranged from a maximum (in a good season) of 7.5 Ml.ha^{-1} (6.0 Ml surface water + 1.5 Ml groundwater) to a minimum (in drought) of 4.0 Ml groundwater with no surface water.[11]

As we saw in Chapter 8, groundwater accounts for roughly 0.6% of the water on Earth, or about 20% of the water not in the oceans. At the very least there is thirty times as much water underground as there is altogether in lakes and rivers. Groundwater is contained within aquifers—formations of rock, the hydraulic characteristics of which depend in essence on their porosity and permeability.[12] In some geographical environments, several different aquifers can be stacked above one another. In general the water is contained within the substance of a porous rock or bed of gravel, but there is at least one important exception to that generalisation. Limestone aquifers are generally impermeable but they can contain large bodies of water within caverns to form underground lakes and rivers. Such formations can be very important, as in the south east of England, but the water they contain is 'hard' because of calcium ions dissolved from the rock.

Unless they are extensively cracked, igneous and metamorphic rocks do not form aquifers. The value of an aquifer to a human population depends in part on the rate at which water can be brought to the surface and, in turn, that depends on the permeability of the rock and the depth of the water table below the surface. For example, Dury[12] cites permeabilities (m^3.d^{-1}.m^{-2}) of up to 5 for sandstones and conglomerates, 500 for loose sands and 50,000 for loose gravels (the depth, and hence the pressure, are not specified). Underground water is not necessarily stagnant; its flow rate depends on hydrostatic pressure, slope and, of course, the permeability of the aquifer through which it flows. Flow rates vary from one or two metres a day to one or two metres a year and, accordingly, residence times of water in undisturbed aquifers can range from a few days to tens of thousands of years.

In some situations, as when a significant proportion of an aquifer extends up a hill or a mountain, sufficient hydrostatic pressure can be generated to force water from a deep aquifer to the surface. The path taken by the water might be natural—through permeable rock—in which case a spring is formed, or it might be man-made to give an artesian well. The adjective, which denotes water under pressure, has its origins in the French province of Artois where the first known examples of free-flowing wells were dug. Artesian water is normally confined within an aquifer supported and capped by impermeable rock, known in the trade as an 'aquiclude'. The world's largest body of artesian water is the 'Great Artesian Basin', occupying about 20% of Australia and situated, roughly speaking, in the north east quarter of the continent.

Groundwater is used very widely. I have already alluded to the limestone aquifers in the south east of England. About a third of the water consumed in the USA is groundwater. At the time of writing, about 60% (by area) of Australia is totally dependent on groundwater to support human activities and, in another 20%, groundwater is the major source of water. Much of this is artesian.

Needless to say, the quality of groundwater varies. In shallow unconsolidated sediments water is subject to pollution from the surface but otherwise it is usually of good quality. Groundwater always contains some dissolved salts, the identity and concentration of which depend on the type of rock and the residence time of the water within the aquifer. Water in contact with any substance, be it the atmosphere, a glass bottle, a metal pipe, a river bed or an aquifer will dissolve some material from that substance. Just how much dissolves will depend on the chemical nature (and hence the solubility) of the substance, the length of time it is in contact with the water and, associated with that, movement of the water. There can also be other factors, of course, such as intrusions of the sea. It should not be too much of a surprise to find that a significant proportion of groundwater is saline.[13] This is the starting point of the salinity of salt lakes and the oceans—but their salinity is also an expression of their evaporation.

Because groundwater systems effectively have headwaters from which they are recharged and, because they flow, they also have discharge areas at a lower altitude than the intake. Given all that, unexploited groundwaters are normally in a steady state (when averaged over time) with respect to total volume of water and its flow rate. The salt concentration increases progressively the further downstream the water travels—as indeed it does also in a river. The total salt concentration is limited, however, by the steady state characteristics of the system — because water is eventually discharged and the time in contact with the aquifer is thus limited.

Quantitative assessment of groundwater resources can be confusing because of differences in terminology, discrepancies between reports, and sometimes because of arithmetical errors. According to a 1987 assessment[14], Australia has a total of 15,921 Gl of groundwater classified as a 'divertible resource'. Of that 38.4%, is 'fresh', 43.2% is 'marginal', 13.1% is 'brackish' and 5.2% is 'saline'. During the period 1983-84, 2522 Gl—or approximately 16% of the total—was used. An assessment of groundwater resources in south eastern NSW and Victoria classified the water differently—as 'low salinity' (less than 1000 mg.l^{-1} total salts) and 'high salinity' (the rest). Estimated volumes in that particular study area amounted to 256,000 Gl low salinity and 756,000 Gl high salinity water — proportions that are somewhat different from those cited above for the entire continent (the omission of Tasmania from the assessment has negligible significance). The 'estimated annual yield' (the 'divertible resource' above?) of low salinity water was 1807 Gl of which 70 Gl, or about 4%, was used in 1984.[15]

Despite some of the uncertainties, a fairly obvious conclusion to draw is that there is a lot of salty water underground. So now let us return to irrigation and salination of the soil.

The mechanism of salination caused directly by irrigation with groundwater is fairly straightforward. It is virtually axiomatic that a heavy dependence on groundwater for irrigation is most likely to occur in arid environments and, in such environments, evaporation rates will be high. For example, in the general vicinity of the Darling River in western NSW, rainfall averages about 150 mm annually—but evaporation rates are equivalent to some 2000 mm.[16] In other words, an open water tank with parallel sides and a flat bottom containing water to a depth of 2 m at the beginning of an average year, would finish the year with but 150 mm of water (always assuming that no-one confused the issue by using any of the water and that the diameter of the tank was great enough to allow normal rates of evaporation).

The consequences of irrigating in such an environment will be something like this. Irrigation with 5 Ml.ha^{-1}.y^{-1} is equivalent to a deposition of 500 mm water over the area for a year; rain adds another 150 mm. This is about a third of the potential evaporation and, depending on just when the irrigation was done and the actual (as distinct from the average) rate at which the water was applied, all could be lost to the atmosphere by simple evaporation—even without the added contribution of transpiration by the plants. Other than the proportion removed in produce, the salts dissolved in the groundwater would remain in the soil which would become progressively more saline with each application. For example, if the groundwater contained salts at a concentration of 500 mg.l^{-1}, irrigation at a rate of 5 Ml.ha^{-1}.y^{-1} would add salts at an annual rate of 2.5 t.ha^{-1}. In some senses that is the simpler of the two major causes of salination. It can also happen with irrigation by surface water but, in that case, the process is likely to be slower and less severe than with groundwater. Moreover, under some circumstances, irrigation with groundwater can produce yet another vicious circle. If water is withdrawn from an 'upstream' segment of an aquifer where the salt concentration is lower than the mean concentration, then that simple act of withdrawal will raise the mean concentration of the remainder.

As well as salts that are relatively benign at low concentrations, groundwater can also contain elements that are toxic in virtually any inorganic combination. One such element is arsenic. There have been zones of serious arsenic poisoning from groundwater in several countries, but to date the worst seems to have been in India, predominantly in West Bengal and Bangladesh. Over a wide area at least 200,000 people have been badly affected by the poison. According to Pearce, in the past the Bengalese farmers grew one rice crop annually. With the advent of irrigation from tubewells up to about 150 m deep, sufficient water became available to grow three or sometimes four crops of modern high yield strains of rice every year. The trouble was that much of the well water contained arsenic at dangerous concentrations, some as high as 2 mg.l^{-1}, a concentration that is 200 times the maximum recommended by the World Health Organization.[17]

There is another type of problem waiting in the wings for communities that are too enthusiastic in their use of groundwater. Withdrawal of water faster than it is replaced can lead to the permanent loss of an aquifer. As Dury[12] points out, porosities "significantly above 25" imply either open packing or very variable particle size or both. If there is a rapid nett withdrawal from such an aquifer, the packing pattern can change suddenly and reduce the porosity. Excessive withdrawals have caused irretrievable collapse of some aquifers, often followed by subsidence. For example, excessive use of groundwater has been responsible for a 7 m subsidence of Mexico City in the course of 90 years.

In coastal regions the sea is only too pleased to move in and replace any missing groundwater. Such a sequence has caused problems on the coasts of California, Florida, Japan and Libya. I shall summarise the Libyan situation. Some 2.64 million hectares of land in Libya is considered potentially suitable for cultivation but, because of a limited water supply, only about 470,000 ha is farmed. In the past, in Libya as elsewhere, the volume of groundwater that could be used was limited by the rate at which it could be pumped to the surface. A pump operated by manpower, animal power or a 'windmill' can supply water at a rate that, depending on the depth of the aquifer, is about sufficient to meet the basic needs of a family and a few farm animals. A portable diesel-powered centrifugal pump, on the other hand, can deliver some 3000 l.min^{-1}. Such pumps came into general use in Libya in the late 1950s.

About 95% of Libya's total domestic water supply is groundwater; at the time of writing, 99% of such water that is pumped goes to agriculture. In the principal farming areas, water consumption exceeds recharge of the aquifers by around 1760 Gl.y^{-1}—about five times the recharge rate. As a result the seawater/groundwater interface has moved 1-2 km inland and, from 1950-1990, the 'average' salinity of the groundwater rose from about 150 to about 1000 mg.l^{-1}. Libya intends to deal with this problem by transferring water from inland aquifers to the coast at a rate expected to reach 6.18 Gl per day. In this way it is hoped to reduce the demand on coastal aquifers below the natural rate of recharge and thus push back the seawater interface. (I understand that the first part of this project, a pipeline of some 1860 km, known locally as the "Great Manmade River", was opened in 1991.) The life of the inland aquifers under this arrangement is estimated to be about 50 years.[18]

In the USA the Ogallala Aquifer stretches south from the southern extremity of South Dakota through the states of Nebraska, Kansas and Texas and, for good measure, its western edge extends into Wyoming, Colorado and New Mexico. Before European occupation this region was predominantly grassland supporting 'millions of buffalo' that, in their turn kept impressive populations of grizzly bears, wolves, vultures and Indians in business. Later white settlers expected it to support, with the same degree of enthusiasm, domestic cattle and wheat. For a while it did—until drought in the 1930s produced the infamous Dust Bowl during which Oklahoma's topsoil migrated, some as far as Norway (see Chapter 10).

As in the grazing areas of inland Australia, farming households usually obtained their water from bores with the help of windmills. After World War II, however, the scene changed profoundly with the advent of diesel-powered pumps. In this case the impact of these pumps was much more extensive and, needless to say, much more dramatic than it was in Libya. Reisner[19] describes it thus:

'The irrigation of the Ogallala region, which has occurred almost entirely since the Second World War, is, from a satellite's point of view, one of the most profound changes visited by man on North America; only urbanization, deforestation and the damming of rivers can surpass it. ... Where one saw virtually nothing out of the window [of an aeroplane] forty years ago, one now sees thousands and thousands of green circles. From thirty-eight thousand feet, each appears to be about the size of a nickel, though it is actually 133 acres—a dozen and a half baseball fields. The circles are created by self-propelled sprinklers referred to by some as "wheels of fortune". A quarter-mile-long pipeline with high-pressure nozzles, mounted on giant wheels which allow the whole apparatus to pass easily over a field of corn, a wheel of fortune is man-made rain; the machines even climb modest slopes which would ordinarily defeat a ditch irrigation system. Wheels of fortune are superefficient but intolerant; they don't like trees, shrubs or bogs. Therefore the millions and millions of shelterbelt trees planted by the Civilian Conservation Corps have come down as fast as the region's fortunes have risen. All that now holds the soil in place is crops and water which cannot last."

By 1975 withdrawals from the aquifer by the main states amounted to 44 Gl per day and exceeded recharge by some 17,300 Gl annually.

'Dryland' is an adjective applied most often to the salinity of soils that are not irrigated. The details of the phenomenon can vary very widely with topography, geology, climate and, of course, with farming practices. The basis of dryland salinity is that the subterranean water table rises to the surface—or very close to it. The groundwater is either already sufficiently saline to be inhibitory to plant growth or, alternatively, inhibitory salt concentrations are reached with evaporation.[20] A common and important component of the process in managed land is the removal of trees. Tree roots supplemented by capillary action in the ground below them can control water levels to as far as 6 metres below the surface. The total rate of transpiration is a direct function of leaf area; a change in leaf area will affect transpiration rate accordingly. In some situations evaporation proceeds to the point at which the solution becomes saturated and salts crystallise. In the following outline I have drawn on examples from Australia's Murray-Darling Basin, a system that has been studied in detail and, among other things, illustrates some consequences of irrigating intensively over a water table that is not far below the surface.[21]

In its entirety, the Basin, which covers an area roughly equal to that of France and Spain, is a major part of Australia's most productive farmland; it produces about half of Australia's total agricultural produce. The Murray, at an age of some sixty million years is no chicken; it has sustained human populations for at least forty thousand years. It is Australia's longest river but, by the standards of major rivers from other continents, it is but a trickle with an average flow rate of about 16% of the Nile's, less than 2.5% of the Mississippi's and 0.25% of the flow of the Amazon. Its flow is now heavily 'regulated' by dams, weirs and the diversion of water for irrigation. It is not only a useful example of general principles but it can also provide an instructive case history of interactions between the management of agriculture and of a water supply.

The natural geological discharge of aquifers is sometimes into the sea, more often a stream or, if it is a closed groundwater system such as the Murray Basin, onto depressions in the ground. The groundwater systems in the Murray Basin have only two such discharge options—directly or indirectly to the Murray River and to depressions in the ground ('salinas'). In areas of high evaporation, salinity of the water discharged into salinas will increase—often to the point of crystallisation.

There are some additional peculiarities of the Murray Basin that exacerbate the problems of salination. One is that the landscape is extraordinarily flat — indeed there is no other continent with such an extent of consistently flat topography as this region of Australia. At Mildura, for example, the gradient of the Murray River is less than 5 cm per kilometre.[22] Under such circumstances, the flow rate of groundwater—as well as the river—is slow, the water table is high (0.5-1.5 m from the surface) and very sensitive to influences likely to raise it further.

The water table can be brought to the surface, or at least within capillary distance[23] of it, by decreasing the rate of withdrawal or increasing the rate of replenishment, or both. Irrigated farming with surface water can cause soil salination by both processes if, as is usually the case in the pioneering stages, trees are removed to provide farmland. The trees need not be on the land that is to be irrigated. The critical thing is that they are growing over the aquifer that runs under the irrigated land. Then, if irrigation applies more (surface) water than is transpired by the crops—as is commonly the case in Australia— the excess must go somewhere. Some will evaporate; depending on drainage arrangements some will run off to a stream, and some will infiltrate the soil and find its way to the groundwater, the level of which will rise accordingly. As always, the proportions of those three components can vary greatly with management procedures and the physical environment but, in situations such as those that prevail in the Murray Basin, they can rapidly bring the water table to the surface—and have done so. Waterlogging and salination are the usual results. Waterlogging of soil is lethal to most plants partly because, in such a context, the water is eutrophic and anaerobic for reasons similar to those outlined in the Appendix. Moreover it is usually another example of positive feedback—another vicious circle. Inasmuch as the rate of transpiration is a function of leaf area, once some plants are killed by waterlogging, the total leaf area—and hence transpiration rate—will diminish, and the rate at which the water table rises will increase.

Diversions from the Murray approximate 3780 Gl.y⁻¹. This amounts to about a third of the 'normal' flow and about two thirds of the flow of the river in a dry year.[24] Ninety four percent of the water so withdrawn was used for irrigating (1990 statistics) 198,000 ha in NSW, 236,000 ha in Victoria and 39,000 ha in South Australia where the annual application rates are equivalent to 7.9, 6.3 and 12.9 Ml.ha^{-1} respectively.[25]

In one relatively small area of the Riverine Plains in the Murray Basin, soil salination became apparent in about 1900—soon after the introduction of irrigation. By 1911 some 80 ha were affected; two years later the affected area had grown to 400 ha. A weir came into service in 1924 and more water became available for irrigation. By the early 1930s more than 300,000 ha were badly salted. Drains were constructed to remove groundwater which was then carried by a small stream, Barr Creek, to the Murray.[26] Subsequently Barr Creek was estimated to discharge some 170,000 tonnes salt annually into the Murray and was recognised as the most significant point source of such discharge.[27] The corrective measures were partially effective and the area of ground degraded by salt was reduced, but subsequently increased again. For example, the areas classified as salinised in 1982 were 4,000, 90,000 and 55,000 ha for NSW, Victoria and South Australia respectively. In 1991 the corresponding areas were 20,000, 90,000 and 224,500 ha. Over the same period, the affected area in Western Australia increased from 264,000 to 443,000 ha—nearly all of which was attributable to deforestation and the resulting dryland salination. The total areas of saline soil for Australia in 1982 and 1991 are given respectively as 426,000 and 790,800 ha.[28] But improved methods of assessment have recently made the picture much worse. Later estimates indicate that more than a million hectares of the Murray-Darling Basin are affected by irrigation salinity and some 500,000 ha by dryland salinity. At least 2,500,000 ha of Australian soil is degraded by dryland salinity, Western Australia being the worst affected.[29]

Responses to salination in irrigation areas have included both the adoption of measures to lower the water table and trying to live with slightly elevated soil salinity by growing *relatively* salt-tolerant crops such as barley. Lowering the water table has been approached through pumping groundwater and through the construction of drains to remove excess irrigation water. The next problem is that, in both cases, the water has to go somewhere. In some situations it is drained to evaporation basins which then become salt pans, but most of it is eventually discharged into the River Murray. In theory saline water can also be piped directly to the sea. Except in some Middle Eastern countries, however, the length of a pipeline constructed for this purpose and the energy needed for pumping the water are each likely to be considerable.

Another approach is to be much more frugal in the use of irrigating water in the first place. Growing rice in a semi-arid climate would not seem to exemplify such frugality. In some regions fruit trees are irrigated by trickle irrigation, as is common practice in Israel, but this method is not so easy to apply to cereal crops nor, of course, to pasture.[30] But, even here, although the technique will not raise the water table, the salts in the irrigating water will accumulate in the soil, albeit slowly, until carried into groundwater or a stream by occasional heavy rain. Finally, of course, an effective way of reducing the damage is to replant trees over the aquifers to replace those that were removed in the first place.[31]

In this and other types of situation, irrigation makes its impact not only on the soil but also on the river in which the result is an increase in salinity the further downstream you go. The Murray is comprehensively monitored from numerous stations effectively along its entire length. I have interpolated the following salinities for three stations from a graphical summary of river salinities for the period 1978-1986.[32] At Jingellic, in the Headwaters, the total dissolved salts (minimum, median and maximum, mg.l^{-1}) were respectively 20, 26 and 47. About a third of the way downstream, at the confluence of the Murray with Barr Creek and the Loddon River, the corresponding concentrations were 53, 160 and 853; at Milang, the last town on the river proper, they were 187, 440 and 920. The proportions of different salts also change with distance downstream. At Jingellic the predominant anion is carbonate; the principal cations are magnesium and sodium in about equal proportions. At Milang the dominant anion is chloride and the dominant cation, sodium. Overall, the withdrawal of water for irrigation, the regulation of its flow by dams and weirs and the drainage of excess irrigation water into the river has raised its median salinity by about 50%.

All of this is significant, not only as an example of some of the consequences of a set of processes, but because the Murray is the principal source of water for the residents along its banks in NSW and Victoria, and for the entire state of South Australia.

Sewage effluent has been used to a limited extent in irrigation since the mid 19th century. The practice is currently being widely reconsidered for its potential to reduce water pollution, because there is increasing pressure globally on water resources and because sewage contains essential plant nutrients. This topic is considered further in Chapter 14 but, for the present, we should acknowledge that irrigation with sewage has its own potential for causing salination by both of the mechanisms already outlined. Moreover, the sewage of large industrial cities usually contains toxic inorganic ions, including heavy metals, and refractory toxic organic compounds.

— A Broader View —

So much for the examples; now let me try to put them and the principles they are supposed to illustrate into some kind of perspective. Dams have two major hydrodynamic effects that are immediately obvious. As well as increasing the proportion of the total runoff from the catchment available to the human population, they smooth out seasonal irregularities in the flow rate of the river downstream from the dam. Some rivers in their natural state normally flood in the 'wrong' season for planting crops and dams can change that—not usually to the point of producing floods in the 'right' season, but rather to regulate the flow so that water is available for irrigation when it is needed.

The effects of a dam on the downstream flow patterns of a river are somewhat different according to the primary function of the dam—whether it is intended for flood mitigation, for generating electricity, or to provide a reservoir for a city or for irrigation. But, whatever the intention, the pattern of flow downstream is different from the unregulated flow and, on balance, is usually less. Floodplains are no longer subjected to the flooding and deposition of sediment that formerly sustained their fertility. And as we have seen, a reduced flow rate is likely to raise the salinity of a river because there is less water to dilute salts acquired from the river bed and from any saline discharges into the river. Moreover, a changed pattern of flow rate will inevitably mean a changed ecology both in the river and on its adjacent land. This might or might not be detrimental—it is not *ipso facto* a value judgment. But when it leads to increased populations of bilharzia in the river and then diminution of sardine fisheries along the coast because their ecosystems are deprived of the nutrients contained in the sediments carried by the Nile before the construction of the Aswan High Dam, it is not too good.

As I have already said, a reservoir for an urban water supply does not necessarily reduce the volume of water reaching the sea (or a lake) from the river if the sewage of the town is returned to the river or directly to the sea. On the other hand, if the reservoir is used primarily to supply water for irrigation, then the total volume of water reaching the sea from that river will be reduced, sometimes to zero. A dam causes a large area of land to be inundated and, since it retains most of the suspended matter normally carried by a river, the reservoir progressively silts up and its storage capacity is reduced accordingly. Moreover, the rising water in the reservoir kills the vegetation that formerly grew on the surrounding hills leaving bare soil. This soil erodes and raises the amount of sediment above that which would otherwise be carried by the river—especially when a drought has lowered the water level to expose the bare soil and the drought is then broken by heavy rain. Lake Burragorang, one of Sydney's principal water reservoirs, was created by construction of the Warragamba Dam. It accumulates about two million tonnes of silt annually—more than twice the rate previously estimated.[33]

The situation is likely to be much worse if forest within a catchment is cut. India's large Tehri dam, situated in the Himalayan foothills, was given a life expectancy at birth of a century; widespread deforestation within the catchment followed by siltation reduced that to 30 years. The record, however, seems to be held by China. The Sanmenxia Reservoir, which had several objectives, came into service on the Huang He (Yellow River) in 1960. By 1964 45% of its storage capacity was lost to siltation and power generation was abandoned. Large outlets had to be opened at the base of the dam to release silt and flush it downstream.[34] My understanding is that the Huang He is known as the Yellow River because of the amount of silt it carries, in which case we might fairly conclude that someone's right hand was not too sure about what his left hand was doing—or vice versa.

The inundation caused by a dam not only kills vegetation—it also kills or displaces all the animals, including hominids, that previously inhabited the affected area. The farms and houses of the hominids are necessarily abandoned in the process. A dam not only has the potential to smooth out the extremes in the flow rate of a river, but it can also change local climate by raising the humidity and dampening temperature fluctuations in its immediate vicinity. In a temperate climate this, together with the amenities offered by the extra water available—especially that used for irrigation—can work wonders for the mosquito population and generally perturb the regional ecology, not merely that of the river. Perturbations of this nature have been associated with outbreaks of Rift Valley fever in Egypt in 1977 and Mauritania in 1987.[35] And sometimes dams burst.

Inasmuch as many rivers cross state or national boundaries, all of these features of dams have tremendous potential for causing political trouble. Among the more obvious regions with such potential are the Middle East and North Africa, but they are not the only ones. Of course it is not only dams that have the potential to cause international trouble. Any water supply that is used by more than one nation, state, or even community, has this capacity—especially in arid regions.[36]

Since the beginning of the 20th century gross water consumption by the human population has increased about 7.5 fold. In 1900 the total volume used for agriculture, industry and domestic/municipal purposes amounted to about 600 km³ of which 90% went to irrigation and 6-7% to industry. By 1950 the total had increased to some 1430 km³ with agriculture and industry accounting for about 82% and 13.5% respectively. Extrapolated volumes for 1995 amounted in total to almost 4560 km³ with agriculture using 66% and industry up to 25% (see also Note 40 and Table A6). A more recent estimate indicated a total annual withdrawal of some 3800 km³ of fresh water from all sources.[37]

By the end of the 20th century there were more than 45000 dams on the planet.[38] Collectively their reservoirs hold more than 5000 km³—or about 10% more than the current annual rate of consumption. By and large droughts are regional rather than global, but regional over-exploitation combined with drought can be serious. From the very outset, constructing dams and pumping groundwater have been subject to positive feedback dynamics, similar in their essentials to those governing the relation between the human population and its food supply. The exploitation of water resources, however, has several components that can make the dynamics of water consumption more complex than those of growing food.

For a start, statistical data of the type listed in Table A6 show that most of the water we use goes to growing food and fibre and is therefore driven by the same forces that drive agriculture.[39] During the 20th century, the demands of secondary industry increased faster than those of agriculture although, so far, industry still uses only about 38% of the volume of water used in irrigation. (Another assessment has consumption by industry equivalent to 28% of that of agriculture.[38]) Some industries also have the potential to use wastewater of a quality that would not be suitable for irrigation. Water is used widely for generating electricity, a use that, in the light of current concerns about air pollution and the 'enhanced greenhouse effect', has the attraction of being clean—and hydroelectric schemes do not preclude diversion of the water for another purpose. Finally, of course, a growing human population needs water for domestic purposes—although that is the least demanding of the three categories listed.

Given all this, together with the small 'safety margin' of the world's reservoirs, it is not surprising that there are plans afoot to build more dams. If implemented, those plans would increase storage capacity by about 50%.[40]

The urge to 'make the deserts bloom' seems to be a sufficiently important part of the human psyche to have become an end in itself, supplementing—perhaps sometimes supplanting—the simpler and more basic need to produce food in an arid environment. When decisions are made to undertake large scale irrigation projects it is probably naive to try to distinguish between economic forces on the one hand and, on the other, a need to provide a growing population with its basic needs. But if one is to try to make the distinction, the realistic—albeit cynical—view is that the economic incentives are dominant and have been so for all of the 20th century. The following quotations give some indication of this type of thinking in Australia in the late 1970s:

"In the agricultural context, water as a commodity has no intrinsic value. It is its agricultural use which provides a return."[41]

"Early advocates of irrigation development believed that the irrigation districts would provide a haven for starving stock in a drought. It is now seen that the best economic use of irrigation water is to produce as much as possible every year, so the option of building up fodder or water reserves is precluded. It follows that these irrigation districts cannot be used to agist large numbers of stock in a drought. However, by maintaining their production in all seasons, they impose a strong stabilising effect on commodity prices."[42]

"The Colleambally Irrigation Area, established adjacent to the New South Wales Murrumbidgee Irrigation Area in the 1950s, ... covers 95,000 hectares, and before intensive development, was a pastoral area with some wheat and oats. In 1978-79 the value of annual production from that area was $23.0 million, and the total capital value some $56.5 million. It has been estimated that had the area been left under the old regime, the present value of annual production would have been $1.43 million, and the total capital value some $6.2 million. Thus the irrigation development has produced a nine-fold increase in property value and a sixteen-fold increase in annual production."[42]

Movement of people into irrigation areas was noted with approval. The population of six irrigated shires in Victoria increased from 29,000 in 1910 to 77,000 in 1977 whereas the population of six adjacent non-irrigated shires remained essentially constant over the same period.[41] Potential problems were not ignored, but they were underestimated.

"On the basis of economic and other studies in the Kerang Region of Victoria, it has been concluded that for every unit of production within that Region left as dry land, with irrigation there are now three units of production. This is despite the spread of soil salinization *to what is believed to be near its ultimate extent.* (My italics) Obviously, under irrigation without the salinity problem, the area could have produced even more; salinity has prevented it reaching its maximum potential, but only by some 20 to 25%."[41]

All of these comments were published in 1980 and all referred largely or wholly to the Murray-Darling Basin. Ten years later perceptions had changed. By then salination was acknowledged to be a major problem; the comprehensive treatise, The Murray, published by the Murray-Darling Basin Commission in 1990, devotes four lengthy chapters to salinity. The problem has since been acknowledged to be much more serious than identified in that publication.

— Some More Examples —

Throughout much of the 20th century large dams have been built in many parts of the world. There have always been clearly stated reasons—ostensible reasons—for these undertakings. Those reasons were inevitably based on a need for more water for agriculture, for the generation of electricity, to sustain a city—or all three. But it would be foolish to assume that these were always the only reasons.

Perhaps it is part of our folklore that big is beautiful. Perhaps engineers, particularly hydraulic engineers, are seduced by size and all that entails. Perhaps they are victims of hubris and seek mastery over nature. But if they are, they are not alone. For as long as I can remember there have been proposals that the Clarence River, one of the larger coastal rivers in northern NSW, should be diverted westwards across a mountain range to irrigate the Murray-Darling basin. I do not know if the proponents of this scheme included engineers. They did include farmers—farmers west of the Divide, not coastal farmers on the fertile floodplain of the Clarence. Fortunately the scheme has not yet been implemented.

When organisations, public or private, are established for the purpose of building dams, then that is what they seek to do—build dams. There might or might not be good reasons why they should build any particular dam, but one reason for doing so is very probably to justify their own existence. This motive is not likely to be stated openly—indeed it might not always be consciously acknowledged within the organisation—but we can be confident that it will influence what should be a dispassionate evaluation of a proposal to build a dam. The Tasmanian Hydroelectric Commission seems to be, or to have been, a good example of an organisation that builds dams for their own sake.[43] Moreover, if there is more than one such organisation, rivalry can be a powerful incentive to build dams. Reisner has described some effects of rivalry between the US Bureau of Reclamation and the US Army Corps of Engineers—and attitudes of politicians—to massive hydraulic projects. Reisner puts it thus:

"In the Congress, water projects are a kind of currency, like wampum, and water development itself a kind of religion. Senators who voted for drastic cuts in the school lunch program in 1981 had no compunction about voting for $20 billion worth of new Corps of Engineers projects in 1984, the largest authorization ever. A jobs program in a grimly depressed city in the Middle West, where unemployment among minority youth is more than 50 percent, is an example of the discredited old welfare mentality; a $300 million irrigation project in Nebraska giving supplemental water to a few hundred farmers is an intelligent, farsighted investment in the nation's future."[44]

In 1950 there were 5270 large dams throughout the world. By 1998 there more than 36,500 and, according to another source, the number had exceeded 45,000 by the end of the 20th century.[45] When dams are proposed for developing countries, and when their construction is dependent on foreign aid, there are probably more opportunities for corruption than occur in the simpler domestic situation. Moreover, it is virtually impossible to build a major dam in any densely populated agrarian country without disrupting the lives of very many people and, of course, inundating a lot of farmland.[46] Over the past several years, a number of proposed dams have received a deal of publicity because of their social, political and economic implications.

The Narmada River flows in a westerly direction, crossing about 1000 km in India at a latitude around 22° N—about the latitude of Calcutta. It discharges via the Gulf of Cambay into the Arabian Sea. In 1985 the World Bank agreed to lend the Indian Government $US450 million towards building the Sardar Sarovar dam, the 'centrepiece' of a $3000 million project on the Narmada. The planned height of the dam is 150 m;[47] by 1995 it had reached 80 m. When full, the reservoir would submerge more than 37,500 ha of land, (including?) 12,000 ha forest, and displace people estimated to range in number from 120,000 to 'about one million'. The aims of the project are to irrigate cropland (40 million ha according to the Indian Government; 1.8 million ha in Gujarat and 75,000 ha in Rajasthan according to another assessment), to generate 1450 MW electricity and to provide drinking water for a population ranging from 8000 villages to 30 million individuals.

There has been considerable opposition to the project, some of it from the residents of the threatened villages, some from outside. The outside opposition has arisen partly from sympathy with the threatened villagers, partly from more general environmental reasons, and partly from reaction against the roles of the Indian Government and the World Bank. A 'Save the Narmada' movement was formed locally; its members threatened to remain in their homes and drown in the rising waters because resettlement sites offered by the Indian Government were unsatisfactory. Other villagers who had been resettled returned to their original villages for the same reason. In 1992 an independent review commissioned by the bank commented that human and environmental concerns had been subordinated to engineering and construction demands.[48]

In 1993 the World Bank withdrew from the loan agreement with $170 million of the loan not yet disbursed. The reasons given were that the Indian Government neither consulted local people adequately nor provided suitable land for those to be displaced.[49] The withdrawal of the bank was followed by celebrations in the affected regions and elsewhere in India. The World Bank has been criticised for its role in the project and for its 'breathtaking ... dishonesty'. The Indian Government continued with the project despite the bank's withdrawal but, in response to a petition from the 'Save the Narmada' movement in 1995, the Indian Supreme Court halted construction of the dam at a height of 80.3 m. In February 1999, the Court modified its verdict and allowed the dam to proceed to a height of 88 m. In October 2000 it gave permission for immediate construction to proceed to 90 m and subsequently, in a series of 5 m increments, to the height originally planned, subject to approval from the 'Relief and Rehabilitation Subgroup' of the Narmada Control Authority. In May 2002 the Authority gave permission to raise the height from 90 to 95 m even though at least 3500 families had not yet been 'rehabilitated'.[50] Because of cash shortages, however, it is doubtful that the dam can be completed before 2010.[51] Nevertheless, it would appear that the Indian government, through its Supreme Court, is determined to suppress criticism of the project, especially when such criticism eminates from a distinguished author such as Arundhati Roy.[52]

A commission established by the Indian Environment Ministry is reported to have found that, of 212 dams under construction or recently completed, about 190 did not comply with conditions laid down by the Ministry. State governments were the common offenders.[53]

The Changjiang, or Yangtze, River rises at an altitude of 5,800 m in the Himalayas and flows across China to discharge into the East China Sea at Shanghai. At 6,300 km it is the planet's third longest river, surpassed only by the Nile and the Amazon. Its average discharge is some 1500 km^3 annually[54] carrying with it 500-600 million tonnes of sediment. The catchment area is about 1,800,000 km^2. The catchment and the floodplain support a population of some 400 million, contain about 25% of China's cropland and produce around 40% of the nation's total farm output—including approximately 70% of the grain crop. More than 700 tributaries join the main river. One of them, the Min River, supports an irrigation system that has been working successfully for about 2,200 years and makes the Chengtu Plain China's most productive area. The Changjiang and its associated canals carry 80% of China's inland water traffic.

The geomorphology of the region and the high population intimately dependent on the river have combined to make some of the inevitable floods catastrophic. In the summer of 1931 a flood inundated 3 million ha farmland and killed 150,000 people. A similar flood in 1954 killed 30,000—the improvement being attributed to diversions and 'the frantic strengthening' of dykes.

About three quarters of the way down the river is a section known as The Three Gorges. The uppermost gorge (Qutang) is 8 km long and the shortest of the three. It constricts to about 100 m making it, not only the narrowest of the three gorges, but also the narrowest part of the river once it leaves its mountain birthplace. The second gorge, Wuxia, or Witches Gorge, is about 40 km and the third, Xiling, about 48 km in length. The entire Three Gorges strip, that is the gorges themselves plus the intervening wider valleys, total about 250 km in length. During various floods within the past century, water levels within the gorges rose by 20-25 m. There is already a low dam, the Gezhouba Dam, at the outlet from Xiling Gorge; a new dam, the Three Gorges Dam, is proposed for its inlet. When completed it will be the world's biggest.[55]

The primary rationale for this dam is power generation, but flood mitigation and improved river navigation are also important. The potential for generating power is put at 17,000 MW, which is some 40% more than the capacity of the present largest hydroelectric dam, the Itaipu in Paraguay. The power so generated will lessen the demand on fossil fuels, according to some estimates by up to 40-50 million tonnes of coal annually.[56] The dam's proponents claim that it will also reduce demand for such domestic fuels as wood and crop residues but, because of the nature of the country and the expected direction of resettlement (see below), that is questionable.

The planned height of the dam is 180 m, the surface area of the reservoir 1150 km² and its capacity 40 km³ (40,000 Gl). By one assessment, the reservoir will inundate 19 cities and counties, nearly 240 km² farmland and 50 km² orange groves.[57] By another it will flood 2 cities, 11 counties, 140 towns, 326 townships, 1,351 villages—and nearly 240 km² farmland.[58] It will enforce the resettlement of a population that is currently about 725,000 but, by completion of the project in 2008, is expected conservatively to be at least 1,130,000. Much of the resettlement will be to higher altitudes in the same general area—that is to say, within the catchment of the reservoir, a movement with a potential for accelerating soil erosion. The general region has already experienced serious erosion, not least because of extensive deforestation.[55] The project was—and is—controversial.

Opposition to or support for the project within China seems to depend on where people live in relation to the dam and its reservoir, and to how they will be affected by it. Which, of course, is not very surprising. But support might well have been amplified because, as Chau[57] points out, "the central government, in anticipation of the project, has refrained from investing in the area for the last 30 years". In other words, there would seem to be more than the usual amount of room for improvement in living standards or, as economists are wont to say, they are starting from a low baseline.

Much has been written about the Three Gorges Dam and a good deal of it is critical.[58] The criticisms are political (the general secrecy of the evaluation and ultimate approval of the project) and environmental—on obvious grounds such as already outlined, and on the inevitable uncertainties of a dam so much larger than any other. A good deal of criticism or scepticism is related to the question of resettlement. As Chau notes, since 1940 about 86,000 "water conservancy projects" in China have caused the resettlement of more than 10 million people. About 30% of these schemes failed for various reasons and, accordingly, there is suspicion about this one. Other critics are concerned in a general sense about changed river ecology. Some believe that the most important needs of the area could be met by a series of small dams on tributaries of the main river. Within this general context Smil[59] has commented that it would be quite possible technically to build more than 11,000 small hydroelectric dams throughout China with a total generating capacity 25-30 times that of the Three Gorges Dam.[60] But, whatever the technical pros and cons of the undertaking, there is little doubt in my own mind that the decision by the Chinese Government to go ahead with the dam was influenced, if not determined, by the psychological/political need for a monument. It was a decision reflecting Man's—in this case predominantly men's—long standing love affair with enormity in engineering projects.

The Pergau Dam illustrates a different type of problem—or at any rate the problem is more obvious than in the two previous examples. In 1988 the British Government secured from Malaysia an arms order worth more than £1,000 million. Some months later Britain announced a loan of £234 million to Malaysia towards the £400 million needed to build a hydroelectric dam at Pergau, near the Thai border. An editorial in *The Ecologist*[61] states that there is documentary evidence that the loan was conditional on the purchase of arms from British companies, a requirement contravening the British Overseas Aid Act of 1966. The same editorial notes that the aid money included a payment of "£60 million to the ruling political party of Malaysia, in addition to £40 million to agents and ruling families in Malaysia". It also claims "close links" between some of the arms manufacturers and the British Conservative Party (in office at the time).

There are many other examples with similar overtones.[62] My point in these three cases is to provide some examples in support of the fairly simple proposition that dams are not always what they might seem. They are not by any means all bad, but they are often built for the wrong reasons. Sometimes those wrong reasons seem like the right ones because the wrong question was asked in the first place. We always—or nearly always—ask questions about natural resources in the form of 'how are we to meet the future food/water/ energy etc. needs of a growing population?'. To take that question at its face value and attempt to answer it in practice is to perpetuate vicious circles of the type linking the human population and its food supply. Until the question is replaced with something along the lines of 'how much food/water/energy etc. can we afford to produce/use without destroying our habitat?' we will continue to accelerate a decline in the planet's overall carrying capacity. That there can be such a decline while the human population is growing might not be immediately apparent, but it is a topic to which I shall return.

In this and the preceding chapter I have given some examples of the ways in which the human population interacts with its water supply. When the population is large that interaction runs the risk of degrading the resource in question—water. When the population is also growing the risk is increased. When the growth of the population occurs in the context of a positive feedback loop with its food supply, the interaction of the population with the water supply also becomes destructive. The all-consuming characteristics of a vicious circle are not confined to the central elements of that circle—population and food. They extend to every other commodity used by the population. And water, of course, is not just 'another commodity'. It is a critical component of that primary feedback loop.

In the next two chapters we will look more directly at the interaction of population dynamics with the soil and with two major nutrient elements, nitrogen and phosphorus.

People and the Soil
Part I – Erosion and Organic Matter

When an industrial product (tobacco, assorted pesticides etc) is accused of being a serious threat to health in one way or another, an apologist for the relevant industry usually claims that there is no 'positive proof' or 'definite proof' or 'scientific proof' to support the accusation. In this type of situation the apologists probably have very little idea of what 'positive proof' etc actually means. Even an experiment whose results are so convincing that there is no doubt in anyone's minds about their significance, is ultimately a statement of probability, rarely of certainty. Failure to appreciate the statistical qualifications inherent in scientific statements is, among other things, a source of misunderstanding and conflict between scientists and lawyers when the former are called as expert witnesses and the law is seeking a simple 'yes' or 'no' answer to its question(s).

'The sun will rise to-morrow' is a statement of probability, albeit one on which you could bet with a fair degree of confidence. But more probable than even that is the statement that the human population will not continue growing for ever. Indeed, unless we get excessively pedantic about the meaning of 'for ever', that is a claim that is about as certain as we can ever reasonably expect.

— People —

Growth of the human population will ultimately be limited by either a decrease in birth rate, an increase in death rate, or both; any of a number of factors could contribute to each of those determinants. But whatever influence might make itself felt first—war, disease, birth control, toxaemia from pollution etc—the ultimate Damoclean sword is the state of the planet's soils—most directly, but certainly not exclusively, its agricultural soils.

A concept that causes more than its share of controversy, at least in academic circles, is that of 'carrying capacity'. Basically it is a simple notion—the

number (or better, the biomass) of animals that can be sustained by a specified area of land. With species other than Man there does not seem to be too
much of a problem, but when we apply it to ourselves there is usually hell to
pay. To the best of my knowledge, Thomas Malthus was not responsible for
the term but its essence was central to his thesis for which, in recent years, he
has been hauled over many a hot coal. Thus Davis' takes the poor old bloke
severely to task for his well known pronouncement that 'population is necessarily limited by the means of subsistence'. What Davis said included:

> "Population is not in fact 'necessarily limited' by a scarcity of the means of
> subsistence. It is limited by whatever contributes to mortality, such as
> wars, epidemics, overeating, alcohol consumption, smoking, and so on."

Strictly speaking, Davis's first sentence is justified, but it would not be if
Malthus had used 'ultimately' instead of 'necessarily'. Indeed 'necessarily' can
be interpreted to mean just that and, in the usage at the end of the 18th century, might well have had 'ultimately' as a primary meaning. In his *Essay on
Population* published in 1798, Malthus said explicitly, "Famine seems to be
the *last*, most dreadful resource of nature" (my italics).

The various factors listed in Davis's second sentence can all affect growth
rate, but the sentence can be generally true only if a population is stationary.
What criticisms like Davis's do acknowledge, consciously or otherwise, is the
broad positive feedback relation between population and the food supply under an agricultural régime—but they do not seem to acknowledge that the
earth is finite. So much, however, for the semantics.

Cohen[1] has listed 66 estimates (with quotations from their proponents and
supplementary comments of his own) of the number of people the earth can
support—the planet's carrying capacity—dating from 1679 to 1994. The first
estimate was by Antony van Leeuwenhoek and deserves a brief digression.

Van Leeuwenhoek was a Delft (Holland) cloth merchant who was also
custodian of the Town Hall, a position sometimes described as a political sinecure. He had little formal education. It is probably just as well that his job
at the Town Hall was a sinecure because he spent much of his time making
small optical lenses, a craft at which he was remarkably skilled—and secre-
tive. No-one was able to match the quality of his lenses during his lifetime.
He made a 'simple' (that is, single-lens) microscope with a magnification of
about 300. There were compound microscopes in existence at the time but
their aberration was so bad that they were virtually useless for a purpose to
which van Leeuwenhoek put his. He was the first to observe individual single-
celled organisms, including bacteria. He first reported his observations in a
letter written in October 1676 and published in the *Philosophical Transac-*

tions of the Royal Society in 1677. That and subsequent letters laid an indispensable foundation for the ultimate acceptance of the 'germ theory' of disease (Chapter 6)—although that foundation was not seriously built upon for another century and a half.

He has an illustrious place in the history of microbiology. It is of no small interest that he has also contributed to demography and has done so with his feet more firmly on the ground than many of his 20th century successors. His estimate of the earth's human carrying capacity was 13.4 billion.

The 'serious' estimates in Cohen's listing ranged from 0.5 to 1022 billion (there was one tongue-in-cheek suggestion of up to 10^9 billion—or 10^{18}). The estimates included two from the 17th century, two from the 18th, two from the 19th with the rest from the 20th century. Of the quotations and summaries compiled by Cohen, only three directly acknowledged the phenomenon of 'overstocking' (see below). One acknowledged a finite life expectancy (about 5 centuries) of 'fertilizers'. In some other quotations the word 'sustain' or its various derivatives was used.

To phrase these observations differently—through all but those four specific examples, and possibly one other, there is an implicit assumption that, by minding its Ps and Qs in some technological directions, in particular energy consumption and pollution, the human population, once it has reached its maximum, can maintain that maximum more or less for ever. There is no apparent practical recognition that a static population is a dynamic entity that constantly makes an impact on its habitat and that the apparent stasis might thus be temporary.

A population grows exponentially when it increases at regular time intervals by a constant proportion of the population existing at the beginning of each interval. Another way of saying the same thing is that such a population will double regularly and thus has a characteristic doubling time. It is a pattern of growth that is commonly associated with single celled organisms that propagate by binary fission—each cell giving rise to two—but it is certainly not restricted to microbes. Under suitable conditions, virtually any type of plant or animal population will grow exponentially. Because populations in this state increase by a constant *proportion* every equal interval of time, it is misleading to say, as is often said of human populations, that such a population is increasing by 'x million' every year. It might be a useful approximation for a few years but, while exponential growth persists, the annual increment needs to be updated regularly.

Another characteristic of populations that are growing exponentially is that, to draw graphs that make any real sense of their dynamics, you need to plot the logarithm of the population—not simply the population—against time. If that is not done you are usually faced with the problems of selecting a scale that keeps the whole range of the population on the same graph—in which case the scale is too small to make much sense when the population is small. Alternatively, if you can make sense of the graph at low populations, you are likely to run off the paper at high populations. With a simple arithmetic plot you cannot determine growth rates in any practical sense because they are constantly changing nor, for any of these reasons, can you usefully compare growth rates of populations that are very different in size from one another. These problems are avoided with logarithmic plots in which the slope of the plot represents the exponential growth rate, regardless of the absolute size of the population. Growth rates are expressed by the proportion (commonly percentage) by which a population increases in a specified time interval or, inversely, by the generation or doubling time (see, for example, Figs 5.1, 6.2 and 10.1). The latter, of course, is analogous (but of opposite sign) to the half-life of a radio-isotope.

It has been quite common for authors of environmental books to give examples of the length of time needed for a population growing exponentially to consume first, half its habitat or principal resource, and then the whole lot. Water lilies growing in a pond have been quite popular in this regard. I used to give students of microbiology the following exercise, partly as practice in dealing with microbial population dynamics, and partly to impress upon them the significance of exponential growth.

A bacterial cell has a mass of about 10^{-13} g; the mass of the earth is about 6 x 10^{27} g. If those numbers are difficult to grasp it is fair to say that the cell is quite small and the earth quite large. If the bacterium grows with a mean generation time (doubling time) of 30 min (quite common under appropriate conditions), how long would it take for the bacterial population to equal the mass of the earth? The answer is about 68 h; if the earth were twice the size, the answer would be about 68½ h (see Appendix). The corresponding question applied to the human population with its present doubling time of about 40 years, population 6 billion and average weight around 50 kg (there is a high proportion of children) gives an answer of about 1770 years. Doubling the size of the earth would add merely another 40 years. Of course each of these situations is physically impossible but they do illustrate a general truth—the 9th 'Law' which says, in effect, that if a population grows exponentially it will eventually consume essential resources faster than they can be replenished. From now on, unless specifically stated otherwise, the term 'growth rate' applied to populations means the exponential rate.

But back to carrying capacity. This concept is related to some of the arguments advanced in Chapter 2, arguments that might stand brief repetition with a slightly different emphasis. In the course of going about its business, any animal population, or for that matter any individual animal, performs two very basic functions. It consumes resources and it produces waste. In the wild, or in any balanced system, the consumption of resources and the breakdown and recycling of the wastes occur at rates that, when averaged over a sufficient period of time, are mutually compatible. In an unbalanced system, such as occurs when a species gets out of control and its population explodes, that species will eventually consume its resources and accumulate wastes faster than can be compensated by natural processes. Then the land supporting that species is overstocked and the system will duly collapse. If the wastes and the bodies of the offending species remain within the affected area and there is no other consequence of significance, such as serious soil erosion, the system will slowly recover but, when it does, it will have a species composition quantitatively different from that which prevailed before the collapse. On the other hand, if the offending animals and/or their wastes are moved out of the affected area, as happens with commercial agriculture, the system will take much longer to recover or, in extreme cases, it will not recover at all within a time scale of any relevance to human history. At least that will be the situation unless, as a minimal precaution, the essential nutrients removed from the affected land are replaced from some other source. And then, of course, there is an immediate question about the life expectancy of that other resource.

These principles apply as much to Man as they do to other animal species. But with modern Industrial Man there is a major added dimension to waste accumulation. This now includes not only biological material, but also prodigious amounts of non-biological refractory objects and substances, many of them toxic, produced by an extraordinary range of industrial activities. All of this stuff has to go somewhere and its disposal usually ensures that some land will no longer be available as a habitat or for farming.

Thus we can have a situation in which purely subsistence farmers, even with the most conscientious management of their land, will exceed its carrying capacity. The dynamics of their ecosystems—their farms—become unbalanced and, as with any other species, their land is overstocked. With commercial agriculture, because produce is removed from farms, the situation is different, more complex, more destructive—and ultimately more difficult to deal with.

It might or might not be the case that we can now identify carrying capacity as the human population supportable by a specified area of cropland or pasture, but to do so in a way that made any sense at all would require taking averages of essentially a stationary population over extended periods. Carrying capacity must refer to the long term if it is to make sense. So far that has not been possible, if only because of the continuing growth of the global population, growth supported in part by bringing pristine land into agricultural production. And that leads to another point. It must be possible for a population to overshoot the carrying capacity of a particular area of land and, by so doing, bring about a deterioration of the soil. Soil degradation need not immediately stop or even retard growth of the population if the rate of degradation is slow. Nor must it immediately affect farm yields if there is some 'reserve capacity' in the soil or the system (see Fig. 4.1). Moreover, for a while, soil degradation can be masked, at least as far as crop yields are concerned, by regular applications of fertiliser. But if a system of agriculture is causing progressive soil degradation, as assessed by virtually any set of objective criteria, then I would argue that carrying capacity *under that régime* has already been exceeded. That is true even if the population is still growing. If the régime persists for long enough, the dependent population will eventually collapse. To say that is not to imply that there is some simple quantitative relation between (rate of) degradation and loss of carrying capacity. Nor is it to deny that management practices might be changed to prevent further degradation, in which case carrying capacity might not necessarily have been exceeded.

It is impossible to quantify these generalisations because the details will vary enormously from one situation to another. And, as I have said, a change in management practices can change the circumstances that determine whether or not soil is being degraded. Moreover, a change in management practices can also change the population supportable by an area of land if the size of that population is a direct function of the yield of food from the land. For example, the yield from moderately degraded land might be increased merely by introducing irrigation, a practice that itself might increase or decrease the level of degradation (however defined), or leave it unchanged.

But, all that admitted, there must be a limit to the number of people a given area of land—any land—can support. It must also be true that, if the soil within that area is being progressively degraded by any means, then the process of degradation must ultimately, if not immediately, reduce the size of the population which that land can support. In other words, it will reduce its 'carrying capacity'. Supporting the affected population by transporting food from somewhere else does not change that basic fact.

In earlier chapters I referred to some aspects of the growth of the human population, the distribution of population between town and country and effects of those various dynamics on public health, farming practice and, by implication, on the state of the soil. The 19th century was marked in the 'West' generally by rapid technological innovation and development, accelerating urbanisation and an increasing overall growth rate that largely reflected a drop in mortality—particularly infant mortality. There was no corresponding increase in overall Asian growth rate, but African rates increased sharply in mid century (Fig. 5.1). The picture changed in the 20th century and, to illustrate that change, I must change the geographical categories.

According to whom you read or listen to, the nations of the planet are divided into 'Developed' and 'Un-' or 'Underdeveloped' or 'Developing', into 'North' and 'South', 'Rich' and 'Poor', or 'First-' and 'Third World'. (The 'Second' World seems not to assume much prominence in this last scheme—rather like Second Class on British railways in the middle of the 20th century.) In Fig. 10.1 and Table 10.1 the source of the material obliges me to use 'Rich' and 'Poor'.

The global picture is dominated by the dynamics of the 'Poor' countries whose population accounted for two thirds of the total at the beginning of the century and more than three quarters by 1990. Rates illustrated in the figure are tabulated in Table 10.1.

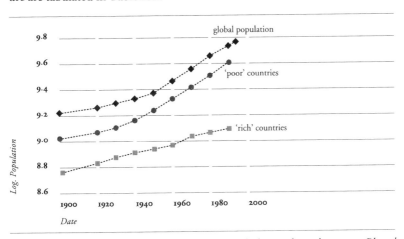

FIG. 10.1 *Semi-logarithmic plots of population growth during the 20th century. Plotted from material tabulated by Livi-Bacci (1992). See also Note 2.*

There are various levels at which population pressure can affect soils. Broadly speaking they can be grouped into (i), the overall physical state of the soil and (ii), the content and availability of individual nutrient elements. In this chapter under (i) I shall restrict myself largely to rates and extent of erosion and to the organic content of soil; under (ii), in Chapter 11, the emphasis will be on the elements, nitrogen and phosphorus. This separation is artificial and I do it for convenience. There are, of course, also many other factors that can influence the state of soil, factors that include availability of other nutrient elements, salinity (Chapter 9), water infiltration rates, accumulation of refractory toxic substances etc. Moreover, because I shall continually refer to soil(s) in a general sense, I should emphasise that, although the technical term covers all the layers ('horizons') of soil, most of my comments refer to topsoil. My second disclaimer is simply to emphasise that, despite my generalisations about 'soil', there are many types of soil whose physical, chemical and ecological characteristics are different'—but I have to draw a line somewhere.

A population that exceeds 6 billion and has a doubling time of 40 years has the potential for an enormous impact on the planet's soils. So the next question is—have we, by the criterion of soil degradation, exceeded the earth's present carrying capacity for the human population?

TABLE 10.1 *Population doubling times and growth rates for intervals selected within Fig. 10.1*

Group	Interval	Doubling Time (y)	Annual Increase (%)
'Rich'	1900-1920	86	0.81
	1920-1940	75	0.92
	1940-1950	150	0.46
	1950-1960	50	1.4
	1980-1990	150	0.46
'Poor'	1900-1920	120	0.58
	1930-1950	55	1.3
	1950-1990	32	2.2
Global	1900-1920	100	0.69
	1920-1950	69	1.0
	1950-1970	35	2.0
	1970-1995	40	1.7

— Soil Erosion —

Apart from special circumstances, such as covering the earth with concrete or bitumen, the ultimate expression of soil degradation is undoubtedly erosion. That is true notwithstanding those situations in which soil can be degraded (*e.g.* through salination) to a point at which it will no longer support the growth of vascular plants—but it does not erode because the physical conditions are not conducive to erosion. On the other hand, soil that is not otherwise degraded can be vulnerable to erosion under some circumstances. Erosion amounts to the removal—by wind, by water (including ice, although the effects of glaciers are very much slower and more specialised than those of liquid water), and by gravity—of soil from somewhere and its deposition somewhere else. It has gone on, if not since the beginning of time, then at least for as long as the planet has been able to support terrestrial vascular plants. To add insult to injury, the soil that is removed usually has a higher content of nutrient elements (about three times higher) than the residual soil.[4]

With the possible exception of arid regions, the commonest erosive agent is water, directly as rain and by the flow of water over or through land. I say with the 'possible' exception because, from time to time, arid regions can be subjected to extensive erosion by water. Water causes three main types of erosion and one or two supplementary types. The main ones are sheet, rill and gully erosion. Water is also usually implicated in 'mass movements' which, in the present context, is not a political term but rather refers to landslides.

Sheet erosion is the removal of soil in more or less a uniform layer by the flow of water over a relatively wide area of poorly protected soil. Rill and gully erosion are mutually related but differentiated by their size. In both cases water flows down a slope, seeks out depressions which it then proceeds to enlarge as it aspires to become a river. Rills are foetal or infant gullies. The distinction between infancy and adulthood—between a rill and a gully—is made arbitrarily at a depth of 30 cm. Notwithstanding any such childhood aspirations, rill erosion is commonly associated with sheet erosion for the purposes of assessment. This is partly because of the shallowness of the rills—and because 'migration' of rills often leads to sheet erosion. Then there is tunnel erosion. This is caused by a fairly vigorous flow of water beneath the surface of the soil which, for a while at any rate, is left apparently undisturbed. From time to time such tunnels collapse and thereby metamorphose into gullies. And of course, there is the ever-present erosion of their banks by rivers and streams.

The soil removed by all of these processes is carried by water and eventually deposited somewhere. Just how far it is carried and where it is dumped will vary enormously in response to many physical variables such as the size range of the soil particles, the volume and velocity of the water, the topography of the land etc. Some of the suspended soil will be deposited on land, some will settle onto stream beds, some will be carried out to sea. Soil that settles on land will normally enrich it; soil that settles on a stream bed or the sea bed is effectively lost as far as terrestrial plants are concerned. Except in arid regions, water erosion is generally more serious than wind erosion—and even in arid regions it can be severe on occasions. Its 'advantage' is that it is easier to measure—both as sediment load in streams and in the runoff from selected or experimental plots.

Many factors affect the susceptibility of soil to erosion either over time or in a single event. They include the physical and chemical characteristics of the soil, its structure, the slope of the ground, climate and season,[5] and perhaps above all, the density and type of plant cover. All of these vary widely and have done so in the past as they do to-day. But once Man took to lighting bushfires and, in particular, took up farming, soil degradation became a problem and erosion accelerated in virtually every region where farming was practised.[6] The term 'accelerated' erosion is used to distinguish the human contribution from the natural process.

Of course there are also natural processes for forming soil; if that were not so, none of us would be here. The question all of this is leading to is whether or not natural and accelerated erosion together exceed the global rate of soil formation. If they do, then that might be reason for saying that we have already exceeded the earth's carrying capacity for *H. sapiens*.

Estimating rates of soil formation is partly a matter of definition. Thus, if we are talking about the so-called 'O horizon'—the surface layer of organic matter in a well-vegetated area—we encounter a situation in which leaf litter etc can be decomposed quite rapidly to humus, provided there is sufficient moisture. (A well managed domestic composting system can produce in a few weeks humus that would do credit to any 'O' horizon.) Some rates given for 'soil' formation under such circumstances in temperate and tropical climates[7] lie in the range of 10-30 t.ha^{-1}.y^{-1}. Another estimate[8] puts rates of formation of 'topsoil' in the USA at between 3 and 4 t.ha^{-1}.y^{-1}. But when we are considering soil erosion, we are rarely dealing with circumstances in which there are plenty of plants to provide the raw material for humus. To make sense of rates of soil formation in the general context of erosion we need to look at the inorganic components of the process—ultimately the weathering of rocks. The speed with which rocks are weathered depends not only on their physical—and biological—environment, but also on the surface area of rock exposed to that environment. Thus, other things being equal, deposited 'stones' are likely to break down faster than parent bedrock, and volcanic ash a good deal faster still. But normally such weathering is very much slower than the decomposition of plant and animal tissues in moist temperate and tropical climates. Rates of soil formation, thus defined, have been estimated to range from 2 to 200 mm per thousand years or 0.03-3 t.ha^{-1}.y^{-1}. Overall, formation of Australian soils is probably restricted to a much lower range[9]—perhaps to a maximum of about 0.3 t.ha^{-1}.y^{-1}.

You do not have to be an expert in this business—and I am certainly not— to appreciate the difficulties inherent in trying to estimate global or regional rates of soil erosion with either precision or accuracy, let alone both. Attempts to do so have been bedevilled by controversy and some serious howlers.[10] There are many obvious inconsistencies in some of the quantitative information that I have tabulated both here and in the Appendix. Table 10.2 lists some estimates of global rates of soil erosion by water. The mean (and standard deviation) of the 'unbracketed' rates is 28.5 (± 10.3) Gt.y^{-1}; the median is 25 Gt.y^{-1}. So, to the extent that there might be a consensus, it is around 25 or 26 Gt.y^{-1}. But a consensus is not, in itself, a guarantee of accuracy.

TABLE 10.2 *Some estimates of total global rates of soil erosion*

Erosion Rate (10^9 t.y^{-1})	Comment	Reference
17.5	'Total'	(a)
(10	'Natural')	
20		(b)
25	Mean for period 1973-93	(c)
26	crop and grazing land	(d)
36	'Total'	(e)
(10	'Natural'. See also Ref. f)	
50	by water*	(f)
25	by wind*	
	(See also Ref. g)	

(a) Dury (1981): (b) Jansen and Painter (1974) cited by Rosewell *et al.* (1991): (c) Brown *et al.* (1993): (d), Brown and Wolf (1984) cited by Johnson and Lewis (1995): (e) Lal and Stewart (1990) cited by Crosson (1995): (f) Myers (1993) cited by Pimentel *et al.* (1995a and b): (g) Crosson (1995). *Pimentel *et al.* cited a total rate of 75 Gt.y^{-1} caused by water and wind. I have split the total into two components on the basis of the generalisation that water erodes about twice as much soil as does wind (see text).

A recent assessment of erosion rates in Australia gives a very different picture from that of Table 10.2. Wasson and associates[11] estimated total sheet and rill erosion for Australia in an 'average year' to be about 14 (13.94) Gt.y^{-1}. Almost 70% of that occurs in those regions that drain directly to the coast.[12] Fourteen gigatonnes is 55% of the median and 28% of the highest global rate in the table. The authors comment "It is surprising that Australia should have an above average erosion rate, given the continent's low gradients and arid climate." But that figure accounts only for sheet and rill erosion. When gully erosion[13] is included the total rate for Australia reaches 28 $Gt.y^{-1}$—about 10% more than the median rate for the whole planet, or 56% of the highest of the tabulated rates.

In 1993-94 Australia was accredited with 469 million hectares of land contained in 'agricultural establishments'.[14] If the sheet and rill erosion occurred only on farmland, then the rates just quoted would be equivalent to a national average of about 30 $t.ha^{-1}.y^{-1}$. Adding gully erosion doubles the rate to 60 $t.ha^{-1}.y^{-1}$ (see Table A7 for various regional rates). The 'continental mean' rate for Australia listed in Table A7 is put within the range 0.27-0.45 $t.ha^{-1}.y^{-1}$. But that is for the whole country. The total area of Australia, including Tasmania, is 7,682,300 km^2; if that is used as the reference then the rill and sheet erosion given above amount to 18 $t.ha^{-1}.y^{-1}$—40 to 67 times greater than the tabulated 'continental mean'—and perhaps at least sixty times greater than the national average rate of soil formation. Adding gully erosion again doubles it.

Part of most erosion is 'natural', part man-made—but estimating the proportions is not easy. Wasson and his colleagues[11] commented on the rate at which ground level is lowered by erosion. They said, among other things, that "it seems that the modern rate of 2.5 m 1000 $year^{-1}$ is about 145 times higher than the average currently estimated long-term rate, even though the modern rate is probably exaggerated". Notwithstanding the very wide regional differences in erosion rates, for the purpose of this discussion it is not unreasonable to treat Australian erosion as almost entirely the result of human activities. A case might therefore be made for considering erosion rates on a *per capita* basis, if only for giving a different kind of perspective.

Using 1995 total population estimates, the highest global erosion rate in Table 10.2 is equivalent to 13 $t.hd^{-1}.y^{-1}$. The Australian erosion rates quoted above give an overall *per capita* rate of Australian sheet and rill erosion of about 750 $t.hd^{-1}.y^{-1}$ and, when gully erosion is included, 1500 $t.hd^{-1}.y^{-1}$. The corresponding rates per *farmworker*[15] are approximately 35,000 and 70,000 $t.hd^{-1}.y^{-1}$ respectively. All in all an impressive performance and, notwithstanding a lack of reliable global statistics for numbers of farmworkers, probably a world record.

But back to the basic erosion rates. Is the situation really so much worse than previously thought or does the disparity between the numbers used above and the 'continental mean' in Table A7 simply reflect the difficulties in obtaining reliable representative measurements in this type of situation? It is a situation in which a disinterested statistical appraisal will probably lead to bewilderment or exasperation. It is a situation calling for judgment. For those such as I, who are not actively engaged in this type of work, that can be difficult if 'judgment' is not to be a synonym for prejudice.

The rates tabulated for the 'continental mean' are old—they are based on measurements made in the late 1960s and early 1970s. They reflect sediment yields measured in rivers—not the material as it is removed from the land surface. On the other hand, the rates obtained by Wasson and associates are recent, comprehensive and obtained by several methods. In this case my own view is to forget the statistics and accept that the recent estimates are likely to be much closer to the truth than the early ones. If that is indeed so, then all but the highest global rate of erosion in Table 10.2 are too low, and the highest rate (50 $Gt.y^{-1}$ by water, 75 $Gt.y^{-1}$ total) is probably also conservative.

In 1991 the Earth was accredited with 1,441,573,000 hectares of cropland and 3,357,292,000 hectares of 'permanent pasture'.[16] If all the eroded soil had come from these two sources, the total of 75 $Gt.y^{-1}$ (water + wind) would be equivalent to an annual loss of about 15.6 $t.ha^{-1}$. But specific erosion rates are usually greater for arable land than for pasture. If we take the rates listed for sediment discharge from the Queensland coast (Table A9) as a rough guide, then 50 $Gt.y^{-1}$ would be distributed on average as 6 $t.ha^{-1}.y^{-1}$ eroded from pasture and 21 $t.ha^{-1}.y^{-1}$ from cropland. If the same guide is extended to include wind erosion, then those rates become about 8.5 and 32 $t.ha^{-1}.y^{-1}$ respectively. This, of course, is grossly oversimplifying—the Queensland coastal distribution is not likely to prevail in other types of environment and, moreover, the global rate also includes erosion from sources such as recently cleared forests as well as erosion from pristine land. (It is worth noting, however, that the sediment discharged from pristine land in Table A9 amounts to 162 $kg.ha^{-1}.y^{-1}$.) Nevertheless, some sort of a yardstick can be useful and, again, that yardstick points to global rates that are at least 3-11 times greater than overall rates of soil formation. (Erosion rates for specific types of situation and for various regions are given in Tables A7 and A8.)

Wind erosion occurs predominantly, but not exclusively, in arid and semiarid regions.[17] In Australia that means where the annual rainfall is less than 375 mm—almost two thirds of the continent. Gravity has an involvement in the dynamics of wind erosion that differs in some important respects from its role in water erosion. Even without considering landslides, water erosion is obviously worse on slopes than on flat ground because of the effect of gravity on the potential and kinetic energies of the water. Most wind erosion, however, occurs on plains, but gravity does have an important say in the size of particles picked up and the distance they are carried by the wind.

Roughly speaking, wind moves soil in three different patterns, according to particle size, although the boundaries between the categories will also be affected by wind velocity. The largest particles (> 0.5 mm) for the most part are pushed along the surface. Their movement is as much the result of being hit by other particles as it is a direct response to the force of the wind. This process is commonly known as 'creep', the participating particles as the 'creep fraction'. The next group contains particles in the range 0.1-0.5 mm. They tend to bounce along the surface, perhaps in their best leaps reaching a height of about 50 cm. When they fall they hit other soil particles, dislodging or disintegrating them. This category, the 'saltation fraction', contains the largest proportion of material moved in the course of wind erosion. It is a major contributor to and component of aeolian landforms, such as sand dunes, and it is usually this fraction that buries fences etc in dust storms. The third class, the so-called 'suspension fraction', contains particles of less than 0.1 mm diameter. This is the fraction that can be carried enormous distances, that provides much of the drama of dust storms and which, when deposited in quantity, produces the type of sediment known as loess. About 10% of the earth's land surface is covered by loess to depths of 1-100 m.[18]

Wind erosion is more difficult to measure than water erosion—there are no streams in which to estimate sediment loads. Such estimates as have been made are often based on apparent loss of soil from affected areas. Losses of up to 10 mm.y^{-1} (130-140 t.ha^{-1}.y^{-1}) have been reported over a period of twenty years from 'Dust Bowl' sites in Kansas.[19] Globally, wind is thought to shift about half as much soil as does water.[17] So let me now turn to this most famous example of wind erosion—the American Dust Bowl of the 1930s. It is important, not only for intrinsic reasons, but also because the types of social, economic and political influences that made it possible have been, and are being, repeated in many other semi-arid parts of the world.

The Great Plains of the United States run roughly north—south down the centre of the continent. They stretch from the Canadian border to about the middle of Texas and, on the way, cover most of North and South Dakota, Nebraska, Kansas and a goodly bit of Montana, Wyoming, Colorado, New Mexico, Oklahoma and Texas. They generally lie at an altitude between 400 and 1000 m. In the past the plains had been grassland, grazed by wild buffalo. Explorers in the early 19th century who had come, of course, from the humid east, saw the area as desert, unfit for farming—especially cultivation. In the second half of the century, however, rainfall increased, perceptions changed and the land was settled by aspiring graziers whose understanding, such as it was, was based on their experience in the wetter eastern states. As in Australia, land was cheap; it was overstocked and grazed with little regard for manuring or applying fertiliser—or for the future. The average annual rainfall is around 500 mm but it is variable, needless to say. Between 1870 and 1890 settlement moved to and fro on the plains in response to variations in rainfall. In the 1750s, 1820s, 1860s and 1890s there were climatic conditions similar to those that were later to cause such devastation, but on those earlier occasions the soil structure was different and there was a grass cover. There were no dust storms. The trouble began early in the 20th century with a move towards the cultivation of wheat. Cultivation was then becoming mechanised; in order to accede to the expectations of tractors and harvesters, the fields were enormous and windbreaks were removed. The change from grazing to cultivation was encouraged by a period of seemingly reliable rainfall, a period lasting from around 1915 into the 1920s.

The 1930s brought a series of droughts that, sooner or later, were inevitable. The first storm hit in May 1934 and topsoil was carried into the Atlantic. The National Resources Board estimated that 14 million hectares had been destroyed, 50 million hectares severely degraded and a further 40 million hectares were in a marginal condition. In March 1935, dust clouds reached altitudes above 3000 m; the soil removed in that storm was estimated to be of the order of 350 t.ha^{-1}. In some regions storms during the 1930s collectively removed more than a metre of soil (*c* 14,000 t.ha^{-1}.) Fences were buried, some houses nearly so. Ducks and geese were suffocated in flight; rabbits were blinded by the dust—an event, as Reisner comments, that provided the human inhabitants with 'something to eat'. Oklahoma experienced 102 such storms in one year; North Dakota had 300 in eight months. Reisner summarises the events thus:

"The Dust Bowl was triggered by the same fatal congelation of hope and drought that caused the plains to empty half a century earlier. The longest severe drought in the nation's history—the one the Bureau of Reclamation planners, ever optimistic, now use as their 'worst-case scenario'—began to descend over the West in 1928. For seven years in a row, precipitation remained below normal. The snow that fell on the plowed-up fields of the Dakotas was so light that the ground, bereft of insulation, froze many feet down; the snow evaporated without penetrating and the spring rains, those that came, slid off the frozen ground into the rivers, leaving the land bare. The virgin prairie, grazed well within its carrying capacity by thirty million buffalo, could probably have withstood the wind and the drought; ravaged by too many cattle and plowed up to make way for wheat, it could not. If not the worst man-made catastrophe in history, it was, at least, the quickest."[20]

The region remains vulnerable. About two million hectares were damaged in each of the winters of 1988-89 and 1989-90.

The Dust Bowl caused emigration from the affected areas, much of it to the San Joaquin Valley which held more than half of California's 'prime farmland'. The San Joaquin farmers were profligate in their use of groundwater for irrigation, an approach prompting Reisner's observation: "Having exhausted a hundred centuries' worth of groundwater in a generation and a half, they did what any pressure group usually does: run to the politicians they normally despise and beg relief". In 1977 the Valley experienced a dust storm that removed more than 25 million tonnes of soil from grazing land in the course of 24 hours. Overall, a total of about 2000 km² land was seriously degraded.[21]

During the 1930s the Canadian prairies suffered dust storms in sympathy with the American Great Plains. The consequences of government plans to cultivate the so-called virgin and idle lands in the north of the Soviet Union from around the middle of the 20th century have been described in some detail.[22] As we saw in the previous chapter, badly managed cotton irrigation in the south west of the former USSR has caused severe soil degradation and allowed subsequent wind erosion—and the effective destruction of the Aral Sea.

Australia has recorded major dust storms since 1840 and detailed meteorological records have been available since 1960. Thirty six 'major wind erosion events' have been recorded since 1840. In many of these storms westerly winds have carried soil from the centre of the continent to New Zealand. One such storm during a drought in the summer of 1982-83 had its worst effects on land that, normally grazed, had been converted to wheat cultivation following good rains a few years earlier.[23] During a drought of generally similar severity in 1994/95, there was less wind erosion in the same region, partly because of better management. Since the early 1970s there has been a significant decline in the number of days of dust storms in Australia, and one of the reasons is paradoxical.

"... since the 1960s, overgrazing has caused a fundamental change in vegetation type in semi-arid environments, with the spread of woody weeds. Because these perennial plants are not palatable to stock, their spread has produced a more complete and persistent ground cover in these wind erosion-sensitive lands, ..."[24]

The drier parts of Africa and Asia are constantly vulnerable, again often because of population or economic pressure's encouraging the cultivation of land that was traditionally grazed. For example, Inner Asia (southern Siberia, Mongolia and much of northern China) contains over 2.5 x 10⁶ km² of grassland—somewhat more than 6% of the grassland of the planet. During the 20th century there has been extensive land degradation in the Chinese and Russian sections of this area, but very little in Mongolia, which is an independent state. More precisely, some 75% of the pasture and about half the arable land in parts of Siberia, and over a third of the grassland of Inner Mongolia (Chinese) has been significantly degraded. In Mongolia (proper), on the other hand, only about 9% of pasture has been classified as degraded. There are perhaps two principal reasons for these differences. One is that, in the Russian and Chinese regions, there has been extensive conversion of pasture to cropland, a conversion accompanied by the use of heavy machinery to plough light topsoil. The other is the general replacement of nomadic pastoralism by the confinement within paddocks of animals at densities high enough to require cultivated hay etc for fodder (see also Chapter 13). In contrast, the Mongolians have largely retained their traditional nomadic habits, but these are also threatened by recent economic developments.[25]

But in addition to extreme cases, there must also be more frequent instances of moderate wind erosion at levels that are much more difficult to measure. For example, about 20% of the dust in the troposphere (the lowest layer of the earth's atmosphere), has evidently been blown from disturbed soils.[26] But the question to which we must now turn is what, overall, is the present state of the planet's soils? Is degradation from all causes—impoverishment, salination, toxicity, erosion, waterlogging etc—serious or not?

The total land area of the planet is slightly more than 13×10^7 km^2. In the period 1989-91, cropland and permanent pasture accounted for 37% of the total (cropland 11%, pasture 26%). Those two categories had increased by 1.8% and 2.4% respectively over the previous decade. Forest and woodland made up 30% of the total but had declined by 7.8% since 1979-81.[27] The increase in the area of farmland, 2.2% over the decade, or an average of 0.22% annually, is about 13% of the global population growth rate over the same period (Fig. 10.1 and Table 10.1). So either the efficiency of farming increased enormously over that time, or we have been over-exploiting the soil—or both. The loss in wooded area over the decade was about three times the gain in farmland, a difference that can be partly explained by clearing tropical forest to replace abandoned farmland.[28]

By 1990 the total area of land considered to be degraded by human activities amounted to some 1.96×10^7 km^2, equivalent to about 17% of the earth's 'vegetated land'.[29] There are various possible definitions of soil degradation; I used my own definition in Chapter 4, for example. If degradation is extended to include 'land' in a broader sense than is normally meant by 'soil' and is defined as 'a reduction in (the land's) capacity to supply benefits to humanity', a different assessment emerges. By this criterion Daily[28] has estimated the total degraded area—land affected by soil degradation, degraded dryland vegetation without (frank?) soil degradation and degraded tropical forest—to amount to about 5 x 10⁹ ha or around 43% of the planet's vegetated surface. All in all this, together with the erosion rates already discussed, does point to significant overstocking with *H. sapiens*.

— Organic Matter —

Apart from plant cover, the content of organic matter within soil is probably the most important single natural source of protection against erosion. Soil organic matter comes in many forms, in many shapes and sizes. It is the stuff of the roots of living plants, of soil fauna and their dung, of the cells of microorganisms, of plant litter in its various stages of decomposition, the dung of surface animals. Organic matter profoundly affects the structure of soil and, with that, the rate at which water infiltrates from the surface and diffuses within the soil. It affects the amount of water that soil can retain—organic matter in general can hold about three times as much water as can the same mass of clay.[30] It is a vitally important source—or, perhaps more accurately, storage depot—of essential nutrient elements. Moreover, elements that are bound organically are usually much less susceptible to loss from soil than the same elements in some of their common inorganic forms. This is especially true of nitrogen and phosphorus. These elements and sulphur are released from organic combination, predominantly by microbial action, at rates that vary widely with circumstances—with water content, temperature, with pH—but usually at rates that keep the soil microflora and the vascular plants very much on their competitive toes.

Other than plant roots, the organic entities that are most important for soil structure are the cells/bodies of the soil biota and the various polymers and aggregates derived directly or indirectly from them. These substances include polysaccharides (of which cellulose is dominant), lignins, proteins, lipids, nucleic acids etc, all in various stages of breakdown, as well as a range of complex molecules, many aromatic, that are formed in the soil and collectively designated as humus. There is no simple method of precisely estimating the organic content of soil; the usual practice is to estimate the amount of organic carbon. But the carbon content of each of these classes of compound is different, and the proportions of those substances can vary from one environmental situation to another.[31] When estimating the organic content of soil in any practical sense, it is therefore necessary to rely on a rule of thumb. One such rule equates the mass of organic *matter* with 1.9 x the mass of organic *carbon*.[32] A content of 2% organic *carbon* (*c.* 3.8% organic *matter*) is considered by some to be the minimum necessary for the maintenance of a satisfactory soil structure; others contend that, with less than 2% organic *'content'*, soils are 'generally erodible'.[33] Another rule of thumb is that '1.16% organic carbon' is the minimum able to supply nitrogen to crops.[34]

In undisturbed soils some soil organic matter is commonly occluded in 'pockets'—usually of clay. When that happens the occluded material is protected from microbial attack and it can have a residence time of up to 5 years in tropical soils and from 10-50 years at high latitudes.[35] But most soil is in a constant state of turnover, provided the physical conditions allow biological activity. It is thus possible to have conditions that approximate a steady state (see Chapter 2) in respect of organic content and, with care and a little bit of luck, to make some broad generalisations about the distribution of organic matter between soil and the biomass above ground. A tropical rainforest, for example, normally has proportionately less organic matter in the soil relative to the 'standing biomass' than does a temperate forest.[36]

Agriculture has the potential in a number of ways to diminish the organic content of soil and, particularly during the 20th century, has realised that potential only too well. Traditional farming did it with the bare fallow. If moisture were retained—one of the objectives of fallowing—then soil biota took advantage of it to consume existing organic matter, since they were no longer being supplied with the fresh stuff. More generally, if pasture is stocked above a critical level, organic matter is consumed and metabolised faster than it can be replaced by the growth of grasses etc and the deposition of dung. For any one set of conditions, the critical level is lower if the product (the animals) is removed from the pasture to be consumed elsewhere. Fertilising such pastures with inorganic fertilisers does not directly replace organic material although it can do so indirectly through plant growth *given sufficient time*. But it is not given time if stocking rates are high.

The same general type of argument applies to cultivation. Traditionally, especially in China and countries with similar practices, the organic content of small intensive farms was sustained by the diligent application of organic manures. With early Australian and North American farmers, on the other hand, there was little such application (Chapter 7). Organic manures have very little place in modern industrial agriculture where the size of fields and paddocks and the heavy dependence on machinery demand, by economic criteria if nothing else, inorganic fertilisers. Some management practices, such as ploughing-in stubble rather than burning it, and 'no-till' cultivation, can diminish the rate of depletion of organic matter in arable land, but there seems to be little evidence that they reverse it.[37]

The organic content of soils throughout the world varies so widely that there is little point in trying to tabulate a few examples. Let me just say that a range of about 0.5-12% organic matter (*c.* 0.25-6% organic carbon) would cover most situations. (In some extreme cases, such as alpine humus, the content of organic matter might reach about 50%, but this has little relevance to the circumstances under consideration.) A range of numbers such as these, however, gives no idea of the geographical distribution of organic matter, nor of what is happening to it.

Here Australian information is again both informative and disturbing. Except for those on the east coast, Australian soils are relatively poor in organic content. Table A12 lists some values for organic matter in various Australian soils, values that fit quite well into the global range. In one case (northern NSW), the table also partially quantifies the lowering of the organic content of soil by cultivation. What the table does not show is that the organic contents of those soils were halved after 15-20 years of conventional cultivation.[38] Nor does it show that the organic content of the listed krasnozems halved over several decades following the clearing of rainforests for pasture, cropping or horticulture[39] or that Mallee soils (see Chapter 9) sharply lost stability after only one year's cultivation[40]—or that losses of organic matter from Australian soils have been put within the range of 0.039-1.244 $t.ha^{-1}.y^{-1}$, depending on location and depth of sampling.[41]

The six southern Queensland soils listed also provide the basis of an instructive story. They included several different soil types and represented six districts with mean annual rainfalls ranging from about 475 to 675 mm. The organic content of the soils increased, as might be expected, with rainfall. The increase was linear for both virgin and cultivated soils (there was one abnormally high content in the virgin soils—the highest of the range in Table A12). The cultivated soils had approximately two thirds of the organic content of their virgin counterparts. With cultivation, organic carbon was lost exponentially with half-lives ranging from 0.6-16.9 years (median 7.75 y).[42] By the 1980s about 75% of all Australian soil contained less than 1% organic carbon (*c.* 1.9% organic matter) in the surface layers.[43] Much of that soil is used for dryland farming.

The same kind of story can be told of most soils in most countries where agriculture is highly mechanised, and in many where it is not but which have high population densities. China is a special case inasmuch as, in the middle of this century, it set out to 'modernise'—to change from a predominantly agrarian society to one in which 'efficiency' and mechanisation were paramount. According to Smil,[44] the 'natural' organic content of 'intensively cultivated soils on the Northeast China plain' was about 9% (matter). By the 1970s it was down to 5% and, by the mid 1980s, 2%.

There is very little room for complacency about the planet's soils. In Chapters 12 and 13 we will look in a bit more detail at some basic reasons for the worldwide decline in soil quality, but first—the two elements, nitrogen and phosphorus.

CHAPTER II

People and the Soil
Part 2 – Nitrogen and Phosphorus

The two elements, nitrogen and phosphorus are essential participants in the biochemistry of every organism on this planet. In their inorganic forms they are 'macronutrients' of plants and many types of microorganism. But the emphasis given to them at the complete expense of a range of other essential elements is not intended to trivialise the importance of those other elements. Nor is it intended to imply that those elements are not subject to patterns of dynamics similar in principle to some of those outlined below. That disclaimer is broadly true whether it is applied to macronutrients such as calcium, magnesium, potassium, sulphur, iron etc, or to any of a range of trace elements. It is with nitrogen and phosphorus, however, that some of the outstanding characteristics of modern industrial agriculture are most obvious and, perhaps, most disturbing. These are discussed, therefore, largely as exemplars. And, as I find myself obliged to say repeatedly—you have to draw a line somewhere.

— Nitrogen —

There is no shortage of nitrogen. Elementary nitrogen accounts for about 80% of the atmosphere—3.8-3.9 x 10^{15} tonnes of N_2. And unlike the situation with some other major essential elements, such as phosphorus, our farming activities cannot suppress its cycling, although they can and do change the dynamics of that cycling. Central components of the so-called nitrogen cycle —it is a complex web rather than a simple cycle—were identified in Chapter 2, but I shall briefly mention some of them again.

Elementary nitrogen is reduced to the level of ammonia by several types of bacteria—symbiotically by *Rhizobium* in the root nodules of legumes, by some free-living bacteria (notably but not exclusively by *Azotobacter* spp and *Clostridium pasteurianum*) and by several genera of cyanobacteria (see Chapter 8). Ammonia is also produced industrially by the Haber process (Chapter 7).

In addition, there is precipitation in rain of some volatile forms of bound nitrogen and, where there is or has been smoke, of some suspended particulate forms. On the other hand, nitrogen is lost from soil by denitrification (reduction of nitrate to elementary nitrogen—see Chapter 2), by the leaching of nitrate to waterways and by the loss of ammonia to the atmosphere. These losses always reduce the 'efficiency' of assimilation by plants of nitrogen supplied in manures and in inorganic fertilisers (see Chapter 7, Table A5, and below).

Because the global functioning of agriculture depends in part on the application of fertiliser nitrogen—that is to say, on the industrial synthesis of ammonia—it would seem that natural processes can no longer supply all the nitrogen needed by to-day's human population. If that is indeed so, it means that the combined processes of biological nitrogen fixation and deposition from the atmosphere cannot supply nitrogen fast enough to meet present demand, a situation exacerbated by the capacity of ammonia or ammonium ion to inhibit biological nitrogen fixation. In that sense it is another vicious circle. Moreover, the synthesis of ammonia from nitrogen requires energy—quite a lot of it—and that has its own implications. So now let me try to attach some numbers to these generalisations.

Table A13 lists some rates of atmospheric nitrogen deposition or, in one case, actual uptake or adsorption by soil. As has been my practice in this type of situation, I have not tried to be exhaustive in listing these rates. My aim is to give some idea of the orders of magnitude of the processes and the range of values that might be expected. The table points to an average global rate of deposition of at least 8 kg.ha^{-1}.y^{-1}.

Rates of biological nitrogen fixation should, by rights, be more difficult to quantify. It is easy enough, relatively speaking, to measure the rate at which a bacterium can fix nitrogen under experimental conditions. It is another matter to extrapolate from that to extensive areas of land in widely different climatic conditions. Even if the attempt is restricted to symbiotic fixation in legumes, fixation rates will depend, among other things, on the density of those plants. In an agricultural context, this is not such a problem but, in the wild, it can be little more than guesswork.

With those reservations in mind, it is not so surprising to find estimates of symbiotic nitrogen fixation in legumes ranging from 17 to 700 kg N.ha^{-1}.y^{-1}. Free-living bacteria and cyanobacteria are not nearly so impressive—they are accredited with 0.1-25 kg N.ha^{-1}.y^{-1} depending on the type of organism[1] and the type of environment. Corresponding estimates for specific regions can be more precise. Thus, in Australia, lupin crops fix nitrogen at 'average' rates of 136-140 kg.ha^{-1}.y^{-1}, peas and peanuts at about 100 kg.ha^{-1}.y^{-1}, soybeans at about 220 kg.ha^{-1}.y^{-1}. The total amount of nitrogen fixed annually in Australia by leguminous grain crops has been put at about 200,000 tonnes.[2] Estimates of total global rates of terrestrial symbiotic nitrogen fixation that I have encountered are (10^8 t.y^{-1}): 0.7-3.6, 1.39 and 1.2.[3]

If we take the total global area of vegetated land as 115.42 x 10^8 ha (Chapter 10), these gross rates of biological nitrogen fixation are equivalent to specific rates ranging from 6 to 31 kg.ha^{-1}.y^{-1}—with indications of an average around 10-11 kg.ha^{-1}.y^{-1}. Total annual terrestrial denitrification has been estimated at 1.07-1.61 x 10^8 tonnes,[4] equivalent to some 9-14 kg.ha^{-1}.y^{-1}. If these estimates are reasonable, then the entire planet is just about in balance as far as the central biological processes, nitrogen fixation and denitrification, are concerned. That is about what you would expect in a pristine environment.[5] But the present total situation is certainly not pristine. A substantial proportion of the globe is dedicated to producing food for human consumption, food that in very large measure is consumed away from the farm on which it was grown, and is consumed under conditions that allow but a minuscule proportion of the nitrogen it contained to be returned to the farm. As a first approximation, then, it would seem that agriculture depends for its nitrogen just about entirely on atmospheric deposition and synthetic fertilisers.

But that might be too simple for at least two reasons. The first is the uneven global distribution of atmospheric deposition, nitrogen fixation, application of fertiliser, denitrification and, of course, losses through leaching and erosion.[6] The second is the very wide range in the 'efficiency' of assimilation by plants of nitrogen from the soil. Trying to quantify something like this can be a messy business, but we can make some fairly simple generalisations.

Plant roots take up nitrogen almost entirely as NH_4^+ or NO_3^-. The main reservoir of nitrogen in soil is organic matter, as already mentioned. Microbial activity releases nitrogen from organic matter predominantly, but not exclusively, as ammonia or ammonium ion. Once released the NH_4^+ is likely to follow each of three primary pathways—assimilation by plants, assimilation by microorganisms and, if the soil is adequately aerated, oxidation to NO_3^-.[7] The extent to which these pathways determine the fate of soil nitrogen depends on their relative rates—and those rates can vary enormously according to circumstances. In a natural situation, however, the release of NH_4^+ and its uptake by plants and microorganisms are likely to be pretty much in balance so that very little nitrogen is lost. Indeed, if there is an initial deficiency, an ecosystem in a state of succession can accumulate nitrogen until it reaches a steady state.[5]

The picture changes, of course, when produce is removed from farmland but, here again, the details depend on the rates of the dominant processes. In some environments, low yields of grain can be sustained over long periods without manuring (Note 5 and Chapter 7), but high yields cannot. In the past, grain yields have been sustained through the application of farmyard manures and/or by crop rotation in which growth of the primary crop was interrupted by a season of a leguminous crop which was duly ploughed in as a green manure. Modern agriculture, however, deals with the problem of high yields by the application of fertiliser nitrogen in the form of an ammonium salt, a nitrate (sometimes NH_4NO_3), urea or, in the case of some highly mechanised enterprises, especially in the USA, by the direct injection of gaseous ammonia. In all of these circumstances there is nearly always a large initial surplus of NH_4^+ (urea is rapidly hydrolysed by soil microorganisms)—much more than the crops can handle—and the efficiency of assimilation is low. For example, in some Californian studies, nitrogen was applied to irrigated plots at an average rate of 679 kg.ha^{-1}.y^{-1} (a rate that includes 126 kg.ha^{-1}.y^{-1} from the irrigation water). The crops assimilated 30% of the total, 37% was leached from the soil and 33% was not accounted for.[8] In a different environment, fertilisation of Queensland sugar cane with urea has been accompanied by the loss, in the form of ammonia, of half the applied nitrogen. In other Australian locations, application of urea to various crops resulted in an average loss, again as ammonia, of about 12% of the applied nitrogen.[9]

Under 'natural' conditions an important part of nutrient cycling occurs through the urine and faeces of animals (see Chapter 2). Some nitrogen is always lost in the process but the quantitative details change in an agricultural environment—and can do so enormously under the auspices of modern 'agribusiness'. For example, in the 1970s-80s, farm animals in the USA excreted somewhere around 4-5 million tonnes of nitrogen annually (*in toto*—not each!). For various logistical reasons amounts equivalent to only 1.4-2.4 million tonnes of nitrogen could be collected. About half of that was lost during handling, transport and application.[10] Human excrement is of no less significance, and will be considered further in due course.

Preparing a nitrogen 'balance sheet' for the world's agriculture is no easy matter. A national balance sheet is bad enough and I shall begin by summarising one such attempt for Australia.[11] Up to about the late 1960s Australia applied less nitrogenous fertiliser and made more use of legumes, especially in 'improved pasture', than most other countries with highly mechanised farm industries.[12] Over the period 1951-1962, combined application rates to crops and pasture increased from about 15 to 38 kt N.y^{-1}. By 1977, however, the rate had risen to 217, and by 1988, to 382 kt N.y^{-1}. The nitrogen budget constructed by McLaughlin and his colleagues refers to the entire Australian land area for the years 1987-88. According to their assessment, total nitrogen 'inputs' from all sources amounted to 3.43 Mt N. y^{-1}; 'outflows' from all causes amounted to 2.28-2.36 Mt N.y^{-1}.

In other words, there is an implied nett annual gain by the Australian land surface of 1.07-1.15 Mt N. On the face of it this is scarcely consistent with the decline in soil organic content referred to in Chapter 10. But not all the components of the balance sheet are directly applicable to agriculture. If the atmospheric deposition (listed as 1150 kt for the whole continent) is adjusted to the total area of 'land with agricultural activities' (see Chapter 10), this entry becomes 700 kt. Nitrogen fixation by 'forest communities' (200 kt) can be omitted. These adjustments bring the total 'inputs' down to 2.78 Mt. The 'outflow' includes an entry (1200 kt) for fire, but again, not all of that can be applied to 'agriculture'. I shall take 75% of the total as applicable to 'farmland',[13] even though I feel intuitively that that might be too high. This reduces the relevant losses through fire to 900 kt N and the total 'outflow' to 1.98-2.06 Mt N. y^{-1}. By and large the other entries do not need changing. But the upshot of these adjustments still leaves us with a nett gain of 720-800 kt nitrogen for all Australian farmland.[14]

It should be apparent by now I am reaching a point at which further assessment will require either an inordinate amount of detailed analysis of statistical material or a lot of guesswork. Given, once again, the nature of this book and the interests of practicality, I have chosen a compromise—some supplementary statistics and quite a lot of guesswork. Applying this approach solely to wheat, and using statistics for the period 1987-89 (see Appendix), leads to some sort of conclusion that about 46% of the Australian wheat yield was dependent on fertiliser nitrogen. The total annual production for that period was about 12.6 million tonnes. If individuals each consume about 100 kg wheat annually, then these numbers imply that 126 million people ate Australian wheat and, of those, some 60 million depended on the industrial manufacture of ammonia for that wheat.

In a somewhat similar vein, Smil[15] has done some calculations for all foodstuffs globally—and doubtless so have others. He has stated some of his assumptions, but not enough to allow a serious appraisal of his quantitative conclusions. But, in the light of the various disclaimers that I have made, both above and in the Appendix, that might not matter very much. His critical conclusion is:

" … global farming without synthetic nitrogen could feed no more than 3.5 billion people, or roughly two-thirds of the current total. Every third, or certainly at least every fourth, inhabitant of the Earth is now thus alive owing to the ingenuity of Fritz Haber and Karl Bosch."

The numbers might or might not be about right. The important thing is that the qualitative conclusion of absolute dependence of part of the population on synthetic ammonia, as identified at the beginning of this chapter, is undoubtedly true. And as we have also noted, it is true not because of any shortage of nitrogen—but because natural processes cannot cycle nitrogen *fast enough* to meet the demands of a human population of the present size. Some feeling for the dynamics of those demands can be gleaned from a simple rephrasing of consumption patterns: a protein consumption of 50 $g.hd^{-1}.d^{-1}$ (see below) by a population of 6 billion is equivalent to 208 tonnes of protein *per minute*. If that protein were supplied solely by wheat, then the wheat would be consumed at an average rate of about 2080 tonnes per minute. If it were supplied entirely by beef steak, the rate would be about 1040 tonnes of steak per minute.

A critical bottleneck, at least as far as crops are concerned, is a need either for mixed crops or, more commonly, crop rotation to take advantage of symbiotic nitrogen fixation in legumes. A 2-course rotation could incorporate nitrogen at a rate in the general order of 140 $kg.ha^{-1}$ every alternate year, but would effectively halve the yield of the main crop—which would be grown only in alternate years. A 3-course rotation (two years main crop—one year legume) could incorporate a similar amount every third year and reduce the average yield of the main crop by one third.[16]

There is another problem, also foreshadowed. It is that the manufacture of ammonia from elementary nitrogen requires quite a lot of energy whether it is done in the soil by a bacterium or in a factory. Fertiliser nitrogen is sometimes referred to by agronomists as 'fossil nitrogen' because of the combustion of fossil fuels used in its manufacture. For example, if we take a third as the proportion of the population dependent on 'industrial' nitrogen and assume they each eat about 50 g protein daily,[17] then collectively they will require energy at a rate of about 845 x 10^{12} $J.d^{-1}$, just in the synthesis of the ammonia. This translates to something like 58,000 tonnes of bituminous coal or 20,000 tonnes of crude oil daily. Those two rates add up to annual discharges of 78,000,000 and 40,000,000 tonnes CO_2 respectively—or about 0.3% and 0.2% of total annual CO_2 production by industry.[18] The proportions might not seem very high; I mention them to draw attention to but one aspect of the energetic ramifications of modern industrial agriculture.

But it might be argued that some of this energy is being spent anyway and, because of that, the picture that I have tried to paint is distorted, at least for the northern hemisphere. In the more densely populated parts of the industrial world, in particular Europe and North America (and I would assume China, Japan and much of India), there is now a problem of excess deposition of bound nitrogen from the atmosphere (see Chapter 8). These substances, consisting mainly of ammonia and several nitrogen oxides, arise from feedlots (see Chapter 13) and the combustion of fossil fuels—predominantly by industry and by motorised transport. Such combustion can produce temperatures high enough to cause significant combination of atmospheric nitrogen and oxygen. A substantial proportion of all these gases is eventually deposited in rain. One estimate is that 60% of the fixed nitrogen, including fertiliser, that is deposited annually on land is produced by human activities.[19] Because this is more than can be assimilated by plants, about 20% of the nitrogen deposited in river catchments is eventually leached into waterways where it contributes directly to eutrophication. Moreover, excess nitrogen in the soil can affect the dynamics of plant metabolism, one result of which is an increased susceptibility of trees to insects and fungi.[20] And, as we saw earlier (if we were paying attention!) excess deposition of nitrogen compounds will not only reduce species diversity in grasslands but will also lower the amount of carbon stored by the biomass so affected.[14]

The remaining implication of this discussion that should be emphasised is that, subject only to minor variations in response to possible changes in management practices, the absolute number of people who can subsist on foodstuffs fertilised with what I shall call 'biological nitrogen' as distinct from fertiliser and pollutant nitrogen, cannot increase on any long term basis.[21] In other words, to use Smil's numbers (above)—if 3.5 billion people can be supported now by biological nitrogen, no more than 3.5 billion could be so supported if the population increases. If the population should double, then, on the numbers I am using, a maximum of 3.5 billion might be supported 'naturally' and the remainder, about 8.5 billion, would depend on nitrogen from fertilisers and man-made atmospheric pollution. Just how serious this might be will depend upon future energy resources and the wider implications (*e.g.* for climate, land degradation etc) of using those resources on a large scale. It is scarcely a prospect to generate complacency—or to suggest the adjective 'sustainable'.

— Phosphorus —

Unlike nitrogen, phosphorus is not obliged to cycle through the atmosphere. Indeed, as we noted in Chapter 2, its vehicles for doing so are limited largely to smoke (which can be significant in Australia)[22] and to small amounts of phosphanes, produced mostly in swamps. Phosphate fertiliser is mined as rock phosphate. It might be applied as such or chemically 'beneficiated' to increase the content and the solubility of the phosphorus.[23] A flow chart for phosphorus—indeed for most nutrient elements[24]—in a modern industrial society should look something like this:

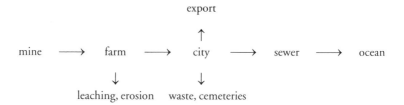

There are some fine details that could be added to the diagram. For example, some phosphorus is returned from the sea to the land through the droppings of marine birds, but the rate of this process is insignificant in relation to the major pathways. Similarly, there is transfer of phosphorus from the sea to the land via fish and other marine foods—but most of that finds its way back again via the sewers. There is also some discharge of sewage onto land. Because most of the phosphorus used in agriculture ends up in the sea (the component that is exported can be expected ultimately to suffer essentially the same fate), some comment is needed about its possible recovery from that medium.

When sewage is discharged into the sea, the options faced by the phosphorus it contains are similar in principle to those outlined in Chapter 8 for the analogous situation in rivers. There are, however, some important differences in detail—differences that derive in large measure from the presence of significant concentrations of various bivalent cations which combine with the inorganic states of phosphate and limit its solubility in the marine environment.[25] The overall consequences of discharging sewage into the sea are that some phosphorus remains in solution and is dispersed throughout the oceans, a substantial proportion is precipitated and becomes incorporated, along with a lot of other substances—many of them toxic— into the sediment. Some is incorporated into the biota. The thermodynamic implications of all this are that the entropy of the phosphorus is now so great that an enormous amount of free energy would have to be spent to recover and concentrate it (see below).

Within the general context of pointing out the abundance of mineral resources and the improbability of exhausting any of them, Beckerman[26] makes the following comment on the amounts of various elements in the sea.

"Sea water has been estimated to contain about a billion years' supply of sodium chloride and magnesium, 100 million years' supply of sulphur, borax and potassium chloride, more than a million years' supply of molybdenum, uranium, tin, cobalt; and so on. Yet who would have thought of including sea water in the list of resources available to us thirty years ago?"

Some implications of this type of argument should be considered a little further. One of the nominated elements is potassium—its concentration in sea water[27] is about 0.4 g.kg^{-1}. In order to obtain 1 tonne of potassium, it would therefore be necessary to process about 2500 tonnes of sea water. Mineral resources are not designated as 'high grade' or 'low grade' without reason.

But our immediate concern is with phosphorus. The average concentration of phosphorus in the sea[25] is about 70 μg.l^{-1}. To the time of writing, the highest global annual application of phosphate fertiliser on record was achieved in the year 1988/89 (Fig. 11.1); it was equivalent to 37.7 million tonnes P_2O_5 (16.5 million tonnes of elementary phosphorus). If this were to be obtained from the sea, water would have to be processed at a rate of 2.4 x 10^5 km^3.y^{-1}. This is more than 70 times greater than the total global annual consumption of freshwater (see Table A6). It is equivalent to a daily rate of 6.5 x 10^5 Gl (or 646 km^3). And it is not simply a matter of pumping the water from one place to another. It would have to be processed in some way—by running over an ion exchange resin, by electrophoresis or electrodialysis or by precipitation. I have absolutely no idea of the amount of energy needed for all of this but I suspect it might dwarf that used in all other human activities combined. And none of this takes any account of the ecological effects of trying to mine the oceans in this way. Arguments such as Beckerman's have at least one foot in fairyland.

An argument, similar in principle to Beckerman's, has also been advanced about the amount of phosphorus in the planet's soils (see below). But, back to the diagram which says, in effect, that if the processes it depicts persist, we will one day run out of phosphorus. So now let me try to quantify that statement.

I shall again begin with the situation in Australia—partly because of accessibility and partly because Australian native soils are generally lower in phosphorus than the soils of other countries with significant agriculture.[28] Tables 11.1, 11.2 and 11.3 represent three attempts to quantify the fate of phosphorus in Australian primary produce. The first (in order of presentation—not chronology) is my own, the second from Melbourne, at the time a student at the Australian National University, and the third from McLaughlin and colleagues. The objectives of each table were somewhat different.

Table 11.1 seeks simply to assess the fate of phosphorus in primary produce. For that reason there is no entry for the amount of fertiliser applied or the amount of food imported. Imported food, however, is automatically included in calculations of food consumed and would have affected the amount of food exported—presumably increasing it by roughly the same amount. No allowance is made for wood chips or timber products in this or the two following tables.

TABLE 11.1   *The fate of phosphorus contained in one year's Australian primary produce*

Item or process	Phosphorus content of item or process (tonnes)	Proportion of total (%)
All produce	107488	
Distribution of produce		
Exports	62325	58
Sugar cane	9698	9
Food waste	8670	8
Sewage	20838	19
Assimilation	86	0.1
Total accounted for:	101617	95

The total phosphorus contents of all primary produce and of exported foodstuffs are taken from Table A14. Sugar cane is listed here but not in Table A14 because the cane is removed for milling and the extracted residue is rarely, if ever, returned to the soil. The amount of sugar cane was derived from Australian Bureau of Statistics (1991a). The amount of phosphorus in sewage was calculated on the assumption of a *per capita* discharge of 300 $l.d^{-1}$ containing 11 mg $P.l^{-1}$ (see also comment in text). Food waste was calculated on the basis of 100 $kg.hd^{-1}.y^{-1}$ (personal communication, van den Broek and Ratton, NSW Environment Protection Authority, 1992) and a guessed phosphorus content of 0.5%. 'Assimilation' refers to the assimilation of phosphorus by a growing population. It was calculated from the difference between deaths and live births for calendar year 1990, an assumption of average body mass of 60 kg and a phosphorus content of 1% in the human body (see also comment in text).

'Assimilation' and 'Sewage' need specific comment. As discussed in Chapter 2, an animal that maintains constant weight does not make any nett withdrawals of nutrient elements from its environment. It excretes or otherwise expels the same amount as it consumes. A growing animal, however, does accumulate nutrient elements. When that animal disposes of its dead in a way that removes the carcase from the essential food web then, in the long run, the retention of nutrient elements within its biomass can become significant. Ideally an attempt to quantify this process for a human population should take account of average growth rate of all the children in each age bracket, the proportion of the population buried in cemeteries and cremated, and the fate of the ashes of those cremated. As it happens, assimilation is of minor significance in comparison with the other processes and I have contented myself with a simpler approximation (see legend to Table 11.1).

The same type of argument, however, leads to an alternative method of estimating the amount of excreted phosphorus in sewage. By rights, what we excrete should equal what we consume minus what we assimilate and what we throw away in the garbage. If that calculation is done, taking 14217 tonnes of phosphorus for the national total consumption of foodstuffs,[29] and the other two quantities as in Table 11.1, we finish up with 5461 tonnes of 'excreted' phosphorus in the nation's sewers. That is less than half the tabulated amount; it accounts for but 5% of the produce. The discrepancy can be explained largely by the widespread domestic and industrial use of detergents, partly by the discharge of other phosphate compounds into sewers, partly by uncertainties in the data and by dubious assumptions.[30]

TABLE 11.2 *A Phosphorus 'balance sheet' for Australian primary produce*

Item	Phosphorus Equivalent (tonnes)	
A *Primary produce*		
Cereals	71305	
Sugar cane, fruit, vegetables	9373	
Livestock and milk	45109	
Total		125787
B *'Inputs'*		
Imported foodstuffs	849	
Fertilisers	198990	
Total		199839
C *'Losses'*		
Exported foodstuffs	58218	
Sewage	12000	
Domestic waste	8528	
Sugar cane	7726	
Total		86472
Total, A+B-C		+239154
Difference, A-B		- 74052

This table has been condensed from Melbourne (1990). Melhourne also had entries for phosphorus recycled within the system.

Despite the uncertainties, there is broad agreement among the three tables on the total amount of phosphorus disbursed—even though the items contributing to the disbursement are not the same in all cases. There is good agreement between Tables 11.2 and 11.3, but not with 11.1, on the amount of phosphorus discharged in sewage. There is very good agreement between Tables

11.1 and 11.2 on the amount of phosphorus in food waste. But it is important to repeat and to emphasise that tables of this type cannot be compiled without resort to some guesswork. They should never be treated as gospel—as a definitive statement of unquestionable accuracy and precision.

TABLE 11.3 *An Australian annual 'balance sheet' for phosphorus*

	'Inputs'		*'Outflows'*
Process	*Mass of P (kt)*	*Process*	*Mass of P (kt)*
Food imports	< 1	Produce export	56
Phosphate rock	326	Urban discharge	11
Imported fertiliser	54	Soil erosion and leaching	9-35
		Fire	4
Totals	381		80-106

The material in this table was taken, with some modifications, from McLaughlin *et al.* (1992). The principal modification was omission from the 'Inputs' of 77 kt P from atmospheric deposition. That item had been attributed to the burning of 'biomass' and would thus have been transferred essentially from one area of soil to another. On the other hand, the loss attributed to 'fire' has been retained since, on McLaughlin *et al.*'s representation, that was presumably a nett loss.

Both Tables 11.2 and 11.3 imply a considerable excess of input over outflow, an excess that persists even when losses from leaching and erosion are included, as they are in Table 11.3. It would seem, therefore, that the phosphorus content of Australian soils should be increasing.[31] For more general reasons this is probably also true of agricultural soils throughout the 'developed' world (see below). Much as in aquatic systems (Chapter 10), when inorganic orthophosphate is applied to soil, its fate is determined by the competing claims of a number of different types of reaction pathways. It can react with cations (*e.g.* Ca^{2+}) to form essentially insoluble salts, or with cations and other anions to form insoluble complexes. It can be tightly adsorbed onto soil particles, it can be assimilated by the soil biota and, of course, by vascular plants. As always, the absolute and relative rates of all these processes vary very widely with circumstances but, suffice it to say, if the crop plants are to do what is expected of them, they need their phosphorus in a hurry—before it is shunted off down some other pathway. Once the phosphorus is tied up in these other forms, it becomes 'available' to the higher plants only through its hydrolytic release from chemical binding or through desorption from soil particles.[32]

Phosphorus released in those ways will face the same demands on its services as when fertiliser is applied, but rates will be much slower and overall dynamics very different. If the process is allowed to proceed without human interference, the phosphorus will cycle within the system which will duly evolve into a complex ecosystem—all other conditions being favourable (see Chapter 2). But that is not exactly what a farmer has in mind when he culti-vates a field. What he wants is usually a single crop that will reach maturity in one season. In the natural situation, there might be plenty of phosphorus in all its forms ('total' phosphorus), to support a very substantial biomass, but the farmer's crop simply cannot get it fast enough. So, although the mecha-nisms are different from those applying to nitrogen, we again have a situation in which the yield of a crop can be limited, not simply by the total supply of an element, but by the rate at which that element can be made available.[33]

There is no counterpart of biological nitrogen fixation for phosphorus. If an area of ground is continually cropped or grazed without supplying phos-phorus in some shape or form, yields will drop—sharply at first, unless the el-ement is present in excess, and eventually more slowly as concentrations of phosphorus in the soil fall and yields approach their minimum. What all of this seems to lead to is that, even though phosphatic fertilisers might be in-creasing the phosphorus content of some agricultural soils, a population of the present size could not be supported under the present régime without their continued use.

Superficially this is a similar type of situation to that described for nitro-gen, but there are some important differences. Perhaps the most obvious is that nearly all of the phosphorus within an ecosystem remains as 'phosphate' ($H_2PO_4^-$, HPO_4^{2-}, PO_4^{3-}, R-PO_4, where R denotes an organic radical, or, after a fire, as P_2O_5) and, as such, is constantly 'available' to organisms, albeit at widely different rates. Loss of phosphorus from an ecosystem nearly always amounts to its physical removal, not its chemical conversion to an unavail-able form. This is different from the situation with nitrogen which is easily 'lost' by denitrification but can be brought back into biological circulation by a very specialised system and the expenditure of energy. If phosphorus is lost, then the continuing functioning of the system requires replacement of the phosphorus either by returning that which has been removed or by obtaining phosphorus from another source. And that, too, requires the expenditure of quite a lot of energy.

A 1971 estimate[34] concluded that a population of only 1-2 billion could be supported solely by manuring with animal dung and plant wastes. There are so many assumptions built into a calculation like this that I shall not try to analyse it. But, whatever the validity of the actual numbers, there are some underlying principles that are true. It appears to take no account of returning either human sewage or, apparently, food wastes to the soil; its conclusion that such a system would support a smaller population than the present one must therefore be valid. In that sense it does no more than assess the consequences of managing farms more traditionally by transferring manure from pasture to arable land (see Chapter 2). It is probably true because of the effect on yields of the rates at which phosphorus becomes available to plants, as discussed above. It is not true, however, if it implies that the 1-2 billion could be stable under such a régime. Simply transferring farmyard manure within the agricultural system would not stop the impoverishment of soil caused by exporting produce from the system.

So, all of this gets back to a comment made earlier in this section—if we continue to mine phosphate and dump it into the sea, we will one day run out of the stuff. How long will that take? Many pitfalls lie in the way of trying to answer that question. Fig. 11.1 shows rates of consumption of P_2O_5 in fertiliser over the period 1961-1998—globally and separately for the 'developed' and 'developing' countries. From 1960 to about 1980 the global pattern was dominated by the developed world. Subsequently rates of phosphorus consumption in those countries declined while consumption in the developing world continued to increase linearly (correlation coefficient 0.934) and overtook that of the industrialised nations.

The figure suggests that global consumption of phosphorus will henceforth be dominated by the 'developing' world. Why did the 'developed' world slow down? A partial explanation presumably lies in a response to accumulation of phosphorus in agricultural soils as discussed above.[35] In attempting to extrapolate regressions of this type there are several virtual certainties that need to be acknowledged from the start and which, once acknowledged, must give pause to attempts to assess with precision the life expectancy of phosphate reserves. The rate of consumption will change, in either direction, in response to variations in the population growth rate, to economic conditions, to farm management practices, to the accessibility of the reserves and to their phosphate content. All or some of these influences are already reflected in the figure. In addition, new deposits will be found and/or 'resources' will be reclassified as 'reserves'[36]—for a while. Estimates of 'reserves' made sequentially over recent years include the following (Mt P_2O_5) and indicate a decline: 33,900, 17160 and 3600-8000.[37] It should also be acknowledged that agriculture is not the only consumer of phosphate, although it is by far the major one. At least since the early 1970s the manufacture and application of fertiliser has accounted for 72-79% of total production/consumption of phosphate rock.[38]

Despite the many obvious uncertainties inherent in the process there have been serious attempts to estimate the life expectancies of phosphorus reserves and resources. Steen, for example[37] made a series of estimates based on various assumptions about rates of application of phosphate fertilisers. Her central conclusion was that "depletion of current economically exploitable reserves can be estimated at somewhere from 60 to 130 years".

The nature of Fig. II.I is such that no extrapolation is possible without a lot of very dubious guesswork. For that reason I shall confine myself to the simple and highly improbable assumption that global consumption of phosphorus remains constant after 1999. In addition I have assumed that agriculture accounts for 75% of total consumption and that 'reserves' of phosphate rock lie in the range equivalent to 3,600-8,000 Mt P_2O_5 (see above). Under those conditions all of the 'reserves' would be exhausted within 85-190 years.

In 1975 Wells estimated the life expectancy of accessible phosphate deposits under a range of conditions some of which, to put it mildly, were very curious.[39] He also estimated the amount of phosphorus in the top 1.6 km of the earth's crust (excluding Antarctica) to be 5.9 x 10^{14} tonnes—and noted that phosphate mining could extend to twice that depth. He then proceeded to calculate that a population of 20 billion using 1.16 billion tonnes of phosphorus annually would take about 500,000 years to exhaust those supplies.

This, of course, is a very handy piece of information and I can only hope that Wells's tongue was in his cheek when he did the calculation. The broader implications of mining the top 1.6 km of the entire land surface of the planet (excluding Antarctica) were not discussed. They should be obvious enough, but some consequences of a similar approach to two very small areas of land nevertheless deserve brief mention.

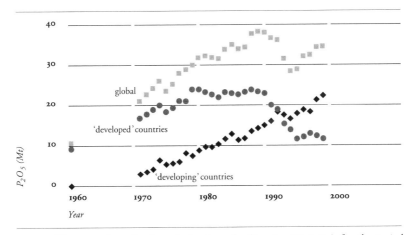

FIG II.I *The annual application of phosphate fertiliser (as Mt P_2O_5) for the period 1961-1999. Squares, global; circles, 'developed' countries; diamonds, 'developing' countries. Source, International Fertilizer Industry Association (http.//www.fertilizer.org).*

Nauru is a small equatorial Pacific island; its total area is about 2000 hectares. It was very well endowed with an extensive cover of high quality phosphate deposits, derived probably from guano or possibly more directly from a marine source.[40] Shortly after the end of World War I Australia obtained rights to mine the deposits which became a major source of its fertiliser—until supplies were exhausted in the late 1980s. By 1966 mining had removed the topsoil from more than 1800 hectares. It was dumped in 'wasteland'. The cost of importing soil to rehabilitate the mined area was put at about $A260 million.[41] This was considered cheaper than recovering soil from the 'wasteland'. In 1993, Australia agreed to pay Nauru $A107 million towards the reparation of the land surface, a process that I assume included replacement of topsoil. The Nauruans lost their soil and the lifestyle that went with it, but they received a very high *per capita* income from the sale of phosphate. A side effect of all this change was that, by 1987, 65% of men and 70% of women were clinically obese and about a third of the population suffered from diabetes.[42]

Ocean Island, about 200 nautical miles to the east of Nauru, fared somewhat differently. It lost so much of its soil in the course of phosphate mining that the inhabitants, the Banabans, were resettled by Great Britain on an island near Fiji. Mining ended on Ocean Island in 1980.

From my perspective, any attempt to quantify the life expectancy of accessible phosphate deposits with a precision of even, say, fifty years would be an exercise in deception—of oneself and/or a reader. All I am prepared to say is that a continuation of present practices will eventually exhaust usable phosphate deposits. The previous calculations suggest that the time needed to exhaust those deposits is quite long in the perspective of the average politician or economist. It is, however, very short on an historical time scale and negligible on an evolutionary scale.

Perhaps I should rephrase the essence of this section of the chapter. Because phosphorus is being used in a manner that is essentially irreversible, that does not mean that the element has been changed irretrievably into something else or physically removed from the planet. It is not locked behind an impenetrable barrier. It is still 'here', not only in the oceans where we are discharging most of what we use, but also in rocks and in deeper layers of the earth's surface—as Wells pointed out. If the human species were to disappear, those sources would be available to relevant ecosystems through weathering, through transport by deep-rooted trees—and by transfer from the sea to the land by marine birds. But the *rates* of those processes are so slow in relation to human demands that, for all practical purposes, they might as well not exist. It is one of the most blatant examples of infringing the 7th 'Law'.

So, what do Chapters 8-11 add up to? I cannot avoid the conclusion that the human population has already overshot the planet's carrying capacity—*under current management régimes*. A continuation of those régimes will support continuing growth for a while, but will eventually lead to a collapse of the population below the present level. I say 'collapse' rather than 'decline' quite deliberately. From time to time the United Nations and other bodies publish graphical projections of the human population over a period, up to about a century. There are commonly three curves based primarily on high, medium and low birth rates with some supplementary assumptions about death rates. In general 'medium' birth rates are considered the most likely—as you might expect (see Chapter 14). It has been generally the case that the highest projection shows the population growing until the end of the 21st century, the medium one levelling off before that and then perhaps declining slightly, the lowest levelling off by mid-century and, after a while, beginning a slow decline. For the most part, recent projections point to a population in the order of 9 billion by mid 21st century.[43] An assumption sometimes implicit in such projections is that, once the population levels off, everything else will more or less look after itself.

What I am saying is, if my contention that we have overshot the earth's carrying capacity is correct, then everything else will not look after itself. Unless we change the way we manage our habitat, the human population will indeed collapse. And since the collapse is likely to be provoked in large measure by soil degradation, and limited access to unpolluted freshwater, it is certain to be accelerated by competition for those soils that are productive and those water supplies that are potable. In other words, the process will be accelerated by wars.

This sort of thing is well enough recognised with other species whose growth rate has outstripped natural constraints imposed by predators. It always leads to degradation of the habitat and, in general, the longer the excessive population growth continues, the greater its eventual collapse. In our own case, the situation has the added dimension caused by the positive feedback relation between the population and the food supply.

In the past, processes of the kind outlined in this and the previous three chapters have led to the collapse of individual civilisations, especially in arid or semi-arid regions. A major difference between then and now, however, lies in the respective sizes of the populations involved. Then the problem was confined to specific populations in specific areas. Now, for all practical purposes, it is global. The primary focus of my attention is, of course, the direct impact of an exploding population on the soil—and water supplies (Chapter 8 and 9). But, as well as that, such a population consumes many other types of resource, the loss of which can also ultimately threaten the survival of the offending species. Three obvious biological examples of this are the global decline in fish populations (see Chapter 14), deforestation (see Chapter 12) and a progressive simplification of the global ecosystem—a loss in biodiversity if you like. The significance of this last, which embraces the other two, is that the planet's ecosystems lose their resilience and thereby become much more vulnerable to stresses.

Could a collapse of the human population under these circumstances lead ultimately to human extinction? The short answer must be 'yes', but I cannot assess its probability. But we need to look now at other aspects of the problem—in particular some of the mechanisms underlying the onslaught on the planet's soils.

CHAPTER 12

Two Related Matters

When we study complex systems quantitatively we are disposed to concentrate on components of the system rather than on the functioning of the whole. In most situations this is unavoidable. It is *relatively* easy to make precise measurements on components, but extremely difficult, if not impossible, to make comparably precise measurements on the whole system—even if you can define just what it is you want to measure. Not that making precise measurements is the same as understanding—but it helps. Broadly speaking, this is true whether the system under consideration is a complex metabolic pathway, the physiological functioning of a cell, of a multicellular organism, the functioning of an ecosystem, a social system—or the weather.

Specialisation or reductionism—for that is what this narrower focus is—is a legitimate end in itself for a very high proportion of people engaged in the study of biological phenomena. It can become focused to the point of absurdity, of course, and give rise to comments along the lines of knowing more and more about less and less. Nevertheless this *modus operandi* has yielded a wealth of detailed knowledge and understanding of natural phenomena covered by virtually all the scientific disciplines. It has provided the theoretical foundations for the impressive scientific and technological achievements of the past couple of centuries.

The intrinsic merits of specialisation are self-evident but its limitations would seem to be less so. If you have enough information, combining all the pieces of many separate specialist studies should, in theory, open the way to a realistic assessment of the functioning of the larger system. In practice it might do so—if you're lucky, have a powerful computer and know what you're doing. When it comes to the management of ecosystems, however, there is precious little evidence to suggest that we do know what we are doing.

If we are persuaded by the methods or the philosophy of specialisation to use a single criterion to monitor the functioning of a complex system, we can be seriously misled as, for example, when an economic measure is used to assess the performance of society or of the world at large.[1]

When ecological or environmental processes are evaluated, one property or group of properties of the system—the 'greenhouse effect' and climate change, depletion of stratospheric ozone, pollution, sanitation, an endangered species etc—is usually emphasised. And, to some extent, that is what I am doing. To be sure, agriculture is itself a very complex system, but it influences and is influenced by all the other processes that affect the global environment. The point of this disclaimer is that, although I believe that the availability of food and water are, as I said in Chapter 10, the ultimate Damoclean sword, my concentration on them at the expense of other environmental problems does not mean that I think those other problems are trivial. Ideally they should all be considered together, but practicality decrees otherwise. Human population dynamics, however, cannot legitimately be omitted from any appraisal of the global environment.[2]

The environmental problems that we have inflicted upon ourselves can be roughly classified into four types or groups, as long as we remember that the boundaries between them are fuzzy, that they all interact and that, as always, the classification oversimplifies. One type—I shall call it Group 1—comprises those conditions and processes that directly threaten human health. It is represented most obviously by poor sanitation and various forms of local or regional pollution. Group 2 also includes pollution, but on a global scale—in particular the accumulation in the atmosphere of the so-called 'greenhouse' gases and those compounds that react with stratospheric ozone. The major impact of Group 2 is likely to be indirect through effects on climate and on the intensity of ultraviolet radiation reaching the Earth's surface. Group 3 amounts to the exhaustion of natural (mineral) resources. Group 4 contains those processes that directly endanger our habitat. In it I include both agriculture and the size and dynamics of the human population. War can exacerbate the effects of all four groups.

I have made this classification primarily for three reasons. One is that commentators about the environment from either point of view—concern or scepticism—tend very largely to focus attention on one or two of the groups. The second reason is that Groups 1, 2 and 3 require different types of response from those applicable to Group 4. In a sense, the first three are symptoms of the fourth. Nevertheless they should be largely controllable through better management practices and improvements in technology. Environmental 'sceptics' (*e.g.* Beckerman[2]) usually take examples from these groups to illustrate the importance of economic growth in dealing with environmental problems. I contend, however, that the dangerous effects of Group 4 cannot be eliminated through improvements in technology, although their impact might be mitigated by such means and by changes in 'management' practices (Chapter 14).

The third reason is that I want to provide what I hope is a rational frame-work for two brief digressions before returning specifically to the modern farm. Both digressions take us into topics that are intimately associated in one way or another with agriculture. One is deforestation. The other is the manu-facture and dissemination of pesticides.

— Forests —

Since the advent of settled agriculture, at least a third (2×10^9 ha) of the Earth's forested area has been cleared.[3] Like agriculture, deforestation is a ma-jor contributor to soil degradation,[4] especially erosion which, in turn, works wonders for the sediment load and the general health of rivers and streams. Forests have a critical role in determining the composition of the earth's at-mosphere—specifically in the rates of cycling, and hence the proportions, of oxygen and carbon dioxide.[5] Their clearing can also add to atmospheric methane, a more effective 'greenhouse' gas than carbon dioxide, when felled timber is metabolised by soil biota, especially by termites. Forests also affect local weather patterns, particularly rainfall.[6]

The effects of deforestation can be extended in a number of ways if the deforestation is associated with burning. Smoke reduces the intensity of ra-diation reaching the troposphere. That can affect regional climate which, in its turn, has the potential to influence global climate as, of course, do the 'greenhouse' gases produced by the fires. About 80% of the combustion of plant material is tropical and, of that some 30% normally occurs annually in South America.[7] In late 1997 forest fires in Indonesia, deliberately lit but aided and abetted by an abnormally long period of drought (which was fore-cast), produced "what is shaping as the most serious environmental crisis in the region's history".[8] Concurrently, Papua New Guinea and Irian Jaya were suffering their worst drought of the 20th century. To the extent that defor-estation reduces rainfall and combustion produces CO_2, the progressive de-forestation of South East Asia may well have exacerbated both the contem-porary el Niño climatic disturbance and 'greenhouse' warming. If that has indeed happened, it is another example of a very dangerous positive feed-back cycle.

Smoke causes respiratory diseases and, by reducing light intensity, it di-minishes photosynthetic activity in areas that are not actually burning. The intensity of ultraviolet radiation is also reduced and, while that might be good news for those susceptible to skin cancer, it is also good news for mosquitoes and for pathogens in surface waters or in the air.[9]

Tropical rainforests are the most complex ecosystems known or, phrased differently, are the most biologically diverse regions of the planet. In some regions, especially in the tropics, forest clearance displaces indigenous peoples and promotes an increase in the incidence of some infectious diseases, including malaria. Overall, forests and their management or mismanagement generate more emotional heat than possibly any other environmental topic. They also generate their share of corruption and political cynicism. Assessing the rate and extent of deforestation in various parts of the world is not easy. As we saw in Chapter 7, even Australia, with but two centuries of European occupation and reasonably good records, has produced conflicting opinions of just how much forest has been lost—and there are still major discrepancies in reported rates of clearing.[10] Part of the problem, needless to say, lies in definitions. Does 'forest' mean native forest, virgin forest, a plantation of native trees, of exotic trees, a monoculture? The definitions are apt to vary with the eye of the beholder. The perception of forest as primarily a means of preventing soil erosion or dryland salination is likely to produce a different definition from that of a conservationist who sees forest as a habitat for a diverse biota or from that of a villager who sees it as a source of food and firewood.[11] And someone to whom a forest is first and foremost a source of timber or wood chips, or someone who sees a forest as an impediment to farming or mining or tourist development may well have his own definition. And, sometimes —just sometimes—governments might find it expedient to suppress information.[12]

With those reservations in mind, let us first consider some bald statistics about forests. Table 12.1 summarises areas of forest and rates of deforestation for various regions. One source (World Resources Institute) does not acknowledge that there is enough reliable information to cite either forested areas or deforestation rates for some regions—and hence for the globe.[13] The other source (Dixon *et al.*) does cite areas for all latitude belts. The current situation seems to amount to a global nett deforestation rate of something like 16 million ha annually.[10]

Most deforestation occurs in the tropics or, more precisely in the terms of Table 12.1, between latitudes 25° N and 25° S. Indeed, as a simple arithmetical balance, the table implies that the low latitudes account for the entire nett global rate of deforestation inasmuch as the rate of reafforestation in the mid latitudes is about the same as deforestation in the high latitudes. In the mid 1990s, the greatest absolute rate of regional clearing (7.4 million ha.y^{-1}) was in the Americas,[14] some 50-70% of which was in the Brazilian Amazon.[15] The greatest proportional rate (1.25%) was in Asia. Africa and Asia had similar absolute rates of clearing (4.1 and 3.9 million ha.y^{-1} respectively[14]).

TABLE 12.1 *Some Regional Rates of Deforestation*

Region	Area of Forest (10^6 ha)	Deforestation Rate (%.y^{-1})		Source
		Range	Median	
By geographical region				(a)
Africa*	527.6	0.5 - 1.0	0.8	
Asia				
South**	63.9	0.6 - 3.3	1.15	
South East**	210.7	< 0.05 - 2.9	1.2	
America*	1374.9	0.1 - 5.3	0.5	
Europe*	140.2	(0 - 1.3)	(0.25)	
Former USSR	755.0		(0.2)	
By Latitude				(b)
Latitude Belt			'Average'	
High (50° - 75°)	1372		0.05	
Mid (25° - 50°)	1038		(0.07)	
Low (0° - 25°)	1755		0.88	
Global	4165		0.37	

The annual rate of deforestation was calculated for the decade 1981-90, for the geographical regions and for the two decades, 1971-1990, for the latitudinal regions. * The range and median are for geographical areas, not for individual states or nations. ** The range and median are for individual states or nations. Data listed for America exclude non-tropical South America; no deforestation rates were available for Canada. Bracked rates denote an increase in area. Areas of forest listed by geographical region represent 'natural' forest only; those listed by latitude include plantations. Some regions are omitted form the first section because of insufficient information. Areas listed under 'Geographical Region' are for the year 1990; those for 'Latitude Belt' are for the period 1987-1990. Deforestation rates for geographical regions are condensed from percentage changes listed in Table 19.1 by WRI (see below); in the interests of consistency, I have converted changes in forested area for the latitude belts to percentage changes of the tabulated areas. (a) World Resources Institute (1994): (b) Dixon *et al.* (1994). But see also Note 3 and Aldhous (1993).

Deforestation has a number of causes. Obviously logging is one, perhaps especially in South East Asia and the South Pacific. This has displaced indigenous peoples and done its share of corrupting governments.[16] Paper production increased from 46 Mt in 1950 to about 270 Mt in 1995.[17] There is some evidence that the most serious cause of tropical deforestation is agriculture.[18] Displacement and sometimes the murder of indigenous Amazonian peoples by well organised, well armed pastoralists and their associated road builders has received some publicity.[19] Much of the meat grown by these pastoralists is exported to the United States. Rifkin puts an interesting perspective on this industry: "Each imported hamburger required the clearing of about 6 square yards of jungle for pasture".[20] Some 200 million 'slash-and-burn' farmers are estimated to be destroying tropical forests, especially in South America. The majority of these farmers have apparently been driven to this because social and economic conditions effectively deny them the opportunity of making a living in any other way.[21] As indicated in Chapter 7, shifting agriculture is not necessarily destructive but, when population pressure reduces the fallow period, it can become very destructive indeed. Of course it is not always as easy to generalise as some of these comments might suggest, not least about Africa.[22] Moreover, a more recent assessment attributes most deforestation and degradation of forests to the timber industry.[23]

There is another important consequence of deforestation—the loss of species numbers, or biological diversity, in the affected area. In most (but not necessarily all)[24] situations a forest can be assumed to provide a habitat for more species of organisms than a treeless ecosystem under similar conditions of climate and topography. The comparison is starkest when the forest is a tropical rainforest, but it is certainly not confined to those conditions.

It is possibly this aspect of clearing land that raises most emotion among environmentalists and generates most sarcasm among their opponents. The sarcasm tends to reach its peak when the defence of a forest is based upon the protection of one species seen to be endangered—be it a small mammal, a lizard, a bird, an ant or a fern. Others, perhaps wishing to avoid this type of argument, adopt what might be seen as a more practical standpoint—they emphasise the capacity of a forest, especially a rainforest, to yield plants that can produce compounds with real or potential pharmaceutical value. Others see the genetic diversity of forests as a potential source of DNA that can be used in the modification—the 'genetic engineering'—of plants for commercial exploitation in agriculture or forestry.[25] The first argument is justifiable ecologically, aesthetically and ethically, but it is not likely to carry much weight in the face of powerful commercial interests. The second and third have some commercial appeal, but they are gambles—especially when weighed against the returns expected from selling the trees and converting the land to agriculture. And they are not without their intrinsic problems.[26]

There are many important advantages of complexity—or a large number of species—in an ecosystem. The ultimate, most fundamental advantage however, even from a hard-nosed anthropocentric perspective, is that complex ecosystems are more resilient than simple systems (see Chapter 2). This resilience stems from the obvious fact that no single species can reach anything like the proportion of the total biomass that is possible in a simple system. It follows that no predator or pathogen, however effective or virulent, can affect the whole system to anything like the extent it can if its target is a major component of a simple system. And that might not be solely because of the relatively low proportion of the target or host species. Complex systems normally include organisms (and viruses) capable of actively challenging in one way or another such potential threats. These challenges are issued, not usually out of the goodness of the defenders' hearts and concern for their neighbours, but rather because they recognise a competitor, potential victim or source of food when they see it.

Another important factor in the general resilience of complex systems is the proximity of individual plants, especially trees, to one another. This affects not only many aspects of the physical environment within a forest—temperature, light intensity, the dynamics of water movement, security of animal habitats etc—it can also affect symbiotic relations and the movement of nutrients within the system. For example, mycorrhizae are fungi that have a symbiotic relationship with plants. They transfer nutrient elements from the soil to the plant's roots and, in return, they receive organic metabolites from the plant. Some mycorrhizae can establish symbiotic relations simultaneously with trees of different species and transfer nutrients between or among those trees.[27] That is a benefit not available to the odd isolated tree.

But not only is the resilience of an ecosystem a reflection of the heterogeneity and population density of its constituent organisms, it is also affected by the total area of land occupied by that system. Complexity is a function of area. If a rainforest is dissected by roads, the contiguous areas may well be too small to sustain significant proportions of all the original species.[28] Fragmentation can also reduce the density of total biomass.[29] Among other things, this exacerbates the decline in the forest's role in regulating the chemical composition of the atmosphere—and everything that flows from that.

So, to summarise in general terms the essence of all this: there is a global nett rate of deforestation that seems to be of the order of 15-16 x 10^6 ha.y^{-1}. Although the fine details of this process vary widely from one situation to another, the basic underlying impetus stems from the growth of a human population that is already very large by ecological criteria—and from increasing demands made by that population. In addition to nett clearance, however, the replacement of complex pristine forests by plantations or 'opportunistic' simplified woodland continues over much of the planet. Both processes contribute to soil degradation, to loss of habitats and ultimately to simplification of terrestrial ecosystems—indeed of the global ecosystem.

The replacement of a complex system by a simple one means that there will be fewer species and that the system will be susceptible to domination. In the extreme case of modern agriculture, a single species is dominant. The system as a whole thus becomes highly vulnerable to any agency, any consumer, any pathogen that can weaken or destroy that dominant species. If the simplified system is used by a human population, it will persist only if it is managed in some way. In this day and age, such management will probably include the use of pesticides.

— Pesticides —

I do not want to digress too far into the area of pesticides—a great deal has been written on the topic since 1962 when Rachel Carlson's *Silent Spring* was published. But modern agriculture and pesticides are currently inseparable partners[30]—so some comment is needed.

Pesticides present a 'Group 1' environmental problem. In one form or another they have been around for a long time. Sulphur was burned for use as a fumigant early in the first millennium BC. Late in the first century AD the Chinese used powdered chrysanthemum flowers as an insecticide (the active ingredient is pyrethrum). By the 16th century they had graduated to arsenic compounds and extracts of the tobacco plant (active ingredient, nicotine) for the purpose. From 1917 to 1942 the pesticides most extensively used in the USA were lead arsenate and calcium arsenate. Mercury compounds have also been widely used and have caused their share of fatalities.[31]

The term 'pesticide' is comprehensive—it includes fungicides, insecticides, herbicides and, strictly speaking, any other ...icide aimed at an order of organisms that you might want to kill. Modern pesticides include both inorganic and organic compounds; they vary widely in their toxicity both to target genera and adventitious victims.[32] There seems to be an assumption held by many users, both domestic and agricultural, that because the primary target organisms are not mammals, the pesticides are not toxic for Man, or at least are relatively non-toxic. And, although in the industrial world, containers of these substances normally carry warnings, I do not have the impression that promotional material gives undue emphasis to the general toxicity of pesticides. The situation is worse in the agrarian countries.[33] The following quotation refers to the so-called Developing Countries, but it is no less applicable to most farmers in the Industrial World in the middle of the 20th century and to a substantial proportion of their counterparts to-day.

"Farmers' misconceptions also contribute to the use of pesticides, which are often viewed as progressive and modern—a legacy of preaching from agrochemical salespeople and agricultural extension agents who paid little heed to the practical limits and substantial risks of the chemicals they were peddling. Many farmers regard pesticides as cheap insurance against the risk of crop loss, one of the few concrete steps they can take to reduce the natural uncertainties of their trade."[34]

Broadly speaking, a substance that is selectively toxic achieves its specificity in one or more of the following ways:
1 It specifically blocks a reaction that is unique to and essential for the target organisms.[35]
2 It is not assimilated or absorbed by non-target organisms.
3 Non-target organisms can rapidly inactivate it.

Commercial insecticides interfere with processes that occur in all, or nearly all, animals and are therefore actually or potentially toxic to all animal species. There are degrees of sensitivity, of course—degrees attributable in part to quantitative differences in Mechanisms 2 and 3 above.

One of the earliest and best known synthetic organic insecticides is DDT (dichloro-diphenyl-trichloroethane). It was first produced commercially in 1939 and was initially seen as a godsend because of its effectiveness in controlling mosquitoes, fleas and lice—all vectors of infectious diseases. Its principle site of action is evidently in the functioning of the nervous system—the transport of sodium across nerve membranes—but it also seems to have a capacity to interfere with hormone function.[36]

Nerve transmission and hormone action occur in all animals, albeit at different levels of complexity and sophistication. The acute toxicity of DDT to mammals is *relatively* low (LD50, 300-500 mg.kg^{-1}),[32] but the relevance of LD50 (the dose needed to kill 50% of a population) is questionable—to put it mildly. People can be, and are, killed outright by insecticides—but rarely by the chlorinated hydrocarbons. The major problems with those compounds lie with their long-term effects, including carcinogenesis.[37] The chlorinated hydrocarbon insecticides are very stable and resistant to biological inactivation; they can therefore persist in soils and elsewhere for a long time. Their persistence inevitably accelerates the development of resistance in insect populations, especially at the edges of treated areas where there are concentration gradients. Under those conditions target organisms can be exposed to concentrations of the pesticide that are lethal to only part of the population; the most resistant organisms survive and multiply. This process continues for as long as the pesticide remains. Quite early in the history of DDT, resistant strains of the malaria mosquito, *Anopheles*, were encountered in regions that grew crops.

Because they are soluble in fats and oils, the chlorinated hydrocarbons accumulate in fatty tissues and in milk. They become more concentrated as they work their way up the food chain, and consequently exert their greatest effect on predators. Their interference with the reproduction of predatory birds is well known, as is their accumulation in the fats of animals as far away from points of application as Antarctica.

The persistence of these compounds has led to their removal from agricultural, but not domestic, use in the industrialised world. On the other hand, they are used widely, and apparently increasingly, in the agrarian world—which has become something of a dumping ground for pesticides that are unwanted elsewhere.[38] A major group of replacements for the chlorinated hydrocarbons are the so-called organophosphates. These substances contain an esterified phosphate group which is easily removed by hydrolysis. Thus they have the advantage of relatively rapid breakdown in soil or water. Their half-lives are normally measured in days rather than the years needed for the chlorinated hydrocarbons. The other side of their particular coin, however, is that they are more acutely toxic than the chlorinated hydrocarbons. This should not be too much of a surprise when it is realised that organophosphate insecticides are chemically similar to the so-called 'nerve gases', and act in essentially the same way as those endearing products of the 20th century chemical industry.[39]

In cropping environments such as exemplified in inland Australia and North America, insecticides are commonly applied from a low-flying aircraft. It is inevitable that the powder or spray drifts away to some extent from the designated track. This might not pose an immediate threat to human health when the aircraft is flying over the centre of a large field, but it has the potential to be very dangerous when the path is close to the edge—and to an adjoining property or a homestead.[40] Sometimes farm workers are sprayed or dusted along with the crops they are attending.[41] Although aircraft—often helicopters—are used in densely populated regions such as Europe, pesticides are more likely than in Australia to be applied at ground level from a tractor. This can be expected to result in less drift, but not to eliminate it. Because farms are generally smaller than in Australia and North America, their perimeters will be larger relative to area, tractors will rarely be far from a boundary —and there are many more people at those boundaries than you are likely to find in the Antipodes. Some of the consequences are not too difficult to guess.[42] In the agrarian countries it is the farm workers themselves who are most at risk because nearly all of the spraying is done on foot, they do not generally wear protective clothing, and they are rarely informed of the dangers.[33] But not only in the agrarian countries.[43]

Spraying or dusting crops is not the only cause of pesticide poisoning in farm workers or those who live close to farms that are so treated. Dipping cattle or sheep to kill external parasites has also made its mark. In the past, sheep and cattle dips in Australia commonly used arsenicals that accumulated in the ground and rendered it toxic for a very long time. Recent evidence has identified another 'unintended consequence' of applying an arsenical pesticide. Bogong moths breed in inland plains of eastern Australia—areas that are used predominantly for pasture and cropping. During summer, adult moths migrate to the alpine region of southeast Australia where they aestivate (enter a state of 'dormancy' during hot and/or dry periods) in caves and crevices. In the past, Bogong moths were an important source of food for aborigines (Chapter 1) and are currently no less important for several native vertebrates, including the endangered Mountain Pygmy-possum. In the spring and summer of 2000/2001 heavy rains washed from a number of caves and crevices in the Snowy Mountains material which killed nearby grass. The death of the grass was attributed to arsenic assimilated in sub-lethal concentrations by Bogong moths and carried over 1000 km from inland plains, where it had been applied as a pesticide. Elevated concentrations of arsenic were detected in the faeces of predators of the moths.[44] A sobering example of movement of a toxin through a food chain and, in the process, its concentration. Progressively arsenicals have been largely replaced by chlorinated hydrocarbons and then by organophosphates.

In 1995 an Australian shearer, one of four seriously affected by the organophosphate, Diazinon, was awarded damages against a NSW pastoral company for its negligence in the use of the insecticide. This was reportedly "the first case of its kind in Australia, possibly the world".[45] Dipping sheep is not confined to Australia, of course. British sheep farmers have also been affected but the National Farmers' Union "has been wary of challenging the Government's claim that 'no firm scientific evidence' links the chemicals to permanent ill health."[46] In general, children are most at risk. The same report cites evidence that children in rural villages contract "fatal brain tumours at twice the normal rate". The danger is not confined to the agricultural scene. An epidemiological survey in North Carolina indicates that children under the age of fourteen "have four times the normal risk" of developing soft tissue sarcomas "if their gardens are treated with pesticides or herbicides."[47] And, of course, pesticides find their way into both surface and ground water.[48]

A pesticide that has generated its share of illness, environmental damage and controversy in recent years is methyl bromide (or bromo methane). It is highly volatile (boiling point 3.6°C) and is used as a fumigant—predominantly for soil but also quite extensively for buildings. It is implicated in the breakdown of stratospheric ozone and thereby in the increased intensity of ultraviolet radiation reaching the earth's surface. It is highly toxic to a wide range of organisms.

Currently the industrial nations account for about 80% of the total amount of methyl bromide used. The USA produces and uses most, disposing of about 20,000 tonnes in 1994—about 40% of the global total. California and Florida account respectively for 36% and 40% of United States application of the substance. Between 1982 and 1993, 454 cases of poisoning by methyl bromide were reported in California.

Moves to ban the use of methyl bromide (because of its toxicity) in California have continued to be postponed largely, it would seem, in response to lobbying by the manufacturers. Under US federal law, the fumigant is classified as a 'Class 1 ozone depleter' and, because of that, was scheduled to be phased out completely by 2001. At the time of writing the manufacturers were seeking to have that classification revoked. It is a situation with extensive political ramifications.[49]

One of the points made by Bartle[42] is that, for several reasons, British records of pesticide poisoning in the human population are unreliable. If that is true of Britain it is likely to a good deal worse for the agrarian countries.[50] Accordingly we find a 1989 global estimate of 20,000 deaths annually and "at least three million cases of acute pesticide poisoning ... mostly in the Third World".[51] A 1994 assessment put human fatalities at 40,000 and up to a million people sickened globally each year by exposure to pesticides. A later estimate, attributed to the World Health Organization (WHO), gives the global annual death rate from such cases as 220,000 with up to three million 'severe poisoning cases'.[52] In a 1990 report, WHO put annual pesticide poisoning of the agricultural workforce in the agrarian countries at about 3%—or some 25 million individuals. The Helsinki Institute of Occupational Health has estimated that, in Africa, there are some 11 million cases of acute pesticide poisoning annually.[53] Another estimate gives an annual global rate of 26 million non-fatal cases of pesticide poisoning (Pimentel, personal communication). The discrepancies among these various estimates can be attributed to failure of victims to seek medical attention, to uncertainties in diagnosis, to inadequate reporting and recording, sometimes to a reluctance to acknowledge causes of symptoms and, at times, a falsification of records.

Pesticides inevitably become incorporated into foodstuffs; the industrial world now has sets of regulations or recommendations limiting the concentrations of specific pesticides in foods.[54] Such regulations led to the rejection of some Australian beef shipments by the USA in 1987 and of Australian sultanas in 1991.[55] The rejections were a salutary experience for Australian agriculture at large, but the benefits of those same regulations for the USA are questionable.[56] The general regulation of pesticides "costs American industry and taxpayers more than a billion dollars a year, but the public health benefits of this effort have been shrinking. ... Growth in reliance on pesticides has offset gains from regulation, and the overall toxic risk of pesticides in use today is as great as it was 25 years ago."[57]

Of course other animals, including bees, are affected.[58] A good many of them, notably spiders and birds, are natural predators of the target insects. They are twice cursed. They too are vulnerable to spraying—and they eat insects that have been hit by the poison. Their vulnerability is not helped by the extraordinary inefficiency with which insecticides are usually applied in agriculture—less than 0.1% of those pesticides that are disseminated by spraying or dusting is estimated to reach the 'target organisms'—99.9% goes somewhere else.[59] Some crops are sensitive to insecticides. About 70% of coffee crops in northern Queensland were effectively destroyed early in 1997 following spraying with an insecticide in an attempt to repel an 'invasion' of the region at large by the exotic papaya fruit fly.[60]

Moreover pest populations are likely to develop resistance to pesticides.[61] That usually stimulates the industry to find substitutes, but then, sooner or later the pests develop resistance to the new agent. A battle might be won, but the war continues. Sometimes defence is sought by modifying a crop genetically, but pests can adapt to that too.[62] An upshot of all of this is that the application of pesticides has had precious little effect on the proportion of crops lost to pest damage. Over the past half century, during which the use of these agents became so widespread, the proportion of crops lost to weeds, insects and pathogens has remained constant at about 30-35%, much the same as it was in the 'prechemical era'.[63]

My purpose in this chapter has been to identify two important activities that, in one way or another, are intimately associated with present day agriculture. Each is ultimately destructive and each is part of a wider set of processes that, if considered objectively, will be seen to deny any valid application of the adjective 'sustainable' to contemporary commercial agriculture. So now let me turn more directly to farming at the close of the 20th century.

CHAPTER 13

The Contemporary Farm

— Industrial Agriculture —

Chapters 10 and 11 looked at some salient characteristics of modern mechanised farming—the dependence on manufactured or mined fertilisers, the decline in the organic content of soils, the increased susceptibility of those soils to erosion. If we are to have a reasonable understanding of why that should be so, we also need to appreciate something of the dynamics and demography of 'agribusiness'.

Table A15 summarises population densities relative to farmland, as well as urban proportions, in a number of selected regions and countries. Needless to say, there are several areas of uncertainty in the tabulated numbers, but they are not of such significance as to hide or seriously distort the essence of the table. One message that emerges is that the planet is at least (allowing for definitions) 10 times more 'urbanised' than was, for example, mediæval Europe (Chapters 5 and 7).

In keeping with its rapid urbanisation in the 19th century, Australia has the highest proportion of town residents and the lowest overall population density relative to farmland. Of the major industrial countries, the USA has the next lowest while Japan has the highest overall population density relative to farmland and the second highest urban proportion. The regions with the lowest proportions of urban dwellers—most of sub-Saharan Africa, much of Asia and parts of South East Asia (not listed)—might reasonably be considered the best representatives of traditional farming. But they are all in various stages of transition and, from my present perspective, they are hybrids. It is therefore easier to start with some unambiguous exponents of modern industrial agriculture. I shall again focus most of my attention on Australia.

One conspicuous feature of commercial agriculture is that, like most contemporary industries, it seeks to employ as few people as possible. In all major industries that policy has very wide economic and social implications. When it applies to agriculture, however, the extent of the ecological and environmental consequences is possibly without equal in any other sphere of human activity.

'Agribusiness' is highly mechanised and the farm 'surplus' is enormous. There are, however, several impediments in the way of calculating that surplus by the simple formula that I used in Chapter 5. For example, the employment of seasonal workers, a practice of some owners or managers to live away from the property and to 'commute' to it, the effect of nationally exporting produce (and, for that matter, importing it) on the basic calculation—and the homogeneous nature of the produce itself. By this last I mean that a large farm that produces but one type of product, be it wheat, meat, milk or oranges cannot supply all the food requirements of its personnel from its commercial produce. An extreme example is a cotton plantation, where the commercial product is presumably not eaten at all—at least by the farmers.

So, those reservations should be kept in mind when I say that the type of calculation that I have already used, but based on the number of farm workers,[1] points to an Australian surplus of about 99%. If, however, the spirit of the earlier calculation is followed and we take into account the farm workers' families (because they live on the farm, they contribute in various ways to the working of the farm—and their wastes should remain there) and assume that the average size of those families is four, then the effective surplus is about 96%. The USA is broadly similar.[2]

In 1993 Australia had, by one definition, 120,656 'farms' or, as the Australian Bureau of Statistics more accurately puts it, 'establishments with agricultural activity'. By another definition there were 151,966 in 1993 and 146,371 in 2000.[3] The total area of these farms amounted to some 469 million hectares—or about 61% of the area of the entire continent plus Tasmania (Chapter 10). The numbers also indicate an 'average' size of Australian farms as 3896 ha but, needless to say, such a simple calculation is misleading.

Meat and cereals together account for about 42% of the total mass of contemporary Australian farm produce. Perhaps more importantly from our point of view, they account for about 76% of the total phosphorus in Australian farm produce (see Table A14). Figure 13.1 shows, for example, a 1993 size distribution of Australian farms producing beef, sheep, and grain—solely or in any combination. Approximately one third (33.8%) of such establishments lie within the size range of 100-500 ha, while between a fifth and a quarter lie within each of the ranges 500-1000 ha and 1000-2500 ha (22.4% and 21.5% respectively).

When they are distributed according to the total area within each size range, however, a very different picture emerges, as shown in Fig. 13.2. One third of the total area is within the range 200,000-500,000 ha.[4]

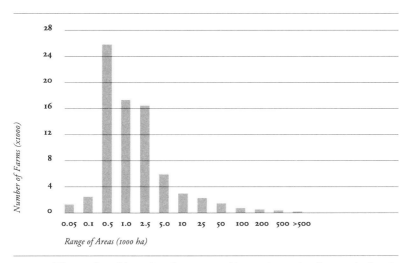

FIG 13.1 *The number of Australian farms engaged in growing grain, sheep, or beef cattle, alone or in any combination. The farms are grouped into size ranges used by the Australian Bureau of Statistics (1993b). Each range is denoted by its nominal upper limit and extends from the top of the previous range to 1 ha below the stated limit (e.g. 100-499 ha). The top range includes properties of 500,000 ha and over.*

The total area of the farms represented in each of these figures is some 443 million hectares[5]—about 94% of all Australian agricultural land. There are 79,952 farms represented in each figure—representing 66% of the number of all Australian farms in 1993 and equivalent to an average size of 5540 ha. This average size is also too simple, of course, but at least the figures give an idea of the extent of the oversimplification.

In Chapter 10 I calculated, with some qualifications, the Australian rate of soil erosion per farm worker. That calculation was made in the interests of a feeling for the problem, rather than statistical accuracy. An analogous calculation (that needs its own set of qualifications) is the amount of 'farmland' per farm worker. The average for all Australian 'agricultural establishments' is about 1632 ha/farm worker.[6] One obvious weakness of that calculation lies in the 89,916 farms[7] with areas less than 1000 ha. So, I must make some more dubious assumptions, one of which is that every farm has the same number of workers. Under those circumstances every farm is allocated 2.4 workers and the properties represented in Figs 13.1 and 13.2 would have, on average, about 2309 ha/farm worker.[8]

The purpose of this exercise is simply to give some substance, albeit rough, to the point that modern industrial agriculture, especially in inland Australia, requires an individual farmer or his employee to manage a very large area of land. The contrast with traditional agriculture in any part of the Old World is obvious. These very low worker densities, have profound implications for the overall management of farms and for their impact on the soil. But now we need to distinguish between the dominant characteristics of cultivating crops and raising animals.

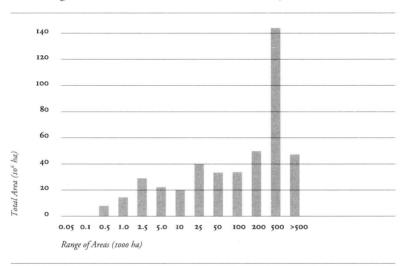

FIG 13.2   *The total area of the farms represented in Fig. 13.1. The mid-point of the 'nominal' range was used to calculate the total area of each group—that is to say, a range was taken as, for example, 100-500 ha, not 100-499 ha as in Fig. 13.1.*

— Cultivation —

Wheat is Australia's biggest crop, occupying some 8.4 million hectares in 1994 and 12.2 million in 1999-2000.[9] The biggest *single* wheat field in Australia reportedly has an area of 16,000 ha. A newspaper article describing a harvest carries a photograph of 17 combine harvesters in strict formation (staggered line abreast) followed by various trucks—in this one field.[10] The landscape is flat and there are no trees. Like the 'wheels of fortune' (Chapter 9), modern farm machines like to travel in straight lines, and they hate trees. As well as that particular fetish, heavy farm machinery compresses soil, diminishing water infiltration and retention which, in turn, affects soil biota. In so doing it exacerbates the effects of declining organic content. Nor can it handle interplanting.

A wheat paddock is a huge monoculture, apart from whatever weeds might be able to get their heads above ground. Like a cotton plantation, it is about as simple a terrestrial ecosystem as you are likely to find outside a laboratory, a greenhouse or a pot-plant. Because it is a monoculture, the entire crop is vulnerable to potential pests or pathogens.[11] It was such a pathogen, rust, that encouraged the movement of wheat cultivation in the 19th century inland to areas of lower rainfall and humidity than prevail on the Australian coast. Perhaps the most famous of crop failures caused by a pathogen was the 19th century Irish potato blight. Although the effects of that catastrophe on the Irish population—and on the ethnic composition of the New York police force—had many political, economic and social dimensions, the fundamental biological cause was the absolute dependence of a large part of the population, the poor, on one crop—a particular strain of potato—and the susceptibility of that crop to the fungus *Phytophthora infestans*.

A monoculture of the scale provided by modern industrial agriculture offers a pathogen not only great ease of transfer from one plant to the next—it also provides a mass of host material to exceed the wildest dreams of any bacterium, fungus or virus. Plants, like animals, develop resistance to infectious diseases. Native plants, like native animals, are relatively resistant to native pathogens—provided the ecosystem is mature (see Chapter 2)—but are likely to be vulnerable to exotic pathogens. Conversely, exotic plants are likely to be vulnerable to native pathogens and parasites. When those plants are grown in monoculture on a large scale, an entire crop can be lost and the soil can remain contaminated for many years.

In the past the response of agriculture to this type of challenge has included intercropping, crop rotation and, within the past century, breeding resistant strains of the cultivar. Most recently, breeding methods have included 'genetic engineering' whereby genetic material from other plants, or even other types of organism, is incorporated into the genome of the crop plant. A problem, of course, is that viruses and microorganisms have high mutation rates. Burning stubble is also used to lessen the likelihood that pathogens will remain in the soil from one season to the next, but the practice has its own problems (see Chapter 10).

Pathogens are not the only ones to rejoice at the sight of a field of wheat—or any other crop. There is no shortage of granivorous and herbivorous animals, both vertebrate and invertebrate, that positively lick their chops at the mere suggestion that someone is going to plant many thousands of hectares with a cereal—or, for that matter, with any single crop. What's more, because there are rarely trees within the cultivated area, they can count on very little trouble from birds. Apart from the regular pests that turn up every year, this type of environment encourages intermittent population explosions of some of them, in particular plague locusts and mice.

Cereal cultivation on farms such as those just described obviously requires the expenditure of a lot of energy. Figure 13.3 summarises the direct expenditure of energy within all Australian farms over the period 1976-1994. By 'within', I mean 'on the farm'—excluding energy spent in the production of fertilisers, machinery etc and in transport to and from the farm. Diesel fuel accounted for 83-88 % of the total. The next most important source of energy was electricity which increased from 10 to 15% of the total over the nine-year period.[12]

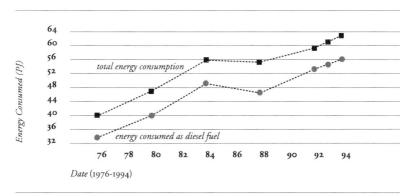

FIG. 13.3 *Energy consumption by Australian farms from 1975/6 to 1993/4. The apparent drop in energy consumption from 1984-88 can be attributed in part to a drop in the number of farms that were officially recognised. This happened because of a change in definition (see Note 3). Squares, total energy consumption; circles, energy consumed as diesel fuel.*

Farms of the type under discussion are mostly a considerable distance from the major cities and ports. In general, large grazing properties are even further away. Transport is thus an indispensable component of the profitable running of these establishments. In the period represented in Fig. 13.3, transport, in its entirety, accounted for 26% of Australia's total energy consumption—a similar proportion to electricity generation and to manufacturing.

By breeding and selecting suitable strains, by the prodigious application of fertiliser, and by the partial suppression of weeds and pests, agribusiness has been very successful in increasing crop yields. Yields of Australian wheat (kg.ha^{-1}), for example, have increased from about 1360 in 1960-61 to 1970 in 1993-94 and 2030 in 1999-2000.[13] In other parts of the world they are a good deal higher (see Table A4). The widespread sense of achievement engendered by the increased yields of the 20th century is not without its corresponding concerns, some of which I have already identified. But there is another long-term problem with monocultures.

Briefly, the total productivity of a grassland community increases with species diversity, the increase being primarily a function of the number of species, not the number of individual plants. In the experiments showing this relation,[14] the importance of species diversity, as distinct from the number of plants, was also reflected in the amount of nitrate in the soil. Under the prevailing conditions, nitrogen was the principal limiting element. The nitrate ion is normally more mobile in soil than the positively charged ammonium ion and is more susceptible to leaching (see Chapter 11). In these experiments the concentration of nitrate in the soil decreased with increasing species diversity. The implication here is that a greater proportion of the ion was assimilated by the plants and therefore less was susceptible to loss by leaching or denitrification. Observations of this type are indicators of differences in the dynamics of complex and simple systems—dynamics that include rates of assimilation, metabolism and excretion by the various species of grass. And, of course, cereals are species of grass. This is another type of reason why complex ecosystems are more resilient and, in terms of their 'sustainability', why they are more stable than monocultures.

Traditionally coffee has been grown in South America under a canopy of shade trees. Among other things, this practice provided a haven for migrating birds, a haven that increased in importance as more forests were cleared. Within the past few decades, however, many coffee plantations have moved to greater 'efficiency' by replacing traditional coffee plants with high-yield strains grown in full sunlight. By 1990 about half of the 5.5 million hectares of plantations in northern South America had been so converted. Since about 1970 the populations of some vulnerable species of birds have dropped by 30-50%. It is not implied that the change in coffee cultivation was the sole cause of this, but it is likely to be an important contributing factor.

By one criterion the results have been impressive; planting density increased from 1000-2000 bushes per hectare to 3000-7000. Yields have increased from around 550 kg.ha^{-1} to about 1600 kg.ha^{-1}. The report from which I have drawn this information[15] puts it this way: "Plied with chemical fertilizer, the sun varieties can out perform traditional varieties by a factor of 3." On the figures quoted individual plants do no such thing; the new varieties yield 230-530 grams of beans per plant compared with 275-550 g for the traditional strains. This is unlike the high-yield cereals of the so-called 'Green Revolution'. In that case the 'harvest index'—the mass of grain as a proportion of total phytomass—was higher than in older varieties. The characteristic that would seem to distinguish the new coffee strains from the old is greater tolerance of sunlight, but even that does not come up to expectations.

The coffee plant evolved in understorey—it is fundamentally a shade plant. The high light intensity and high temperatures encountered in the new plantations have caused water stress, leaf damage, and changed the dynamics of carbohydrate metabolism in the plant. A high proportion of the plants die within a year. The intensive coffee plantations inevitably need a great deal more fertiliser and some pesticides, especially herbicides. There are also the usual problems of loss of soil organic matter and increased erosion. A number of large plantations in Guatemala are putting back the clock and returning to traditional methods.[16]

Another problem is inherent in an agricultural system that exports produce from the farm and replaces the nutrients in that produce with fertiliser from somewhere else. The fertiliser might contain toxic impurities that accumulate in the soil and duly find their way into foodstuffs. This is not likely to be a problem with synthetic nitrogenous fertilisers but it certainly occurs with mined minerals such as phosphate rock.

The elements of potential significance in this respect are arsenic, cadmium, chromium, fluorine, lead, mercury, strontium, thorium, uranium and zinc. Zinc is an essential trace element—presumably for all organisms—but it becomes toxic for mammals if accumulated in more than 'traces'. 'Traces' of fluorine protect against dental caries because of the toxicity of the element for bacteria. To the best of my knowledge, neither it nor any of the remaining elements has any essential role in mammalian biochemistry or physiology. Unlike zinc therefore, none has an optimal rate of assimilation—there is no benefit known to be derived from ingesting any of them. Of these elements, cadmium is generally considered to be the most dangerous in soil, partly because of its innate toxicity, both acute and cumulative, and partly because, when it is in soil, it remains 'available' to plants for longer than the other metals and arsenic. The role of fertiliser in raising the cadmium content of food crops has been acknowledged since 1963.[17]

Regular application of mineral fertiliser to a plot of ground will add a contaminating element. Some of that element will be removed in the crop and duly assimilated by its consumer. Some will remain in the soil where its concentration will increase with every subsequent application of fertiliser. And, of course, some will be leached out from time to time. As the content of the contaminant in the soil increases, so will the amount taken up by the crop. The relation between those two variables is unlikely to be linear, but it will be positive. Selecting a crop or a strain of crop with a lower rate of uptake of the element will improve the situation for consumers—but only in the short term. Because the new crop removes the toxic element more slowly than the old one, that element will accumulate faster in the soil for a given rate of application of the fertiliser.

McLaughlin and colleagues[18] point out that the content of cadmium in phosphate fertilisers used in Australia has been 'historically high' but we have been able to get away with it (so far) because of relatively low rates of application. They present what is, in effect, a cadmium balance sheet for two Australian crops, wheat and potatoes. The essence of that balance sheet is as follows:

Rate of addition of Cd (g.ha^{-1}.y^{-1}): wheat, 6.0; potatoes, 20.0. *Removal of Cd in crop* (g.ha^{-1}.y^{-1}): wheat, 0.12; potatoes, 2.50. *Doubling time of Cd in top 100 mm soil* (years): wheat, 45; potatoes, 15.

The doubling times in the soil of F, As, Hg, and Pb were estimated to range from 100-7500 years for wheat and from 25 (for F)–1800 years for potatoes. Fluorine, mercury and lead were considered to "pose negligible risk of accumulation to toxic concentrations in agricultural food crops".

— Livestock —

In 1961 there were some 941 million cattle and 1342 million sheep and goats in the world. In the early 1990s the corresponding numbers were 1285 million and 1785 million respectively. By 2000 these numbers had changed to. 1331 million cattle and 1774 sheep and goats. In the period 1990-92, the USA had about 99 million cattle and about 13 million sheep and goats.[19] In 2000 Australia had some 27.6 million cattle (including 3.1 million dairy cattle), 119 million sheep and about 0.5 million goats.[20] To put these numbers into some sort of perspective, the global cattle population is about 22% and the sheep + goat population about 30% of the global human population. The corresponding proportions for the USA are about 40% and 5% respectively. In Australia, however, cattle outnumber people about 1.4 fold while there are some 6 times as many sheep as people. A very rough comparison of masses gives a different perspective again. Globally, the total mass of cattle is about 1.7 times that of the human population, in the USA it is about two and a half times as great, while in Australia it is about ten times greater. Pimentel (personal communication) advises that, in the USA, the livestock biomass is some 5 times the human biomass. Australian sheep also outweigh the human population about tenfold.[21]

For a number of reasons these and other animals are going to have a very considerable environmental impact. Collectively they weigh more than the human population, the growth rate of individual animals is considerably faster than individual human growth rates and, although growth does not continue for as long as human growth, once it slows appreciably the adult animals are replaced by young ones—at least when the animal is grown primarily to be eaten as distinct from exploiting its capacity to produce milk or wool.

The numbers I have used above suggest that, in some respects, Australia is once again an extreme case. As we saw earlier, Australia's cattle population comprises about 11% dairy and 89% beef cattle. As Figs 13.1 and 13.2 show, most of the area of the Australian continent is used for the commercial production of beef and wool/mutton, separately or in combination with each other and/or with grain. Maps of the distribution of these activities[22] indicate that about half of that area is given to raising beef cattle only. The largest properties—those of 100,000 ha or more—are used solely for grazing. As a rough approximation, about 20% of all grazing land is in regions of less than 200 mm annual rainfall, perhaps half is in the 200-400 mm belt with the remaining 30% or so in wetter areas. For the most part, grazing is an activity of the arid or semi-arid regions of Australia—and that is one of its problems.[23] Apart from minor differences between different strains of cattle, the animals have a basic daily need for water. As with any other homeotherm, that need will increase under hot dry conditions. That is apart from the water needed to grow their feed. Rifkin comments "Nearly half the water consumed in the United States now goes to grow feed for cattle and other livestock."[24]

Grazing animals that are culled for human consumption remove nutrient elements from the soil. Obviously, as with high yields of crops, high stocking rates will remove those nutrients faster than will low stocking rates. General aspects of this were discussed in Chapter 2 but I should now like to give the process some dimensions for the element phosphorus. The same general principles apply, but with different numbers of course, to all the other essential nutrient minerals, be they trace or major elements. Nitrogen, the other major element that received attention in Chapter 11, is not necessarily so critical here. Pasture or rangelands are rarely monocultures and symbiotic nitrogen fixation is often possible—but grazing can affect the chemistry and dynamics of nitrogen in the soil (see below).

In the year 1999-2000 some 8.6 million adult cattle and calves were slaughtered in Australia to give a total body weight of about 3,440,000 tonnes[25] which, in turn, amounts to some 34,000 tonnes of elementary phosphorus.[26] Corresponding statistics for sheep indicate an annual rate of about 15,500 tonnes for the removal of phosphorus from the soil. If all the cattle used in this calculation had been raised at a stocking density of $0.01.ha^{-1}$ (impossible because it would require a total area of more than 800,000,000 ha— see Figs 13), their despatch would have removed about 40 g $P.ha^{-1}$ annually.[27] If they had all come from stocking rates of $0.5.ha^{-1}$, they would have removed 2 kg P.ha^{-1} but, for several reasons, those estimates are likely to be conservative. For example, direct measurements put phosphorus losses from high country tussock grasslands in New Zealand at about 5.5 kg.ha^{-1}.y^{-1}; the land had been subjected to grazing, or grazing with intermittent burning.[28] By way of comparison, the 1996 average Australian wheat yield of 1970 kg.ha^{-1} would remove about 7 kg P.ha^{-1} in the grain and leave some 2 kg P.ha^{-1} in the straw.[29]

So, in a very simple sense, grazing might be seen to be a less demanding activity than cropping, at least in terms of its capacity for depleting the soil of nutrients. And, up to a point, so it is—which is one of several reasons why most farmers in the past and some of them today include pasture in patterns of crop rotation. There are many qualifications, however, especially in comparisons of grazing in arid or semi-arid rangeland with virtually any form of cultivation.

I have already noted that an animal at constant weight causes overall neither a nett removal nor addition of nutrient elements from or to the soil —with the possible exception of stimulation by excreted phosphorus etc of nitrogen fixation. Animals remove nutrients when they are growing, pregnant or lactating and, of course, that means most farm animals. One obvious reason why pastoralism, at least in the Australian context, removes less phosphorus (and other elements) per area than cultivation is that most grazing takes place in regions of lower rainfall than most cultivation. Plant growth is therefore poorer. Another—and this applies everywhere—is that herbivorous animals are one step up the food chain from the grasses and, other things being more or less equal, would contain only about 10% of the energy and nutrient elements contained in the plants that support them.

But the animals have other effects, the significance of which can vary widely with climate, season and stocking rates. It is often stated that exotic livestock compress and adversely affect Australian soils to a degree never approached by native animals. The excessive impact of the exotic animals is associated with their weight and, more specifically, the pressure on their feet or hooves, their numbers, their propensity for remaining relatively close together in herds, their patterns and speed of movement etc. By and large this seems to be fair comment. For example, in semi-arid north Queensland, soils that are relatively heavily grazed have fewer species and lower total populations of termites and some other arthropods than their lightly grazed counterparts. The rates of infiltration and flow of water within the soil were also reduced in these areas.[30]

Extended grazing on native pasture can be expected to reduce the number of species of grasses—especially native grasses—and, intentionally or otherwise, to introduce exotic species that are less resistant to drought. Grazing tropical savannas, which account for about 37% of the area of Australia, increases their vulnerability to invasion by various woody weeds that, among other things, progressively decrease the area of available pasture. The impact of livestock on soil, like the corresponding effects of cultivation, is exacerbated when forested land is cleared for that purpose. Of course, grazing by large animals is not of itself necessarily destructive—if it were there would be no large herbivores. When large grazers are free to roam over a wide area and are essentially in balance with their habitat, they can have complex effects on soil chemistry and ecology, effects that can ultimately be beneficial to themselves.[31]

Probably the most serious effects are associated with drought. It is in this situation more than any other that the concept of overstocking asserts itself. From time to time Australian graziers are criticised[32] for stocking for good seasons and then calling for help in a drought. The critics usually point out that droughts are inevitable and that the determination of stocking rates should take that into account—which it does on well-managed properties.[33]

In a severe drought there will be no growth of a crop without irrigation—no nutrients will be removed from the soil by the crop, although much could be lost by wind erosion. In a real drought there is no growth of pasture either, although hardy plants will survive. But the animals continue to eat the dormant, dying or dead grasses, as well as the leaves of any palatable shrubs or trees to which they might have access. Many animals die. Some are moved to other properties for agistment—and thereby increase the pressure on those properties. Some are sustained by the purchase—sometimes the donation—of hay. And that too increases pressure on other properties. Animals that remain on pasture under these conditions are living on borrowed time—they are eating their capital. The carrying capacity of land depends ultimately on the duration of the drought. If the drought persists for long enough the carrying capacity is reduced to zero.

When animals remain on land under such conditions, even if they are given supplementary feed, it is common for extensive areas of bare soil to be exposed—especially in arid regions where ground cover was thin to start with. Such an environment is inevitably susceptible to erosion by wind or, when the rains come once again, by water. Moreover, the damage done in this way is largely irreversible within a time scale of immediate relevance to human beings. As we have already seen, overstocking with grazing animals has made its contribution to the prehistoric degradation of soils and the genesis of deserts where formerly had been fertile soils. A doubling of reindeer numbers over the past 40 years (current density c 0.2.ha^{-1}) has degraded tundra in the Norwegian Arctic and led to extensive soil erosion.[34] The essentially irreversible effects of severe overstocking have been apparent, if not always acknowledged, for quite some time.[35] Although cultivated land is generally more vulnerable than pasture to erosion, the extent of grazing land in Australia, and the location of much of that pasture in arid or semi-arid regions where ground cover is normally sparse, make it the major contributor to the high total erosion rates indicated in Chapter 10.

There are other dimensions to the livestock industry, dimensions that in part have their genesis in the destiny of most livestock to be slaughtered, whatever functions they might perform during their lifetimes. But the largest of these other dimensions undoubtedly stems from the pervasive pressure of economic 'efficiency'—and frequent misunderstanding of what might be needed to be efficient. It shows up in a number of ways. It is reflected in those grazing properties with no trees or where trees are few and far between. Apart from their many other attributes, trees provide shelter for animals. The sight of a herd of cattle trying to get into the shade of a solitary tree in the middle of a hot summer's day can be very disturbing. The sight of a herd of sheep or cattle without any shade on such a day is worse. On well managed properties, trees or shrubs that are retained or planted as windbreaks[36] also provide shelter in winter. The complete clearing of land for pasture or a reluctance to plant a few shade trees is, in most instances, presumably an indication of 'efficiency'—a determination to put all available land to good use. A similar determination to exploit all available land by allowing cattle to graze to a river bank is an important cause of the erosion of stream banks.[37]

Economic considerations influence, if not determine, what livestock eat. Cattle are the archetype of what, in Chapter 2, I called 'professional herbivores'. They are ruminants and are about as efficient at digesting grasses as any vertebrate you are likely to meet. But agribusiness too often decrees that they should eat something else. Thus, in Britain, cattle were fed meals made from left-over bits of sheep that could not be sold for human consumption. The prion[38] disease, scrapie, has existed in British sheep for many years. One result of feeding cattle with tissue from infected sheep was the well publicised outbreak of 'Mad Cow Disease' (bovine spongiform encephalopathy, BSE). In consequence entire herds of cattle were slaughtered, the British beef industry experienced severe economic damage—and a number of people who had eaten infected beef or, in at least one case only dairy products, contracted Creutzfeld-Jacob Disease, which is fatal. The problem was evidently exacerbated by unwillingness of the British Ministry of Agriculture, Fisheries and Food to make relevant information of the problem available to scientists.[39]

I do not know if any Australian cattle have been fed sheep remainders, but some have been fed poultry dung. In 1990, yarded cattle in Queensland were fed a meal comprising pieces of poultry carcases and 'chicken litter'—a mixture of hay, sawdust and fowl dung. At least five thousand animals succumbed to botulism.[40]

The notion of economic efficiency has spawned the practice of 'factory farming' with virtually all types of livestock raised for commercial ends. It is not my purpose to review this industry—that has already been done to good effect.[41] But I shall make some limited observations, mainly in relation to beef cattle.

The highly intensive raising of livestock, or production of animal products such as eggs or milk, commonly involves confining the animals within a very small area and denying them exercise. There is often insufficient space for an animal to turn around—especially in the case of poultry and sometimes with pigs and dairy cows. They are fed materials very different from their natural foods. They are commonly treated with antibiotics—in some cases partly to stimulate growth but normally to lower the incidence of infectious diseases that are readily transferred under such conditions. Often the animals are also given growth hormones.

Both the antibiotics and the growth hormones have the potential to affect human health—the antibiotics by generating resistance in human pathogens to related antibiotics, the hormones by direct physiological action. It is difficult to know which species of animal is most cruelly affected by imprisonment under such conditions. The answer may well be different in different countries. Suffice it to say that any animal confined for all or most of a lifetime indoors within a small cage on a floor of mud, or concrete or steel bars is treated cruelly indeed—despite official protests to the contrary. The air in such establishments is sometimes so foul that both the livestock and their primate attendants are sickened by it.[42]

Perhaps in some respects the beef cattle feedlot is not quite so bad. A primary objective of the feedlot is to hasten the growth of animals, an objective that is achieved largely by fattening. Coats[41] has commented that, in the USA, cattle formerly reached 'salable weight' in about three years. Now they are slaughtered at 10-11 months of age. Much of the additional fat that is deposited under these conditions is intramuscular. It produces an effect known as 'marbling' which, despite the medical problems associated with eating excess animal fat, leads to the meat's classification as 'high quality'. This result is achieved by denying the cattle exercise and feeding them materials other than grass. The predominant feedstock is grain.[43] Under such conditions the animals are prone to digestive disorders.[44]

A feedlot is essentially a series of yards, each with a feeding trough running the length of a boundary fence. The largest American feedlots are reported to have capacities of the order of 100,000 animals.[45] In Australia the lots are usually classified in groups with holding capacities ranging from less than 50 (seasonal or 'opportunistic' feedlots) to over 5000. In 1992 there were 140 lots in the smallest category and 22 in the largest.[46] The largest commercial establishments have capacities in the order of 30,000. Australian cattle enter feedlots at around 12-14 months and remain there for periods ranging from about 70 to 300 days. Their feed consists of grain (about 70%; see Note 43), fibrous material designated as 'total roughage' (about 14%) and 'other concentrates' (16%) (see Note 40).

In 1992 the total Australian pen capacity was approximately 500,000 animals, the 'throughput' 700,000-800,000 (presumably per year). By 1996 the total pen capacity had increased to 850,000.[47] Feedlots rarely provide shelter for their inmates. Coats[45] shows some photographs of American feedlots. One is of cattle in a mid-western compound in mid winter. The animals are standing in a mixture of mud, slush, their own excrement and wet snow—and they have no other option. At the other extreme, more than 2,500 cattle died from heat stress at an Australian commercial feedlot in February, 1991. Paragraph 7.52 of the Senate Standing Committee's report says:

"The Committee considers that the need to provide shade and shelter should be based on the health and welfare of cattle in feedlots and not on cost. The Committee notes, however, that individual feedlots in several States and regions have decided to install shade or shelter because of the economic benefits gained from the enhanced well being of the cattle."

A feature, if you can call it that, of a feedlot is that the small area in which the cattle are confined accumulates their excrement and they are obliged to stand in it. How serious this is depends on a number of factors—the density of the cattle (or its inverse, the area per animal), the slope of the ground, climate (particularly rainfall and evaporation rates), drainage arrangements and the frequency with which excrement is actively removed. Let me again quote the Senate Standing Committee—this time paragraph 7.80.

"The Committee notes that animal welfare organisations, industry and government bodies have proposed varying minimum requirements for pen size ranging from 9m² to 25m² per animal. The Committee's view is that the minimum pen space of 9m², recommended by the Australian Lot Feeders Association, is too small and considers that a minimum of 15m² to 20m² per animal, as recommended by the National Consultative Committee on Animal Welfare, will enhance the welfare of cattle in feedlots."

Elsewhere in the report, a 450 kg steer is accredited with producing about 29 kg of wet manure and urine each day. At 15m^2 per animal, this is equivalent to a daily rate of about 2 kg.m^{-2} or 20 t.ha^{-1}. A feedlot of 30,000 would receive some 870 tonnes of manure daily—roughly equivalent to 8,500 and 1400 kg of nitrogen and phosphorus respectively.[48]

Excrement deposited at such a density will, among other things, stink. Odour was considered by the Senate Committee as "one of the most difficult issues facing feedlot managers". It was the substance of complaints made from Queensland, NSW and South Australia. A Queensland submission commented that the odour from a feedlot was 'so vile' as to cause complaints in the town of Dalby, 20 km from the source. A NSW submission referred to similar complaints 10 km from a feedlot.

To its credit, the Senate Committee emphasised the importance of the general welfare of cattle in feedlots and the term 'behavioural requirements' occurred frequently in the report. Clause 7.74 says:

"The Committee is surprised that the *National Guidelines for Beef Cattle Feedlots in Australia* do not reflect the importance of animal welfare as a consideration for site selection for a feedlot and recommends that the Guidelines be revised accordingly."

Various other recommendations to improve animal welfare were made but, in the end, the Committee rejected "the view that lot feeding of cattle is unnatural". It would be interesting to know how the concept of 'natural' might have been defined, consciously or otherwise.[49] A situation in which cattle have, at worst 9 m^2, at best 20 m^2 of ground space per animal, where they cannot escape their own excrement which can accumulate in such quantities as to be offensive to people 20 km away is, to say the least, a curious definition of 'natural'. But the ultimate *raison d'être* for feedlots is economic; it is reflected succinctly in the following comment given in evidence to the Committee:

"One of the most simple yet effective ways of value adding from the grain industry is to get an animal to eat it. We convert vegetable protein into animal protein. With the processing of that animal protein, we get into high value adding."[50]

But now let me turn to the more direct environmental implications of this industry. There are several levels at which cattle feedlots, as distinct from the same number of animals on open pasture, can have a significant environmental impact. One is that the bulk of their food must be cultivated. As we have seen, cultivated land is normally more susceptible to erosion than is pasture. A minor mitigating factor is that, at least in the Australian context, the area of land needed to cultivate the grain will be smaller than that required to support the same number of cattle on pasture—but not by much.[51] Cultivation requires a substantial amount of energy, as does the subsequent transport of the grain to the feedlot. Neither of those criticisms applies to conventional pastoralism under normal circumstances, although there can be a related need for energy during a drought if supplementary feed is required. Energy is also needed to transport the cattle to the feedlot in the first place—but whether or not that is significantly more than taking them directly to an abattoir will obviously depend on the locations of all the relevant establishments.

The smells from a feedlot are objectionable, but they are likely to be associated with a significant ecological threat only under some circumstances. Those circumstances arise when there is enough deposition of atmospheric nitrogen to cause eutrophication of waterways or sensitive land communities. It is a problem associated with frequent rainfall and relatively high overall densities of farm animals. It is not so serious around most of Australia's large feedlots but it can be a significant problem in parts of Europe and North America (see also Chapter 11).

It is at the wider level of nutrient flow, however, that the feedlot's environmental impact can be most serious. As I have already emphasised, grazing cattle remove nutrients from their pasture only if, in effect, they are growing, and then only the amount that is actually incorporated into the animal, its foetus, or its milk. The rest is returned to the soil in dung and urine. At this level a feedlot, or for that matter any other type of 'factory farm', is an analogue of a city—or, perhaps more accurately, a prison. Food is grown somewhere else and transported to the feedlot. Something must then be done with the inhabitants' excrement for essentially the same reasons as a city needs a sewerage system. As with sewage, the manner of disposal of that excrement is the most important single determinant of the overall environmental impact of the feedlot. What we can be confident about, as we can with a modern city's sewage, is that little of the waste material will be returned to the farms that grew the cattle feed.

So, quantitatively, how does a feedlot compare with its human counterpart in terms of waste production? Conversion of statistics summarised in the Senate Committee's report indicates that a feedlot with 30,000 cattle is roughly equivalent to a city of 300,000 in terms of nitrogen excretion and of 270,000 for phosphorus. A total feedlot cattle population of half a million equates in those terms to human populations of about 5 million and 4.4 million for nitrogen and phosphorus respectively.

Feedlot manure escapes a problem that besets urban sewage. It does not (or should not!) contain the toxic wastes of secondary industries. Australian feedlot wastes have enhanced the fertility of nearby farmland, especially in semi-arid regions of relatively poor soil—as is usually the case. Like sewage, the wastes contain and can provide water, organic matter and plant nutrients—all of which have great potential value for farmland. But there is a limit to the rate at which feedlot waste—or sewage—can be applied to land. That limit is determined, most obviously, by soil type, by topography, by the level of any groundwater, by climate—especially rainfall, and by the purpose to which the land is put. It is all too easy to exceed the limit when effluent from a feedlot is not properly controlled and flows, unrequested and unauthorised, through adjoining property.[52] So other options are needed.

According to the Senate Committee's report, some manure is "occasionally" sold "further afield". It would seem, however, that there is little recycling to the farms that grew the cattle feed. Because the manure is part of a chain that merely transfers nutrients from one part of the country to another, there is no basis for the claim made by two witnesses to the Committee that the sale of feedlot manure might reduce "the need for imported fertilisers". It may well reduce the need for the recipients of the manure to buy imported fertiliser, but not that of the farms that grew the cattle feed.

So, to summarise: industrial agriculture employs very few people in relation to the area of farmland. Accordingly, there is an extreme dependence on machinery—and hence on inanimate energy—most of which is supplied by fossil fuels. Surpluses are enormous and, for that reason and because of the size of farms, *direct* recycling on the farm of the nutrients contained in the produce is impossible. Transport of farm produce to points of consumption also requires a great deal of energy, again mostly supplied by fossil fuels. For the most part, cultivation produces extensive areas of monocultures which, if left to their own devices, are extremely vulnerable to pests. The industry's main response to this has been the extensive application of pesticides. It is generally characteristic of the industry that its criteria of success are economic rather than biological or social. But before we consider the wider implications of these practices and how they might be modified to advantage, we need another bird's eye view—this time of the Agrarian World, the Third World or the Developing World—according to which term might have the most appeal.

— The 'Third World' —

I had originally intended to allow a chapter for this topic but duly came to realise that the whole subject is so complex that a chapter would not be nearly enough to do it justice—even in the general terms of this book. So I shall go to the other extreme and try to summarise the essentials in a way that might well be described as cavalier. With due apologies.

The agrarian countries contain most of the world's population and the proportion of that population that is growing fastest (Fig. 10.1). They cover geographical and climatic regions that include the wet tropics, arid, semi-arid and alpine conditions. Overall, about 80% of "the farm families of the Third World may be regarded as small-scale peasant farmers".[53] For several centuries, but especially for the past two, the societies, politics, economies, and the environments of these countries have been affected by foreign exploitation.[54] The principal exploiters were—and are—European nations and the USA. At first the medium of exploitation was colonisation. Goldsmith[55] quotes the following statement, made late in the 19th century, by Cecil Rhodes:

"We must find new lands from which we can easily obtain raw materials and at the same time exploit the cheap slave labour that is available from the natives of the colonies. The colonies would also provide a dumping ground for the surplus goods produced in our factories."

Much of the exploitation of these countries in the second half of the 20th century has been at the hands of large corporations with some assistance from the International Monetary Fund and the World Bank—and quite often from corrupt governments. In the early 1990s, L. H. Summers, then the Bank's Chief Economist and its Vice-President for Development Economics, wrote an internal memorandum that was leaked to the press. Its essence was that the bank should encourage more migration of the 'dirty industries' to the less developed countries. His argument for this policy included the following:

"The measurement of the costs of health impairing pollution depends on the foregone earnings from increased morbidity and mortality. From this point of view a given amount of health impairing pollution should be done in the country with the lowest cost, which will be the country with the lowest wages. I think the economic logic behind dumping a load of toxic waste in the lowest wage country is impeccable and we should face up to that."[56]

The agrarian countries support a range of types of traditional agriculture that are compatible with the prevailing terrain and climate. There is shifting agriculture, wet rice cultivation, fixed-field cultivation on small allotments (with emphasis on interplanting and crop rotation), montane cultivation, nomadic pastoralism. Dickenson and colleagues[53] comment that, in many of these situations, peasant farmers can obtain "high yields from small plots of land with little equipment, *but only at the cost of intensive labour*" (my italics).[57] This is one of several factors that limits the size of peasant farms —"normally no more than 2 or 3 hectares in much of Africa". North of the Sahara, 95% of Moroccan family farms are less than 10 ha; in Algeria the proportion is 85% and in Tunisia, 75%.[58] A result of this type of situation—perhaps the most important ecological result—is that surpluses are very small and that most of the produce is consumed on the farm or, through exchange, within the community. In that sense it is predominantly subsistence farming.[59] This, of course, is anathema to economists and to those looking for 'efficiency'.

The ecological and environmental consequences, however, are along the lines discussed in earlier chapters. There is little need for frequent application of fertiliser and, although soil degradation can and does occur, it is more likely to reflect poor husbandry than an inherent weakness of the system— unless it is a direct consequence of excessive population pressure.[60] But it is not generally such a serious problem as when produce is removed from the farm. Crop rotation, which can follow quite complex patterns,[61] intercropping and constant human attention together with, in many cases, moderately complex surrounding ecosystems, means that pests are usually less of a problem than they would be on an industrial farm in the same location.

I said earlier that the agrarian countries are in a state of transition. More accurately, they are in various stages of transition from 'traditional' to commercial farming. This transition accelerated in the 1960s with the 'Green Revolution' in which traditional strains of cereals were replaced by dwarf varieties giving higher yields—and thus requiring more fertiliser and, in many situations, more water. For the most part these varieties have been grown in monoculture leading to the high levels of pesticide poisoning encountered in the Third World. The change has been engineered largely by foreign land owners and by policies imposed by the World Bank, the International Monetary Fund (IMF) and by the Food and Agriculture Organization of the United Nations (FAO).[62]

Table A15 lists population densities relative to farmland for the world and selected regions. 'General' population densities (based on total land area) for 1993 are (individuals.km^{-2}): World, 43; Africa, 24 (range for individual nations 2-546); Asia, 123 (range, excluding Singapore, 1.5-939); North and Central America, 19 (range 3-266); South America, 18 (range, 3-41); Europe 109 (range, excluding Iceland, 14-450); former USSR, 13 (range, 6-129); Oceania, 3 (range, 2-41).[63] Whatever other problems they might face, some agrarian countries—especially India, China and some African states—are susceptible to the combined effects of high population densities and insufficient money to lessen many of the direct impacts of those densities. Elsewhere, such as South America, other effects are of greater significance.

Growing populations make increasing demands on farmland. In situations such as those of the peasant farmer in China, India or some African nations where more land is not likely to be readily available, 'excess' population is often forced to leave the farm and seek employment in a town. As well as that, the three organisations mentioned above, FAO, IMF and the World Bank, have persuaded and financially assisted the agrarian countries to 'get real' by bringing their farming practices into the late 20th century—to mechanise, to plant high-yielding crops, to export their produce—and to buy fertilisers and pesticides. Many social and environmental consequences of these changes are widely recognised.[64] Mechanisation makes some of the farming population surplus or redundant and accelerates the movement to cities. This process is not confined to the 'Third World'.[65] By 1994 there was a major migration in China of former farm labourers seeking employment in cities, a movement expected to continue for some years.[66]

All of this means that nutrient flow in these countries is changing from localised cycling to extended linear processes. In other words these countries are moving towards the dominant pattern of the industrial world—and the rate of that transition is accelerating. Needless to say there are other problems associated with this process, especially in countries such as India, Pakistan and China which have very high overall population densities. As their cities and economies grow, so does the level of urban air pollution. In turn this can drift over nearby farmland where, at least in regions of India and Pakistan, it is implicated in reducing crop yields by as much as two thirds.[67]

The production of food for export, either across the national boundary or merely to a town within it, clearly means that nutrient elements removed in that produce must be replaced if the soil is to remain productive. Inasmuch as produce is sold rather than eaten by the farmers, replacement food must be bought. If the produce is changed from a range of foodstuffs to a single crop such as tea, coffee or cotton, then the farmer must buy virtually all his food. That might be manageable if he has enough money—but that is not often the case. Of course he should obtain money from the sale of his produce, but, if he has only a few hectares of land, he will not have much to sell. If all the other farmers in his district—or his country—have converted to growing the same things, the prices of those commodities are likely to drop. And, of course, under this new 'efficient' system, the farmer now has to buy fertiliser and possibly farm machinery as well as food. If he cannot buy enough fertiliser his soil will degrade and become increasingly susceptible to erosion. All too often the upshot is that he can no longer feed his family.

This type of situation contributes to and is exacerbated by the acquisition of small farms by companies or wealthy individuals, and the subsequent incorporation of those farms into large plantations. It is also exacerbated by international monetary policies, commonly implemented through the IMF or the World Bank. The loans made through those agencies are just that—loans. They have to be repaid. Repayment of debts often dominates the economies of many 'Third World' countries, especially in Sub-Saharan Africa. A recent report from the United Nations indicates that about 1.3 billion people are trying to exist on incomes of less than $1 a day.[68] Or peasants are displaced so that rice fields can be converted to a golf course.[69]

One should not be too astonished to learn that such changes have made their contribution to the widening gap in living standards between rich and poor. In Brazil, for example, 1% of landowners are reported to control 44% of 'productive farmland' while 53% of farmers 'eke out a living' on 2.6% of the land. Such inequity has provoked not only political protest. It has led to the effective enslavement of peasants as well as the murder of resistant or protesting peasants by police and by hired gunmen.[70]

Yields of the major crops are generally acknowledged to have increased. For example, FAO statistics[71] indicate that 'agricultural production' for the period 1970-1988 increased by about 35% in Africa, within the range of 55-62% for Latin America, the 'Middle East' and South and Southeast Asia, and about 80% in the 'Far East'. How do these changes in yield relate to changes in population? FAO statistics,[71] again for 1970-1988 show a *per capita* decrease in production of about 15% for Africa, changes from about 0 to +12% for the 'Middle East', Latin America, South and Southeast Asia, and an increase of about 34% for the 'Far East'.

It is not clear from the citation just what is measured for these statistics.[72] Agricultural statistics for the agrarian countries are, to put it mildly, not always reliable. One obvious reason for questioning their veracity is that the output of small subsistence or peasant farms is unlikely ever to be estimated reliably, let alone accurately measured. The time required for a few FAO staff to undertake such assessments would be prohibitive and I doubt if the majority of these nations have well established, well funded, trustworthy 'Bureaus of Statistics'. There might also be an ideological reason for ignoring or underestimating subsistence production. By that I mean that an *objective, comprehensive* appraisal of agricultural production in these countries would possibly produce results that contradicted the FAO's, and the World Bank's, basic assumptions and guiding philosophy.[73] It is conceivable, therefore, that the statistics for agricultural production reported by the FAO reflect *the area of land* converted (essentially) to mechanised monoculture more accurately than they indicate actual increases in total production.

Moreover, where a single crop replaces a mixture of crops, grown either concurrently or in rotation, the yield of that single crop will normally be higher when it is grown as a monoculture than when it is grown in conjunction with others—other things such as soil conditions being more or less equal. But it might not be as great as the combined yield of all produce obtained by traditional intercropping and crop rotation. Indeed it is likely to be less. The point here, of course, is that the crop in question provides the whole yield when grown in monoculture but, with intercropping or crop rotation, it contributes only a fraction of the average total annual production from a field. This possibility should be borne in mind in relation to published yields of specific crops. For example, yields of wheat in China are reported to have increased from less than 700 kg.ha^{-1} in 1950 to about 3200 kg.ha^{-1} in 1991.[74] It is pertinent to wonder if, in the early years of the designated period, what proportion of Chinese wheat was grown in monoculture and whether or not the 'official' yields reflected only that wheat.

These are not the only areas of uncertainty in statistics of this kind.[75] In traditional farming some plant material is used as animal feed, some is used as manure, some as domestic fuel. Official statistics normally reflect only the yields of grain—or whatever is the relevant component of the plant in question. Official statistics rarely, if ever, acknowledge the by-products.

The many environmental, economic and social consequences of the 'Third World's' transition from traditional farming to modern agribusiness are exacerbated by the pressures of free trade and globalisation.[76] One of the many predictable results of the modern displacement of peasant farmers and the incorporation of their land into large plantations, is that the plantations are likely to be owned by foreigners, as we have already noted. Much of the profits from these plantations leave the country, a migration that does little for its health—in any sense of that word.[77]

Central American bananas illustrate some of these consequences tolerably well. The major plantation owners in Costa Rica, Guatemala, Honduras and Panama are the US companies Chiquita, Del Monte and Dole. Costa Rica is the largest exporter of bananas in Central America and, after Ecuador, the second in the world. It is a heavy user of pesticides, at least eight of which are banned in the USA. Some 250-300 cases of pesticide poisoning of agricultural workers have been reported annually. Pesticide run-off from banana plantations has killed most of the coral reef off the Caribbean coast. Pesticides have also been implicated in the killing of thousands of fish in the Gulf of Nicoya on the Pacific Coast.[78]

Grenada is a small island in the Caribbean. Its economy also depended predominantly on the export of bananas which, because they were grown largely without pesticides and were therefore given more tender loving care, were more expensive than those from the mainland. Nevertheless Grenada sold its bananas profitably because, in the 1970s, the European Union established tariffs to enable them to compete with their Central and South American counterparts. With this tariff in place, Grenada and the four Windward Islands supplied some 7% of the bananas consumed in Europe.

But this was evidently too much for Chiquita. It seems that effective lobbying by its boss together with a reported donation of 'more than $500,000' to the Democratic Party's 1996 election campaign funds might have achieved something. The US Trade Representative apparently persuaded the World Trade Organisation to force the European Union to remove the tariff. As a consequence, Grenada is reported not to have exported a single banana since January 1997 (at least until the following October). Now, it seems, the "youngsters ... make money by growing marijuana in the hills and trafficking in cocaine".[79]

There are several lessons in this of which the one about the number of eggs in one basket is probably the most fundamental. The conversion of agriculture from traditional farming to agribusiness—be it in the industrial or the agrarian world—does just what this lesson says it should not. It greatly increases the size, and correspondingly reduces the number of baskets—of individual farms or of a nation. And when that happens, the eggs become much more vulnerable, not only to the various ecological threats to extensive monocultures as we have discussed, but also to the slings and arrows of international trade, and its political handmaidens.

So let me recapitulate. The 'Third World' countries have a higher proportion of farmers than their industrial counterparts. Some of those farmers still farm traditionally, some are converting to mechanised farming, some have completed that conversion or, as is probably more often the case, been displaced by others who undertook the change. The conversion exposes the land and the soil to the types of pressures outlined earlier. But, for the most part, the environmental and social consequences of those pressures are worse than in the industrial countries. Overall their populations are greater, population growth rates are higher and the highest population densities are to be found in the 'Third World'. They are financially poorer than the industrial countries[80] and less able to afford the responses needed to lessen soil degradation. Corruption seems to exist on a grander scale, to be more blatant, and bring with it a degree of ruthlessness and violence rarely matched elsewhere.[81] All of these factors combine to degrade the lives of the displaced peasants and to assure them of a bleak future. By and large, the various changes imposed upon these countries would seem to have achieved something more than a little different from their stated objectives.[82]

PART III

CHAPTER 14

Where To From Here?

At this point I should summarise the essence of my argument and the processes outlined in the previous chapters. There is a finite limit to the total biomass that can be sustained in any ecosystem. In major terrestrial systems (as distinct from microbial subsystems), the primary determinants of that limit are the growth rates and total biomass of green plants. For any one physical environment the proportions of total plant and animal components in a mature system will remain roughly constant within limits imposed by intrinsic fluctuations and seasonal variations. The proportions of individual species are affected by the physical environment, competition, predation etc. Nutrient elements are cycled essentially within the system.

Total animal biomass is limited ultimately by the biomass and growth rates of edible plants. In other words, the system is controlled by negative feedback in which consumption of plants at a rate faster than that of regrowth will lead to interrupted reproduction and/or death from starvation by relevant animal species. That process will also interact with competition and predation among species. Mature ecosystems are relatively stable dynamically but they can be disrupted by new species, plant, animal or microbial—especially if the last is a serious pathogen—or by a significant change in the physical environment. Such disturbances, however, usually lead ultimately to a new steady state.

This situation changed with the advent of human agriculture. From that time onwards the interaction between the human population and its food supply moved from negative to positive feedback. In other words, if the human population increased beyond the point at which it could be supported 'naturally', it had the capacity to produce more food and hence sustain a larger population which could, in turn, produce more food. Depending on how far it is able to proceed, such a process is potentially or actually self-destructive—as stated in the 2nd 'Law'. Every major impact that Man has had on the natural environment for the past 10,000 years has its origins in that particular feedback loop.

The ecological consequences of the change have been complex and extensive. The ecosystems immediately involved in the conversion to agriculture became simplified and thus lost resilience. In other words they became more vulnerable to further disruption from any of a number of causes, a vulnerability that has led, among other things, to the extinction of many species.

With improvement of farming techniques the farm surplus increased, allowing or obliging some people to do something else for a living and, in due course, live somewhere else. Those processes generated a new range of skills including making tools which, in turn, further increased farm surpluses and thereby generated another positive feedback loop with a major impact on the distribution of the population. A progressively increasing proportion of farm produce was consumed away from the farm, resulting eventually in the essentially irreversible use of some nutrient elements. The rates of other processes, once natural but now managed, increased to levels at which natural cycling or natural maintenance was no longer possible. These changes were accompanied by an astonishing range of technical and technological achievements. In some ways those achievements might be included among the most ecologically dangerous components of human activity, not simply because of their direct environmental impact, but also because of the complacency and over-confidence that they generate.

Throughout all this the human population grew, with intermittent setbacks or reversals attributable (in the broadest sense) partly to drastically changed living conditions and a failure of understanding to keep pace with either the changes or with human manipulative ability. The present size of the human population (c 6 x 10^9 at the time of writing) is such that, apart from its many direct impacts on the natural environment, it cannot be supported without the expenditure of a great deal of energy for obtaining and moving nutrient elements within and produce from the global agricultural system.

Chapters 1-7 outline processes involved in the genesis and subsequent evolution of human societies dependent on agriculture. Chapters 8-13 summarise characteristics, demographic consequences and environmental impacts of modern commercial agriculture. But here I should give a word of warning. Those chapters contain various types of quantitative information. Given the time involved in writing and publishing a book of this nature, many of the statistics cited will inevitably be out of date by the time they get to a reader. This is not intended to be a review. The primary purpose of the statistical information is to give some feeling for the nature and dimensions of the various processes discussed. With time the quantitative details of those processes will change but the principles they are intended to illustrate will not.

What does all this mean? At this point ecologists and environmentalists who foresee serious trouble and feel that the public should be warned are usually faced with two difficulties. One is that there have always been prophets of doom—people who derive satisfaction from walking down the street, ringing a bell and calling 'The end of the world is nigh!'—or running down the street announcing that the sky is falling. So far their track record has been somewhat less than perfect.

The other problem is that, if one's assessment is seen to be objective and reliable, but deeply disturbing, publicising that assessment might lead simply to apathy. For reasons such as these it is common for an author of a disturbing assessment of some environmental situation to conclude—albeit with gritted teeth—on an optimistic note. I shall try to avoid such pitfalls and try to state objectively what I consider the evidence points to. My problem, however, is that the remedies I consider necessary are of such a nature that they are unlikely to be taken seriously—at least in the short term. But we will come to that in due course. At this stage I must also emphasise that, although my comments are perforce confined largely to 'Group 4' phenomena (Chapter 12), processes from Groups 1-3 should not be forgotten. As well as influencing the dynamics of soil degradation etc, any one of them can easily dominate in some locations or at some period. Moreover, virtually every one of them, as well as those of Group 4, consumes energy. For example, over the period 1860-1991 the human population increased approximately fourfold, but consumption of 'inanimate' energy increased more than ninetyfold.[1] That sort of thing brings its own set of ramifications at many levels, including those affecting the global climate—but they are not levels that I can pursue.

The basis of my argument is this: we have succeeded in producing a system that, by some criteria, has worked very well for a limited period but cannot be 'sustained'. No matter what else we might do, there are two fundamental processes which, if allowed to continue, will *certainly* lead to the widespread destruction of our habitat, the collapse of civilisation(s), and perhaps the extinction of our species. If the last should happen it will be an impressive achievement—to the best of my understanding a world first. Plenty of other species have become extinct, usually because of changes in the physical environment and/or competition from or predation by other species. Extinction caused by a species, not only having the capacity consciously to modify the global environment so profoundly as to make it uninhabitable for itself, but actually going ahead and doing so, has all the essential ingredients of a particularly diabolical Irish joke.

The two basic processes to which I refer are: (1) The 'Vicious Circle', *i.e.* the positive feedback interaction between population and food supply, and

(2) the use of some essential nutrient elements in a way that, for all practical purposes, is irreversible.

If the system is to 'collapse', what exactly does that mean, when is it likely to happen, and how will it proceed? In trying to answer those questions, perhaps the first thing I should say is that the two processes exert their effects differently and on different time scales. So let us first look at the positive feedback mechanism.

— The Vicious Circle —

The environmental implications of a large and expanding population are widely acknowledged, but it is my impression that the underlying significance and implications of the positive feedback between the population and its food supply are not. They are certainly not given the attention they deserve. The 'Vicious Circle' has been accelerating human impact on the planet since the beginnings of agriculture. The increase in the extent and, more importantly, in *rates* of soil degradation,[2] of deforestation, of depletion and pollution of water resources, and the accelerating impacts of the various components of Groups 1-3, can all be laid ultimately at the door of the aforesaid vicious circle. Does that mean that the system has been 'collapsing' for the past 10,000 years? In the light of the growth of the human population—to say nothing of its achievements—not many people would think so. But the writing was on the wall. Destructive processes were part and parcel of the human expansion. Early farms have been abandoned and civilisations have 'collapsed' because of soil degradation (a few examples were given in Chapter 4).

From the beginnings of agriculture, human ecology has developed into a complex dynamic expression of 'productive' and 'destructive' processes. In some senses that might be said of all ecosystems, but when a system is essentially in balance, describing any of its processes by these adjectives would be artificial and misleading. One cannot, for example, point to a diagram such as Fig. 1.1 and say this or that pathway is intrinsically productive or destructive.[3] But any pathway might become destructive if the system gets out of balance and that pathway becomes dominant—something that will inevitably happen if an ecosystem is simplified considerably and/or a positive feedback circuit becomes firmly established. The positive feedback is central to the phenomenon and to the argument. I pointed out in Chapter 2 that the arrival of a new species in an established ecosystem will change the balance of that system and might cause the extinction of some pre-existing species. The original human settlements of Australia and North America were followed by the extinction of large mammals[4]—but, while that was hard on the said mammals, it was not *ipso facto* destructive of the ecosystems of which they were part. There was initially no positive feedback involved and a new balance was established on both continents.

The passage of time has seen the human population grow and the global ecosystem become greatly simplified, a few of its 'reaction pathways' become dominant and, in consequence, highly destructive. Some of these destructive processes are most easily associated with Groups 1-3. If the system does collapse because of the vicious circle, the immediate cause(s) could be any of a number of possible processes operating separately or in concert over a significant period of time. Moreover, ecosystems that have been simplified enough for resilience to be significantly reduced are also vulnerable to 'catastrophic shifts' to a different state in which species composition is very different.[5]

Chapters 8-13 have attempted to summarise and quantify some of the processes currently in train around the globe. What they tell us is that there is a serious shortage of *clean potable* water in much of the world and the problem is getting worse. Depletion of groundwater supplies in some regions is, for most practical purposes, irreversible. There is extensive nett soil degradation resulting from human activities and, associated with that, continuing deforestation on a wide scale. The implications of those processes are that the most basic requirements for survival—production of food and access to water—are threatened by a number of mechanisms that stem basically from a large human population, an increasing *per capita* consumption of energy, and the continuing growth of that population. Grain production is a relatively simple indicator of some consequences of this type of interaction. Over the course of 40 years world grain production increased almost threefold—from 631 million tons in 1950 to 1780 million tons in 1990. But in the late stages of that period and beyond, production *per capita* declined as follows (kg.hd^{-1}): **1984**, 346; **1990**, 336; **1996**, 313; **2000**, 308. A related statistic is the area of grainland *per capita* which declined from 0.23 ha.hd^{-1} in 1950 to 0.12 ha.hd^{-1} in 1995.[6]

Fish provide another important source of food that those chapters did not discuss. It is clear that virtually all the world's fisheries are being exploited at a rate that cannot be maintained for much longer. For example, total oceanic fish catches in 1950 amounted to some 19 million tonnes, by 1988 they had risen to 88 million tonnes[7] but they have not risen since 1989—and that is not for want of trying. The over-fishing is aided and abetted by global subsidies amounting to some $22 billion annually.[8] Depleting fish stocks is not simply a matter of over-exploiting a particular group of organisms under circumstances that would automatically allow a rapid recovery of their numbers if the exploitation ceased or slowed substantially. It is another example of sim-

plifying an ecosystem, and it has far-reaching consequences that arise in part
from removing excessive numbers of those fish that operate at a high trophic
level.[9] Moreover, statistics of catches take no account of the incidental damage caused by nets on the sea bottom or the consequences of using cyanide to
kill fish in and around coral reefs.[7]

To put it simply, information of the kinds discussed in Chapters 8-11, and
to some extent in Chapter 12, indicates that the human population is already
demanding too much of its habitat. If the demand continues at that level *under present systems of management and husbandry*, there will be two consequences of profound significance for the future of very many species, including our own. One is that the global ecosystem—our habitat—will be simplified to a level at which it becomes dangerously vulnerable to any of a number
of types of stress—both physical and biological. The other is that the habitat
will be physically degraded to a point at which it can support but a fraction of
the present human population.[10] I should again emphasise that this immediate effect is basically a dynamic function of the absolute *size* of the population
(see below) and its behaviour—not of *growth rate*—although a growing population will obviously exacerbate and compound the problem.

I make this emphasis because there is a feeling abroad that, if the population levels off, then everything will be all right. My argument is that it will not
be all right under present systems of management. Under these systems we
cannot 'sustain'[11] the present population—let alone a bigger one. As I implied
in Chapter 10, we are already 'overstocked'. The consequences of that overstocking are similar in principle, but much more complex in detail, to those
of overstocking an area of land with any other species of animal. But as I also
indicated in Chapter 10, this is not necessarily to say that a population of the
present size could not be maintained if we did some things differently. In the
light of my emphasis on population, it is perhaps a paradox that, of the industrialised countries with 'significant agriculture', Australia—with the lowest
population density—is probably most at risk from general degradation of its
soil and water. There are several contributors to this situation, of course. A
fragile environment is one but, combined with that, is an ethos of exploitation and the fact that more than half Australia's primary produce is exported,
a process that increases the effective pressure of population on the land.

But what about population growth? From time to time an organisation—
often an arm of the United Nations—publishes estimates of likely population changes over periods ranging from several decades to about a century.
For example, 1992 projections by the United Nations and the World Bank
put the global population at about 11 billion by 2100.[12] A UN projection to
2025 gives a 'high fertility' estimate of 9 billion, a 'medium fertility' (most

probable) estimate of about 8.4 billion and a 'low fertility' estimate of about 7.8 billion.[13] Another assessment based on a different set of premises leads to a high fertility estimate for 2100 of 22 billion, medium fertility (most likely) of about 10 billion and low fertility of slightly over 5 billion. In this case the medium estimate levels off at around 2080 while the low fertility estimate peaks at 7-8 billion near 2055.[14]

Estimates are constantly changing, of course, as new statistics become available or different assumptions are made. To the best of my understanding, all these projections are based on 'conventional' demographic processes—likely birth rates, life expectancies and death rates within the context of expected social and economic conditions. They are not allowing for any ecological breakdown which might become significant during this century. So—if the 'most likely' projections are about right, the positive feedback relation between population and its food supply will last for somewhat less than another century. But, even then, if the population does level off for demographic reasons, it does not follow that it will remain constant. As with other species it is likely to fluctuate in response to any of a number of influences and, if environmental circumstances permitted, it could enter another period of exponential growth.

Given the widespread and growing awareness of at least some of the implications of a very large population, there is a good deal of speculation, calculation and experimentation done under the banner of 'how are we to meet the future food needs of a growing population?'.[15] There is, however, a disturbing corollary to the 'vicious circle', a corollary with its own set of singularly unpleasant derivative questions. The corollary is that, while the relation between the population and its food supply remains in a state of positive feedback, *the only point in trying to increase the food supply is to buy time to get the population under control.* Without that condition, any open-ended increase in food production would succeed in little more than adding fuel to a very dangerous fire.

The 'Second Law' says that any system in a state of positive feedback will destroy itself unless a limit is placed on the flow of energy through it. So how could such a limit be placed on the system involving the human population and its food supply? There are two options and they are not mutually exclusive. One is to place a firm limit on the rate of food production. The other is to place an equally firm limit on the size of the population.

The first option will necessarily cause hardship but not such severe hardship as allowing the system to continue out of control. It will invoke questions such as "is it better for 50 million people to starve now or for 100 million to starve in 30-40 years' time?" The second option, placing a cap on population growth, can be achieved humanely—but not necessarily easily—through limiting birth rate. It is to be hoped, therefore, that the projections showing a cessation of population growth within the next century are not too optimistic.

But even if they are right, we are by no means out of the woods. A population might be 'stationary' in a demographic sense, but it is nevertheless part of a dynamic system. As I have argued already, its existence depends on the continuation of countless processes each of which has a critical range of rates within which it must operate. Increasing the *size* of a population will increase proportionately the *rates* of various processes that, in broad simple terms, include food consumption and waste production. But, if that increase should exceed the matching capacity of other reactions or processes within the larger system, then the system is out of balance and 'overstocked'. It is headed towards some kind of collapse. All of that is true whether the population is steady at its new level or continues to grow. A population that exceeds the 'carrying capacity' of its ecosystem can survive only by consuming reserves and, for that reason, it is temporary.

The trouble, as I have said, is that we are already overstocked. Any species that is degrading soil and polluting its water supplies to the extent indicated in earlier chapters and, in the process, is rapidly reducing the genetic heterogeneity of its habitat has, by my reckoning, exceeded carrying capacity. If those demographic projections are right, stocking density will have doubled by the time the population stabilises and, under current management practices, that will double the demands made on agricultural soils—and a whole lot of other things. And it could well be that the end of the 21st century is too late to prevent essentially irreversible[16] damage to the global ecosystem.

So, at this point, I should address my earlier question of what I mean by 'collapse'. Basically I mean that progressive soil degradation, accompanied and exacerbated by deforestation, depletion and pollution of freshwater supplies, and destruction of fisheries will place growing numbers of people under severe stress, especially in the poorer countries—at first. The stress is likely to be worsened by other factors from Groups 1-3, including increased air pollution, a decline in public health and a range of climatic consequences of the so-called enhanced greenhouse effect. Depletion of basic resources, together with the social and economic ramifications of soil degradation and water pollution, would almost certainly provoke wars which, in their turn, would accelerate general environmental degradation—and probably generate another vicious circle. The demographic consequences of these and other factors would include an increase in mortality and probably a decrease in fertility. In other words the human population would decline.

I cannot begin to calculate when this process would become well established—or even merely apparent. Nor can I calculate how far it would go. All I can do is guess. Guessing might not be the best way of trying to arrive at a conclusion, but it is not necessarily much worse than the deception of an inordinately complex computer model working with some reasonable data, some very uncertain data and a lot of assumptions.[17] So my guess is that, if current practices remain in place, the decline will be obvious by the middle of the 21st century. I shall not try to guess how far it might go but, as I said earlier, extinction of our species must be acknowledged as a possibility.

Before considering what we might do about all this, I must turn briefly to the second challenge, the 'irreversible' use of phosphorus and, by implication, other nutrient elements that do not cycle through the atmosphere to any significant extent.

— The Exhaustion of Phosphorus Deposits —

Chapter 11 dealt with some quantitative aspects of the consumption of phosphate rock and the uncertainties inherent in estimating the likely duration of those reserves. Roughly speaking the estimates point to the effective exhaustion of known reserves within 1-2 centuries. This would seem to be longer than the 'most likely' time for the human population to level off (above). It is also longer than my own guess of the time remaining before environmental degradation begins obviously to affect survival prospects. So, although current practices have the potential to exhaust phosphate deposits within what, on an historical time scale is a relatively short period, it would seem that the wider problems arising from soil degradation etc are more immediate.

Moreover the dynamics of phosphate depletion will affect agriculture differently from those of soil degradation. Whereas the latter will diminish yields etc more or less continuously, depletion of phosphate resources will have virtually no effect until they are close to exhaustion—or, perhaps more realistically, sufficiently depleted to push costs up. What I am saying here is that a well-managed farm can maintain its yields and conserve its soil—even though all about it are losing theirs—for as long as phosphorus (and, by implication, all other essential elements) is available. When phosphate fertiliser is no longer available the farm will rapidly decline if it attempts to remain a commercial operation. An analogy here is with a motor car which can continue running normally while depleting its fuel supply. It will stop suddenly, albeit with some coughing and spluttering, when the fuel tank runs dry.

None of this is intended to imply that we do not have to worry about exhausting deposits after all. What I am saying is that, on my assessment, general soil degradation etc and analogous problems with supplies of freshwater are the most immediate threats, but phosphate depletion is waiting in the wings—perhaps next in line. Thus, if we are to adopt effective measures for dealing with the problem, those measures must be capable of dealing with the whole problem, not just part of it.

— Some Responses—In A Theoretical World —

I have listed below three primary conditions that I consider must be met if we are serious about 'sustainability' and wish to avoid a major ecological collapse. They are minimal conditions—others are also needed—but without them no attempts to deal with other environmental challenges are likely to be of much relevance in the long run. I have also added a fourth condition applicable to semi-arid regions—which include much of Australia. Those conditions are:

1 End the positive feedback interaction between the human population and its food supply.

2 Change the structure and dynamics of agriculture to ensure that the flow of nutrients between the soil and the human population is wholly reversible.

3 Manage the global ecosystem to ensure that there is no further reduction in the genetic heterogeneity of terrestrial or aquatic ecosystems.

4 Withdraw from commercial production all land where the growth of crops depends absolutely on irrigation, and all grazing land in regions of less than 300 mm annual rainfall.

How might these conditions be met? I shall begin trying to answer this question by making two unrealistic assumptions. The first is that the population is brought under control. I must repeat emphatically that, while the population and its food supply remain in a state of positive feedback, no improvement in the management of the world's farms—or, for that matter, in dealing with the various other challenges of Groups 1-3—can correct the problem.[18] All it can do is buy a little time. There are no 'ifs' or 'buts' about that. The second assumption is that people are the equivalent of pieces on a chess board and can be moved about at will. I shall try to translate the conclusions from this section to the 'real world' in due course. Perhaps a third assumption should be that we, the human population, do care to some extent whether or not our ecosystem collapses.

I should also add one more disclaimer. The proposals I am going to advance might seem conceptually simple. Of course they are not. Even with the underlying assumption that people are the equivalent of chess pieces, my conclusions should be seen only as a preliminary approximation. To achieve greater precision would require a lot more information and a great deal of time for its interpretation.

As I have implied, the most destructive aspect of contemporary commercial agriculture is that the major portion of the flow of nutrient elements through the larger system is fundamentally linear rather than cyclic. This has the practical effect of turning some essential nutrient elements into finite resources. It is also a major contributor to general soil degradation, partly through the depletion of organic matter in soil, a process that, in turn, is exacerbated by mechanisation, high yields, enormous surpluses and a virtual absence of trees over much arable land. In the 'Third World'—or at least that part of it that has changed from subsistence to commercial farming—an important supplementary cause of soil degradation has been inability to afford sufficient fertiliser to prevent general impoverishment of the soil (Chapter 12).

I have referred several times to 'management' or 'husbandry'. Changing the way a farm is run certainly has the potential to reduce erosion. For example, crop rotation, terracing or contour ploughing of moderate slopes, the exclusion of steep slopes from cultivation, planting windbreaks,[19] the use of organic manures, 'no-till' cultivation etc can greatly reduce erosion rates.[20] Except for the 'no-till' cultivation, all these approaches lie at the core of 'organic' farming which, in general, has a much better record of soil conservation than its highly mechanised counterparts. Moreover, judiciously applied manures will normally result in very much less leaching of phosphorus (and other elements) from soil than is the norm with a soluble inorganic fertiliser such as 'superphosphate'. (Unmodified phosphate rock is better than superphosphate in this respect.)

But this addresses only part of the problem because there are two systems to consider. One of those systems is the farm. If it is a subsistence farm, for most practical purposes it is the entire system. It will survive or not according to its internal management. Once it becomes a commercial farm, however, it becomes part of a very much larger system—which must also be managed competently if it is to survive. At present that wider management is anything but competent. All too often appraisals of food production focus attention predominantly on the management of farms. The wider ecological system of which those farms are a part is given little attention—although the economic ramifications are often considered at length.[21]

The internal management of commercial farms does not address the underlying problem of the linear flow of nutrient elements from the mine to the sea. To be sure, an 'organic' commercial farm that fertilises with animal manures and observes all the other precautions will conserve its soil and, because of that, it will slow the rate of phosphate application and thus extend the life of phosphate deposits. But its effect on the general dynamics of the removal of nutrient elements in produce will be merely to add another step—another link in the chain—from the mine to the sea. That other step is the farm or the paddock that supported the animals that provided the manure to fertilise the organic farm. The nutrients in that manure will have to be replaced. In the case of phosphorus, that means ultimately from the mine. The overall effect would be a substantial reduction in general soil degradation, especially erosion and leaching, and an associated reduction in the rate of consumption of nutrient elements. There would be little effect, if any, on the loss of phosphate in urban sewage and food wastes.

Is it possible to identify in simple terms any basic goals that must be met if global agriculture is to become genuinely sustainable by the two criteria of soil and phosphorus conservation? We saw in earlier chapters that European farms during the first millennium AD and into the early stages of the second had an apparent surplus of up to roughly 5%. That proportion, which was assessed on the distribution of the population, was assisted by some recycling of urban wastes. If they were competently managed those farms seemed to experience very little soil degradation. It would thus seem that natural processes could compensate for the small proportion of produce that was sold or otherwise removed from the farms (see also Chapter 11, Note 5). So we might reasonably conclude that, under some circumstances (see below), a nett removal of a very small proportion of produce from farmland, if accompanied by good management in all other respects, might allow real sustainability.

I commented in Chapter 13 that the common response to the unintended consequences of the 'First Law' is to find a remedy for those consequences. This is perhaps most obvious in the field of medicine when attention is given to possible side effects of medication. Sometimes those side reactions are treated with additional supplementary medication. This, of course, complicates the system further—especially if the additional treatment produces its own side reactions. Another approach is to find a substitute for the original medication.

The same type of thinking is evident in many environmental situations. For example, a physicist once commented to me that, if controlled hydrogen fusion becomes possible on an industrial scale, there would be no problem in meeting the energy requirements of recovering phosphorus from the sea

(Chapter 11). In the more general context of exhausting phosphate deposits, an engineer once asked if 'we' could not find a substitute for phosphorus. I felt obliged to point out that the element whose chemical properties are closest to those of phosphorus is arsenic. I have also encountered more than once the *serious* comment that, by the time this planet becomes over-populated, we will be able to go and colonise another one. And presumably start the same process all over again.

What is rarely, if ever, considered in this context is to 'put back the clock' and abandon the process responsible for the problem in the first place. Returning sewage to farmland is putting back the clock a little. It has been done to some extent throughout history but, since the introduction of modern sewerage systems and the explosive growth of cities, the proportion recycled has diminished. It is currently being attempted on an increasing scale in various ways and places—perhaps especially in Europe. But it is probably fair to say the motivation for doing so lies at least as much in limiting or reducing water pollution as it does in fertilising soil. Globally, however, the scale of the undertaking is minuscule in relation to the dimensions of the problems it seeks to address. Apart from the matter of the toxicity of urban sewage (see below), the general logistics of using sewage to complete a nutrient cycle effectively are daunting. Let me illustrate this with a very simple and modest example.

The population of the city of Sydney is about 3.8 million. The apparent annual *per capita* consumption of wheat in Australia is about 100 kg. Averaged over several years, Australian wheat yields amount to about 1500 $kg.ha^{-1}.y^{-1}$. Thus about 2500 km^2 is needed just to supply Sydney with its wheat. At its nearest point the NSW wheat belt is some 300 km from Sydney. So, if Sydney's wheat were to be grown 'sustainably'—that is with complete recycling of its elementary components—the equivalent amount of sewage would have to be piped or otherwise transported at least 300 km from the city and then distributed over 2500 km^2. Obviously the area is a function of yield; a higher yield would implicate a smaller area, a relation that puts yield on the other side of the fence from its role in influencing tolerable loss (see below). When other foodstuffs—meat, dairy products, fruit and vegetables—are taken into account, the logistical barriers in the way of fully recycling the sewage from just one major city becomes obvious. And none of that takes into account the material lost as food waste through a garbage system (Chapter 11) or the primary produce that Australia exports. The corresponding problems for huge cities such as Mexico City, Jakarta or Los Angeles or a city state such as Singapore scarcely bear thinking about. It is not surprising that the concept of a city's 'ecological footprint' has arisen.[22] But that is still only part of the picture.

Chapter 8 gave some values for the concentrations of nitrogen and phosphorus in sewage. And of course there are also other nutrient elements, as well as organic matter. But the major constituent of sewage is water which, if applied in substantial volumes to cultivated land or pasture, can be anything from a godsend to a scourge, depending on the prevailing conditions.

There are recommended rates of application of sewage effluent to land[23] but they can be no more than very general guidelines. Application rates would need to vary widely with topography, type of soil, climate, rainfall patterns, proximity of groundwater and type of crop—with particular regard to the requirements and tolerances of the plants for both nutrients and water. Transpiration rates of crops and other plants should also be taken into account.[24] The discharge of sewage onto land necessarily affects soil ecology[25]—for better or for worse depending on circumstances. And, as with any other form of irrigation, frequent irrigation with sewage runs the risk of inducing salination.

All of this means that, apart from the barriers imposed by distance from a city and area of farmland, any system for the disposal of sewage onto land would need to have a great deal of flexibility if it were not to be ultimately self-defeating.

What about sludge? Sludge is separated from sewage in the course of conventional secondary treatment (Chapter 8). It consists predominantly of solid organic matter and it contains about 40% of the phosphorus of raw sewage. It can be dried, transported and distributed easily. It is potentially a very valuable manure and, to a limited extent, is used as such in many parts of the world.[26] It is also possible to purify the effluent to such a degree that it is potable. If that is done, then it follows that the colloids and solutes are separated from the water, and further purification of phosphorus compounds should be possible. The cost of doing so, however, is likely to be prohibitive in many parts of the world.

There are two other problems in all this. As I have foreshadowed, one is that urban sewers act as receptacles for more than human excrement. They receive, legally or otherwise, industrial wastes and a range of synthetic substances that have been used domestically and, through ignorance, indifference or lack of a practical alternative, are thrown out via the sink or the lavatory. A substantial proportion of industrial discharges include heavy metals, all of which are toxic. They and the illicit domestic discharges also include synthetic organic compounds, some of which are both refractory and toxic. A proportion of all these substances—inorganic and organic—is incorporated into the sludge. Thus the repeated application of whole urban sewage or sludge to soil will lead to the accumulation of heavy metals and toxic compounds in that soil, in crops, and especially in the organs of animals sustained by it.[27] (Indeed, as we saw in Chapter 13, repeated application of inorganic fertilisers can also cause the accumulation of some metals, most notably cadmium, in soil.) For this type of reason there are moves in Australia and elsewhere to irrigate tree plantations directly with sewage or fertilise them with sludge.[28] But, while that might avoid the immediate problem of contaminating farmland, those metals will accumulate and pose long-term problems for the soil to which they are applied and, in many cases, for neighbouring soil, for groundwater and for rivers. The practice of discharging raw sewage and abattoir waste onto farmland has also predictably raised concerns about pathogens finding their way into the human food chain, especially via livestock.[29] That, however, is a problem that can be dealt with by appropriate methods of treating the offending material.

What all this amounts to is that changes in the overall management of farms and in the interaction between cities and farms, *vis à vis* the flow of nutrients, can effect some improvements and, to a limited extent, slow rates of soil degradation and depletion of phosphate deposits. It would not be enough to warrant the adjective 'sustainable'—so what else needs to be done?

Part of the problem is, of course, the size of cities. The logistics of returning sewage to farmland would be simplified if cities were smaller and distributed more uniformly throughout farming country—in other words, if the population were 'decentralised'. Concentration of the populace in urban centres is particularly acute in Australia which, as we have seen, is the most urbanised of those nations with 'significant' agriculture. It is unlikely to be entirely coincidental that rates of Australian soil erosion are so very high (Chapter 10). Small population centres would mean less sewage to take from each town or city. Moreover, if the decentralisation were undertaken judiciously, the distance between a city and its neighbouring farms would be much less than at present—especially in countries such as Australia, Canada and the USA. The area of farmland needed to support any one city would be less than that needed for the current set of megacities. Of course, neighbouring farms could not supply all the types of food required by their associated city, and problems of transport of primary produce would not necessarily be lessened by decentralisation. But the logistical problems of distributing the sewage and food wastes onto farmland would.

Inasmuch as an active programme of decentralisation would require establishing new cities and/or expanding existing towns, fresh opportunities would arise for excluding those secondary industries with a predisposition to dump toxic materials down the drain. Decentralisation would not address the problem of the nutrient elements lost in exported primary produce. It would

not, of itself, deal with the loss of nutrient elements and organic matter in domestic garbage—but it could provide an opportunity for addressing that
problem afresh. And, no matter how skillful and well intentioned the engineers, decentralisation would be unlikely to lead to the recycling of *all* the
sewage produced by each city and town. The overall process could, however,
slow the rates of both soil degradation and consumption of phosphate deposits. It should also be acknowledged that decentralisation is not an option that
would be available everywhere. Singapore, for example, would find the challenge more than a little daunting.

At this point I should address the question of just how long a period is supposed to be covered by 'sustainable'. Ideally it should mean 'for ever'—whatever that might mean precisely. An Australian government report[30] says:
"Ecologically sustainability (sic) is essentially about ensuring that each generation does not compromise the potential well-being of the next." In other
words, for as long as the human species continues to reproduce.[31]

In practice, however, the goal seems to be rather less ambitious. Even
though exhaustion of phosphate deposits is probably not the most immediate
threat to the viability of our habitat, it is real and serious. There are obvious
uncertainties in estimating the life expectancy of these deposits with present
indicators, as we have seen, pointing very roughly to 1-2 centuries. So—to
take the more optimistic number—does a period of two centuries amount to
'sustainability'? It is a very short period on the time scale of history and negligible on an evolutionary time scale. But it is quite a long time in relation to
individual life spans and, perhaps most significantly, it is a very long time indeed in the perspective of politicians, economists and captains of industry.

Any reaction to the proposal I have just made about decentralising is likely
to be influenced by one's preconceptions, conscious or otherwise, of the duration of sustainability. Decentralisation might be acknowledged in official
circles as having some merit for various other reasons—including its potential
for reducing traffic congestion and its attendant air pollution. But its implementation would take both time and money—and the time needed is likely
to be quite long in relation to the terms of office of those making the decisions. Moreover, decentralisation is unlikely to be able to deal with more than
part of the recycling problem.

What I have said so far about recycling sewage has been almost entirely
qualitative. I have made no attempt to put any numbers on the proportion of
urban sewage that should be returned to the soil. What is certain is that it
would not be the whole lot. It is highly probable that the best that could be
achieved would account for less than half of the total amount of nutrient elements consumed in farm produce. There are natural processes for forming

soil, so perhaps 100% recycling of sewage etc should not be necessary for real sustainability. While that is doubtless true in theory for some locations, we have already seen in Chapter 10 that natural rates of soil formation are very slow in relation to rates of accelerated erosion. The discrepancy between complete and effective recycling will therefore be small and will vary from one type of situation to another. To take extreme examples, farmland on a coastal flood plain is likely to require less manuring than its counterpart on a slope or in a semi-arid interior.

Clearly then, real sustainability needs more than partial, albeit greatly increased reuse of urban sewage and food wastes. But what? The short unhelpful answer is 'do something else to recycle the rest of the stuff'.

A definitive characteristic of modern industrial farms is that the surpluses are enormous, generally in excess of 95%—depending on definitions (Chapter 13). The dominant contributor to such a proportion is a very high degree of mechanisation. It is this mechanisation that, in turn, engenders enormous fields of crops and hence enormous areas of bare soil after ploughing—perhaps the most indecent of exposures. It encourages—or even demands—monocultures which, in turn, would rarely be possible without the repeated use of pesticides. Fertilising such fields or pasture, especially pasture of the kind covering much of Australia, is also effectively impossible without machines.

What all this is leading to is my next conclusion: real sustainability of agriculture will not be possible without getting rid of most of the machines and changing the scale of the whole operation. That would have the effect of reducing the farm surplus because it would require more manual labour—and it would demand smaller farms. To what extent should the surplus be reduced? For a population of a given size, the process of lowering the surplus will, of itself, reduce the number of urban dwellers and hence the total amount of urban sewage—but the answer to the question will nevertheless depend on a number of factors, not least the proportions of urban sewage and food wastes that are recycled.

For the sake of argument I am going to assume that, with good management, genuine sustainability can be achieved if the nutrient elements and some organic matter derived from 95% or 98% of farm produce are returned ultimately to farm soil. Of course the real situation is much more complex than that,[32] but my aim here is to illustrate some of the dimensions of the problem rather than prescribe an absolute cure—even in the present theoretical context.

The critical determinant of success is likely to be the proportion of urban wastes that can be recycled. The larger that proportion the smaller need be the amount recycled within the farm, and hence the larger the manageable surplus. That is obvious in principle, but the statement needs amplification.

For example, let us say that a city returns to its neighbouring farms[33] sewage and food wastes equivalent to 60% of the primary produce consumed in that city. If we work on the basis that a continuing nett loss of 5% of harvested nutrient elements is manageable, then the 40% of the city's wastes that is not recycled should not exceed 5% of the total produce of the relevant farms. In turn, this would mean that the city consumes but 12.5% and recycles to its supporting farms the equivalent of 7.5% of the total produce of those farms. The remaining 87.5% of the produce would remain on the farms and, we trust, be fully recycled within them.

These numbers, which incidentally take no account of the disposal of human dead,[34] represent a surplus of only 12.5%. They also imply that if 'real' sustainability is to be achieved with even a moderate (proportional) urban population, then the proportions of urban sewage and food wastes that are recycled must be very high. The relation between the extent of such recycling and the amount of food that could be consumed sustainably within a city is, however, not linear—as Fig. 14.1 shows. That figure represents two assumptions—'acceptable' losses of 5% and 2%.

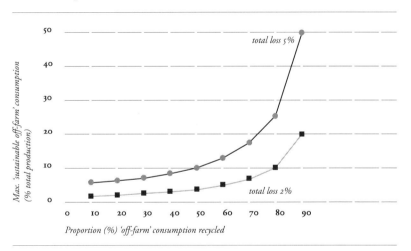

FIG. 14.1 *Two relations between the proportion of urban sewage and food wastes recycled to the supporting farms, and the maximal proportion of farm produce able to be consumed 'sustainably' in a city. The upper curve assumes that a continuing total loss of 5% of produce can be tolerated; the lower curve assumes a loss of 2%.*

A system that shifted the distribution of the population from cities back to farms would, of course, encourage self-sufficiency and thus discourage monocultures. It would reduce the size of farms but it need not eliminate some degree of specialisation if there were exchange of foodstuffs among farms —as undoubtedly there would be.[35] Theorising such as this—to say nothing of the associated calculations—is complicated enormously if food is exported beyond national boundaries—particularly overseas. Indeed if food is exported in the absence of a balancing import of food (perhaps to provide variety) and/or arrangements to receive sewage and food wastes from countries to which food had been exported, it would be virtually impossible to achieve 'sustainability' in any real sense. A similar type of argument can—and should—be applied to the management of feedlots.

A possible physical impediment to the combined processes of decentralising and returning most of the population to the land might be seen in the availability of water for domestic purposes—especially in semi-arid regions. Yet it need not necessarily be so. Small towns do not experience the type or intensity of air pollution that occurs in large cities. In many cases the water requirements of residential buildings could probably be met by rainwater tanks filled by runoff from the roof—as indeed they would on farms. Apart from anything else this would ensure a water supply more or less in proportion to the population without building dams or diverting water from elsewhere. Another question, however, must consider the direct impacts on farmland, soil, and ecosystems generally, of moving most of the world's population from cities to rural areas. That impact would be extensive and serious, but probably not as serious as the effects of present practices. If it were, then that would be clear evidence of the human population's exceeding the 'carrying capacity' of the planet, regardless of the system of management.

Of course all of this amounts to a case for 'putting back the clock' with something of a vengeance—but not quite back to the dungheaps of yesteryear. Redistributing the excrement of a labour-intensive farm can now be facilitated by the relatively modern septic tank, the more modern oxidation tank or the so-called composting toilet. Effluent from the first two can be used directly for irrigation. If properly used, these devices can avoid the danger of gastro-intestinal diseases previously associated with the use of human excrement as manure. Food wastes are easily composted and subsequently distributed.

It should be evident, even from a brief simplified argument such as this, that the 'sustainable' distribution of the human population between town and country will depend primarily on the proportion of urban sewage and food wastes that can be returned to the relevant farms. In other words, the ball is firmly in the court of managing urban sanitation. Obviously, the smaller a town or village and the closer it is to farmland, then the easier that task will be. A movement of population from urban to rural settlement would reverse trends that are current in many parts of the world.[36]

An immediate criticism of this approach—even in the present theoretical context—is likely to be that yields would be reduced and therefore more land would have to be brought into production in order to meet the needs of the existing population. It remains to be shown that this really would be so. In Chapter 13 I mentioned some of the uncertainties of assessing yields in the agrarian countries, partly from the point of view of the reliability, or otherwise, of the relevant statistics and their possible failure to compare like with like. By that I mean comparing the yield of a grain when grown in monoculture with the same grain grown as part of a mixed crop or, more importantly, comparing the yield of the grain in monoculture with the total yield of all produce from a system of intercropping and/or crop rotation. It would be important, of course, to ensure that the content of nutrient elements in the soil was at an adequate level at the beginning of this conversion, and there is no reason why that should not be done with conventional fertilisers—with due attention to trace elements. Thereafter, however, those levels should be maintained by the recycling process and/or, in the case of nitrogen, by biological nitrogen fixation, if necessary.[37]

Throughout history, irrigation has caused its share of problems, predominantly of salination, especially in those regions that are too arid to allow effective cultivation or growth of pasture without this additional water. In response to the demands of a growing population, irrigation has caused and is causing the irreversible depletion of some groundwater supplies and has stopped the flow of some rivers (Chapter 9). Accordingly, if we are serious about sustainability, it would seem that irrigation should be restricted to a role of sustaining 'normal rain-fed' farming during abnormal dry periods. It should not be used to grow crops or pasture in zones where growth would be impossible without irrigation, unless the source of the irrigation water is sewage—and that still leaves the danger of salination etc. Moreover, I would argue that the volume of water made available for irrigation should be limited to a specified proportion of the average annual rainfall.[38]

An extension of this argument can be applied to rangeland. A large proportion of Australia's grazing land lies in regions with an average annual rainfall of less than 300 mm where stocking rates range from about 10 to more than 100 ha per animal. The areas of these properties are so large that applying fertiliser is a daunting process and, in any case, it would be effective only if rain fell. The cattle are removed and sold so the soil is progressively impoverished, albeit slowly. But the process of impoverishment starts from what, these days, would be called a low base. The soil was not much good to start with. Much of the soil is bare for a lot of the time and is vulnerable to erosion—by wind and, on odd occasions, by water (Chapter 10). Such land should be taken out of commercial production.[39]

I discussed in Chapter 12 and mentioned briefly in this one some aspects of deforestation. Clearing forests, even if not done for the specific purpose of making land available for farming, has two characteristics in common with agriculture. One is that it can expose bare soil to erosion. The other is that it will remove nutrient elements and organic material from the area. The details of both of these processes will depend on many factors including the type of forest, the reason for clearing and, associated with that, the fate of the felled trees.

Trees contain more than cellulose and lignin. If they are burnt the elements they contain will be distributed over an area the size and location of which is determined by the extent and intensity of the fire, topography and wind velocity. In the short term this is likely to benefit neighbouring soils at the expense of the soil that supported the forest. If the logs are used, the process is analogous to cropping and will remove nutrients from the ecosystem—which includes the soil. Because of the depth of roots and the *relative* infrequency of commercially logging the same area, however, the dynamics of nutrient removal will be very different from those achieved by agriculture. There will be more time for replenishment by natural processes which, depending on the immediate physical environment, might or might not be significant (*cf* Chapter 7 and swiddening).

None of these comments addresses the impact of clearing forests on genetic heterogeneity. The present situation is that the global ecosystem has been greatly simplified, particularly during the twentieth century. This process might not yet have progressed to the point of serious danger, although some regions, including much of Australia, are probably close that point. Meeting 'Condition 1' (above) would contribute to meeting 'Condition 3' but it would not, of itself, be sufficient. A supplementary requirement would be a complete cessation of clearing or logging virgin forest—with the corollary that logging for timber be confined to plantations or natural forests already in use and maintained specifically for that purpose.

The foregoing requirements do not address the topics of fishing and fish populations. Since most of the fish consumed are wild, although that proportion is declining, their numbers would be affected negatively by increases in the size of the human population unless constraints were imposed. Nor does the (ir)reversibility of the flow of nutrients between the food source and the human population (Condition 2) apply to fish stocks under present conditions or any that are likely to arise for some time. In these terms, the threat to the sea is eutrophication and pollution, not impoverishment. That could ultimately change in the ideal theoretical world.

The arguments advanced in this chapter are applicable most obviously to the industrial world—perhaps above all to Australia. Some of the 'West', notably western Europe, is managed tolerably well by the criteria of soil degradation and relatively low rates of population growth—but population densities throughout Europe are very high. Asia is threatened by enormous populations, high population densities and a destabilising transition to mechanised agriculture (but see Note 35). A 1993 assessment[40] gave Africa the highest population growth rates of any continent, an overall population density about 22% of Europe's and, especially in Sub-Saharan Africa, an approach to food production that is increasingly dominated by commercial interests with the inevitable displacement of traditional cultivators and pastoralists. (Since then the growth rate of Africa's population has declined in response to an epidemic of AIDS.[41]) South America, like Southeast Asia, suffers from the displacement of indigenous traditional farmers by large commercial pastoralists and plantation companies. So they, too, fall within the scope of the arguments.

Before summarising what I regard as imperatives for sustainability, I should put into some sort of perspective the probabilities of my major generalisations about threats to the global ecosystem.

– The statement that a dynamic system in a state of positive feedback will eventually destroy itself unless a limit is placed on the flow of energy through that system is certain.

– In the context of the positive feedback relation between the human population and its food supply, the meaning of the term 'eventually' is uncertain.

– The statement that a continuation of the present system of commercial agriculture, in which most sewage finds its way into the sea, will eventually exhaust accessible phosphate deposits is highly probable. The degree of uncertainty implicit in 'probable' stems from two other variables. One is the possibility that techniques will be developed for recovering phosphates from the sea and from ocean sediments on a scale sufficient to meet the needs of the global population. As I indicated in Chapter 11, my personal view is that such a possibility, although finite, is very remote. The second variable is whether or not this practice will be overtaken by other events such as widespread land and soil degradation. As in the case of the positive feedback loop, the meaning of 'eventually' is also uncertain, although some indication of its significance was given in Chapter 11.

So now, with the assumption that the human population has become stationary, let me restate in more specific terms some of the underlying practical implications of Condition 2.

Full implementation would mean a balance between two extreme approaches. One extreme is technological, requiring virtually complete recycling of urban sewage, food wastes—and possibly human dead. This could not be achieved, especially in highly urbanised nations such as Australia, without extensive decentralisation to give a greater number of smaller towns more evenly distributed throughout farming districts. In all its aspects, it would require very comprehensive sophisticated planning. It would also require a system of generating and processing sewage that ensured that both the sludge and effluent were effectively free of chemical contamination and that the effluent was purified sufficiently at least for industrial use but preferably for irrigation and for domestic consumption.

The other extreme represents a major regression to 'traditional' farming in which farms are small (less than 50 hectares?) and most of the physical work is done by people and draught animals. These farms would provide a range of crops and some types of animal produce. Ideally they would be self-sufficient but that should not exclude the exchange of produce among farms in order to extend nutritional choices. They would support a 'significant' number of trees. They would not depend absolutely on irrigation but they might have access to it during droughts.

Because they would depend heavily on manpower, their surpluses would be very small by present-day commercial standards. Consequently, there would be movement of people from cities to the country—specifically to farms. Cities would thus be smaller, the total amount of food they consumed would be reduced as would the logistical difficulties of recycling sewage and food wastes. As indicated by Fig. 14.1, the actual surplus—and hence the distribution of population between town and country—would ultimately depend on the proportion of its food consumption that each city could return, uncontaminated, to its supporting farms. If 'putting back the clock' is seen as the less attractive or plausible of the two extreme options, then primary responsibility for striking a manageable balance lies with getting in place *effective* technology for recycling.

Finally, I should emphasise the obvious. In one chapter I have advanced in very simple theoretical terms some enormously complex proposals. The implementation of these proposals, even for a single relatively small region, would require a great deal of careful quantitative planning—especially if it were to be done with minimal hardship to those involved. The basic measures that I have proposed have been advanced with the unstated assumption that the theoretical country—or continent—under consideration is self-sufficient in food. Obviously, trans-oceanic trading in food introduces another level of complexity in efforts to achieve genuine 'sustainability'. My justification for all this lies not only in expediency, but also in the conviction that, if we are serious about sustainability, we have to start somewhere—and that 'somewhere' needs to be at a very fundamental level. In the next chapter, however, we will look at a few of the barriers that lie in the way of these proposals.

CHAPTER 15

The Real World?

The two fundamental changes that I identified in Chapter 14 are minimal requirements for 'sustainability'. Many others are also needed but, unless those two basic conditions are met, the most that legitimately can be expected from other responses is a slowing of the overall process of destruction of our habitat. Moreover, in outlining those requirements, I have reduced a highly complex problem to a simplified statement of a few underlying principles. In the present context that is unavoidable. And, simplified or not, there are enough obvious obstacles in the way of my proposals for them not be taken seriously in many quarters.

Broadly speaking, there are five major types of hurdle. One is the preoccupation of somewhere around 99.99% of the world's population with more immediate matters. Another embraces the enormous logistical, social and economic impediments to the redistribution of the population, most conspicuously in the industrialised countries, that I argue is essential for genuinely sustainable management of the planet's soils. By and large, these impediments are both so obvious and so complex that I shall do little more than acknowledge their existence—and their dominance. A third is deliberate cynical opposition arising from vested interests—interests that are usually financial. A fourth barrier stems from a number of peculiarities of human behaviour. This is the most nebulous, and perhaps the most fundamental of the conscious reasons for opposition. To varying degrees, it underpins the other 'rational' objections. The fifth obstacle is the intrinsic inertia or resistance to rapid change of virtually any complex dynamic system, regardless of whether or not human beings are involved.

Among those who do consider the proposals seriously, there is likely to be some disagreement about which of the two basic requirements—breaking the population/food supply feedback loop, or redistributing the population to enable effective nutrient recycling—is the more difficult to implement. In a sense the former might seem more approachable but, however it is done, breaking that positive feedback interaction will raise some very sticky ethical questions. The barriers to the second requirement are more tangible but possibly more intractable.

The conventional view of the problem of excess population is that an improvement in the economic status of those countries with the highest population growth rates will eventually produce a 'demographic transition' leading, more or less, to a stationary population.[1] I assume the economic component of the demographic transition influences the general philosophy and the policies of the World Bank and the IMF towards the agrarian countries.

So far the results of those policies have fallen somewhat short of that objective. One problem is that most of the money provided by these organisations, and to some extent by the governments of industrialised countries, is lent and has to be repaid with interest. Thirty three of the 42 countries that are most deeply in debt are in sub-Saharan Africa. In 1962 this region owed $3 billion; in 1999 it owed more than $200 billion.

A United Nations Human Development Report estimates that about 1.3 billion people have incomes of less than $1 a day, about 840 million are chronically hungry and have life expectancies of less that 40 years. Infant mortality is very high. For example Niger, ranked the poorest country in the world, has a child mortality (before the age of five) of 318 per 1000. Some of the diseases responsible for that statistic are particularly gruesome but, given reasonable nutrition and hygiene, and simple medication, are easily prevented. Niger spends three times as much repaying international debts as it does on health and education. Zambia's repayments to the IMF are ten times greater than its expenditure on education. Uganda's debt repayments exceed its expenditure on health by a factor of five. Africa as a whole pays its international creditors some $10 billion annually—more (by an unspecified amount) than its total expenditure on health and education.[2] One might be forgiven for wondering just what are the social benefits of acceding to the dictates or the admonitions of the IMF.

There are arguments that, even without the substantial economic changes conventionally associated with the demographic transition, birth rates can be reduced by providing women with better access to education and to contraceptives. Bangladesh, for example, provides evidence in support of this viewpoint.[3] But a reduction in birth rate, although important, does not necessarily mean that a population has ceased growing. It need mean only that it grows more slowly than previously. China has adopted another approach by introducing a policy of one child per family, at least in cities. This has had a demographic impact and, if it is fully and consistently implemented, it should eventually lead to a decline in China's population. It has also caused resentment within China, criticism from without, and an increase in infanticide, mainly of females.

But, if the forecasts of a cessation of population growth by the middle of the 21st century are right, and the reasons for those forecasts are as I have assumed, then the vicious circle will have been broken—or at least interrupted. It does not necessarily follow that the breaking of that feedback loop would be permanent. It may well be just a matter of declutching, not disengaging a gear—let alone switching off the motor. Another complication in the whole situation—population and the management of the planet's soils—is that the logistical dimensions of my 'cure' for the latter are likely to act, or be seen to act, as powerful inhibitors of the socio-economic changes considered important for the demographic transition in the agrarian countries.

Those dimensions are indeed extensive and, as I have noted, some are fairly obvious. They include a major redistribution of the population, an increase in the payment of wages by farmers, a decrease in the manufacture and sale of machinery. There would possibly be an overall increase in employment, but also a disappearance of some industries with a resultant loss of some luxuries and, more importantly, amenities. The 'standard of living' in the industrial countries would almost certainly fall. In accordance with Fig. 14.1, the extent, nature and significance of those changes would depend, among other things, on the proportion of urban sewage and food wastes that could be recycled—and hence on the distribution of the population between town and country. The whole transition would take quite a long time. All of this would directly confront, and be confronted by, current economic theory and practice. Moreover, in accordance with the 'First Law', there would be a range of unintended consequences.

There are also major barriers imposed by very basic aspects of human behaviour, perhaps especially by some of those aspects that are innate. Of course, merely by using the word 'innate' I shall infuriate some people. That in itself can be quite entertaining, but the word does need some supplementary comment. It is probably true that in any culture, in any ethnic group, one can find individuals distributed over behavioural spectra ranging from sadistic cruelty to saintly self-sacrifice, from pathological competitiveness to supreme relaxed indifference, from narrow-minded bigotry to open-minded receptiveness—and so on. Some individuals will go to enormous lengths to save a stranded whale. In the name of 'sport' others will shoot almost anything that moves. Generalising about human behaviour in terms of individuals is indeed difficult. Collectively patterns do emerge, however, although here too there is no shortage of anomalies. At great trouble and expense human societies will rescue an individual in trouble at sea or on the side of a mountain, but those same societies might massacre all and sundry in a village or in bombing raids on a city. Of course there are cultural differences and, in biological terms, it would be extraordinary if there were not also ethnic differences in behaviour—even though such differences are much more difficult to identify. But it is not the differences that I am concerned about. The point that I am trying to make is that, despite all these differences, there are patterns of behaviour that are characteristic of our species—as there are of every other species of animal. But more of that later.

In the previous chapters I based my arguments on material evidence. In most cases there were inconsistencies in the evidence and, in some cases, the inconsistencies were substantial. That is scarcely surprising given the practical difficulties in taking the relevant measurements. Nevertheless, it was nearly always possible to draw objective conclusions provided those conclusions were not dressed up with unattainable claims to precision. What I am going to do now is express my personal opinions about a number of matters, mostly related to human behaviour—indeed I have already begun to do so in the preceding paragraph. Given the nature of this book so far, this digression might present itself as a *non sequitur*. Its rationale is my contention that, while some readers will accept or reject my earlier arguments—especially those in Chapter 14—on their merits, many will accept or reject them for other reasons. It is those 'other reasons' that I want to point to, however imprecisely.

I have the distinct impression that, throughout history, explanations offered for human behaviour have been very long on imagination and very short on evidence. But if they were offered with an air of authority they were likely to engender a cult following. Sigmund Freud's musings, fabrications and pontifications are one example. The same criticism can be made of the many sweeping generalisations made from time to time about the behaviour of other species—such as:

"Where the human mind is independent, free to roam over memories, plans and imaginings, the animal mind is dependent, permanently shackled to the present and the flow of experience. Animals are always doing, never reflecting."[4]

Apart from its implication that Man is not an animal, that statement, in its totality, might or might not be true. I have serious personal reservations about it—indeed I think it is nonsense. My purpose in quoting it, however, is to offer an example of a sweeping claim that, for all practical purposes, is untestable. It is beyond the reach of anything that would currently pass for science. Had it been prefaced with 'I think that...' or 'In my opinion ...' there could be disagreement with its substance but no valid objection to stating that opinion.

Now I am about to do the same sort of thing—make some generalisations about human behaviour, the explanatory parts of which are also largely untestable. So, almost everything I say in this regard should be seen as prefaced by 'It seems to me that ...'—or something along those lines. It would become very tedious were I to repeat such a clause on every occasion.

The essence of my thesis is something like this. Human behaviour in a general sense must be determined in part by genetic makeup ('nature'), in part by experience ('nurture') and in part by purely rational thought processes. The proportion contributed by each of these components is likely to vary widely with circumstances along lines illustrated elsewhere.[5] One consequence of such a pattern is that there is a very high probability that any argument advanced about virtually any topic is likely to be publicly assessed by a range of criteria of which purely rational appraisal is but one—and very often one of minor significance.

At the present time, especially in the industrialised world, we think of ourselves overwhelmingly in social, political and economic terms and we have largely forgotten our biology. We do not even seem to acknowledge that we are a species of animal. The quotation above is by no means unusual in distinguishing Man from 'animals'—as if we should be classified in a different taxonomic kingdom. During the early stages of human evolution Man developed various patterns of behaviour that were important, if not essential, for survival. Many of those patterns have been retained, but our environment is now very different from that of our formative years. Accordingly much of that residual behaviour is now either irrelevant or potentially dangerous for the survival of the species in the longer term.

Man is certainly an unusual species in a number of respects—most obviously mental capacity, an ability to speak in a very complex way, and an extraordinary manual dexterity. It is sometimes claimed that the mental capacity of dolphins is quite close to that of human beings. But even if that should be so, dolphins could never affect the planet in any way that even remotely approached the human impact. Apart from the constraints imposed by an aquatic habitat, they have no hands. These human attributes have enabled our species to change the environment in directions and to an extent that eventually gave Man effective dominance over the entire globe. That dominance, however, has produced both delusions of grandeur and a system of managing the global ecosystem that is self-destructive. Removing the destructive elements is a daunting task, partly because of the momentum of the system itself and partly because of deeply ingrained systems of belief. Their removal is also likely to be impeded by some types of innate behaviour.

Most vertebrates—especially homeotherms—that I have met are curious about unusual objects or events. Human curiosity, however, goes further than that of other species inasmuch as it is disposed to produce complex verbal explanations. In the past, natural phenomena that could not be accounted for in direct physical terms commonly spawned myths and then religious explanations. I assume that all religions arose essentially in this way—indeed the expression 'God in the gaps' tacitly acknowledges such a process. Supernatural explanations were usually the prerogative of witchdoctors and their ilk—they held the copyrights. As communities grew the religions became more sophisticated and developed organisational structures.[6] These structures gave some individuals authority and power over others. The increasing complexity of the community at large did much the same. But the authority of the religious officials was not confined to administrative control within the church. It also extended to the area of 'expertise'—they were able to make pronouncements about a range of matters and they expected those pronouncements to be taken seriously. In more recent times technical expertise has come to replace religious omniscience to varying degrees.

It is common among social animals for individuals to seek dominance. Konrad Lorenz puts it this way:

"All social animals are 'status seekers', hence there is always particularly high tension between individuals who hold immediately adjoining positions in the ranking order; conversely, this tension diminishes the farther apart the two animals are in rank."[7]

The same can be said of *H. sapiens*. Any human community—be it of hunter-gatherers, swidden farmers, nomadic pastoralists—will probably have an 'elder', a leader, a chief, a 'big man'. At a different level, a corporate structure provides a framework for a hierarchy which, in turn, enables a dominant individual to have a chain of command. Among other things, the chain of command increases that individual's power and insulates him/her from the populace at large. Mostly, but certainly not invariably, that individual is male. Indeed, there are some long-standing, exclusively female environments, such as a nunnery or a community of concubines, where the same sort of thing happens. I also have the impression that, of the women who enter the nursing profession, a small proportion does so because of the power they will have over others—their patients. The power or authority available within a corporate structure provides a cogent stimulus to ambition and to rivalry between individuals. This is functionally similar to the forces that, in social birds and mammals, drive individual males to seek domination of their herd/flock/pack and thus to have first call on the females. Similar access is also often a fringe benefit of domination in human organisations.

There is another attribute that typifies almost every type of human community or organisation, large or small. It will adopt some kind of collective attitude, most probably suspicion, towards another clan, tribe, or group with broadly similar beliefs or terms of reference, but with some differences of detail. On prolonged acquaintance that suspicion is likely to harden into rivalry, antagonism or outright aggression—especially if the two groups compete for a resource. In some circumstances, most obviously when both groups are antagonistic to a third, their mutual interaction might evolve into collaboration.

As a form of 'light relief' (?) while writing this chapter, I read two books more or less concurrently. They are John Pilger's *Hidden Agendas* and Andrew Riemer's *Sandstone Gothic*. Their juxtaposition emphasised quite dramatically the ubiquity of some basic patterns of behaviour. Whereas Pilger writes about corruption, lying and dissembling, competition, ruthlessness, hypocrisy, and personal ambition in corporate and international dealings, Riemer gives an account of the "fundamental nastiness, vanity and the display of ruthless ambition" among the staff of Sydney University's English Department in the 1960s.

To the best of my understanding, all species of animal are aggressive under some conditions. Terms such as 'mob psychology', 'mob behaviour' are almost clichés. They imply that people are apt to behave differently when they are in a group, a gang, a mob from the way they do when alone. This is scarcely surprising, but it is particularly true of aggressive behaviour. It is a characteristic that we share with dogs and other pack animals. For example, it sometimes happens that a domestic dog, who is thoroughly well behaved when at home, will join a pack of other domestic dogs. Given the chance, that pack may well attack and kill other animals, including sheep or cattle, many of which they might simply mutilate without bothering to eat. Under some circumstances many vertebrates and arthropods, in groups or individually, will kill others of their own species. Chimpanzees, our closest relatives, will send raiding parties that will attack, mutilate and often kill one or two members of a neighbouring community, a practice that has very obvious human counterparts. In other circumstances, however, the implementation of human aggression seems to be unique—at least among vertebrates. So far as I know, Man is the only vertebrate willing to line up two opposing teams that proceed to hack one another to pieces—or otherwise dispose of each other, depending on the weapons available. This practice has been followed in various forms throughout all the historic period, and doubtless before that. With some differences in technical detail, it would seem to be a characteristic that we share only with ants.[8]

Any of numerous characteristics can serve as a common factor that binds a community together, distinguishes it from and shapes its attitude to another community. The most fundamental is doubtless ethnicity. Another is allegiance to a particular football team. It has been argued[9] that soccer is a form of tribalism with its counterparts of tribal hierarchies, rituals, organisations and so forth. Of course the argument has been criticised but, in its basic form, it makes very good sense and can be applied almost equally to other forms of team sports. In March 1996 the captain of the Pakistani World Cup cricket team was threatened with death by his fellow countrymen after their team was beaten by India. Almost as fundamental as ethnicity, however, is adherence to a particular set of beliefs. This last is perhaps one of the most extraordinary, yet most characteristic, peculiarities of our species.

It is probably not unkind to say that an organised religion is a system of beliefs that arose in the first instance from attempts to explain natural phenomena, evolved to provide rules or guidelines for behaviour, offered an administrative structure that enabled individuals to acquire authority, and persisted despite the lack of any tangible evidence for its basic tenets. Jesus Christ was executed in a singularly unpleasant manner essentially because he held beliefs and advocated practices—benign practices—with which the Romans largely disagreed. Subsequently quite a number of Roman citizens were fed to lions because they had accepted what Christ had to say. This practice served several purposes; it helped sustain the lions, it entertained 'the masses', and it warned them not to get ideas above their station and start thinking like the Christians. Ideological dictatorships, such as Germany under Hitler, the USSR under Stalin, or China under Mao, are apt to think along similar lines, but not necessarily with the same emphasis on entertainment.

Islam and Christianity share the same god but differ in the details of their beliefs, their organisational structures and in their practices. Those differences, however, were the underlying rationale of the 'Crusades'—two centuries of war by Christians against Muslims. At the time of writing, Christians and Muslims are killing one another in Indonesia. 'Heretics' have been executed by the Christian and Islamic churches basically because they held unorthodox opinions. In many cultures and societies, 'witches' have been burnt or otherwise disposed of.[10] The warring between Catholics and Protestants in Northern Ireland has been scarcely less fanatical. Although these two communities evolved to develop social, economic and political disparities, the starting point for their separate identities and their mutual antagonism was a few minor technical differences in their respective religions. The mutual aggression of some Islamic sects is even more extreme. The depth of the antagonism of the 'West', especially the USA, towards communism has many of the same elements. Indeed the fundamental basis of the nuclear confrontation between the USA. and the USSR—a confrontation that, if expressed, would probably have ended civilisation—lay in different opinions about the way an economy should be run. The same fanaticism has embraced the USA's attitude towards Cuba. And, since 11th September 2001, fundamentalism on all sides seems to have dominated international relations—especially relations involving the USA, Israel and a number of Islamic countries.[11]

Adherents to any religion presumably believe that their religion is better than others. Moreover, zealots within a religion are predisposed to convert others to their way of thinking. They have been encouraged in this by their church and—especially when those they have been seeking to convert are the 'natives'—by their government. Quite clearly their system of beliefs was better than that/those of the natives, otherwise they would not be in a position of power, telling the natives what to do. Conversion of the natives from their own barbarous/heathen/primitive beliefs would obviously be good for them and prepare them to be better citizens of what was now a colony. As better citizens they could be more useful in promoting the economic objectives of their colonial masters. It does not require an enormous leap of the imagination to see similarities between missionaries and the process of religious conversion on the one hand and, on the other, the role of the missionaries of the IMF and WTO in changing the societies and the farming practices of much of the Third World.

I have argued elsewhere that, although human beings can be quite rational under some circumstances, for much of the time we are not. For much of the time our arguments or analyses are rationalisations invoked to defend a preconceived position of some kind. That position can be instinctive, emotional, or the defence of a point of view which we accept or with which we have been indoctrinated. In many cases that point of view is likely to be the contemporary conventional wisdom. I said in that essay:

"A recognition of deterministic components of human behaviour is built into our language in words such as 'personality' and phrases such as 'face-saving formula'. Personality recognises that different individuals will respond differently to the same environmental situation for reasons that are built in with the bricks. A purely rational response would, in most (but not all) cases, be the same for everyone. The face-saving formula recognises the game-playing and deeply rooted competitive component of negotiation. But all this recognition is confused and bedevilled by an equally widespread realisation that we can also be quite rational for some of the time and by the paralysing uncertainty about just when that might be."[5]

Not long after taking office, Lyndon Johnson said publicly that he was not going to be the first American President to lose a war. This is a very revealing statement. It says that Johnson's primary concern here was not the fate or well-being of the American or Vietnamese armies or nations. The important thing was his own perception of his place in history. It was an interesting example of rationalising inasmuch as a hidden motive had surfaced and been enunciated.

Early in the 19th century the English criminal code identified more than 250 capital offences—for which the penalty was hanging. Children aged 7 and upwards were hanged for petty theft. Shop-lifting goods valued at 5 shillings was punishable by hanging. In 1808 the House of Commons passed a Bill reducing the number of capital offences, but it was another 30 years before the Bill got through the House of Lords. The delay was the result of 'rational' argument advanced by judges who were members of that House. In 1810, the first time the Lords debated the Bill, Chief Justice Lord Ellenborough spoke against it in the following terms:

"I trust your Lordships will pause before you assent to an experiment pregnant with danger to the security of property, and before you repeal a statute which has for so long been held necessary for public security. I am convinced with the rest of the Judges, public expediency requires there should be no remission of the terror denounced against this description of offenders. Such will be the consequences of the repeal of this statute that I am certain depredations to an unlimited extent would immediately be committed...

My Lords, if we suffer this Bill to pass, we shall not know where to stand; we shall not know whether we are upon our heads or our feet..."[12]

If an intelligent, well-educated, sophisticated forum such as a house of parliament of a civilised nation will accept a supposedly reasonable argument for hanging 7-year-old children, then we may expect acceptance of an equally reasonable argument about absolutely anything, especially if that argument is in defence of the status quo.

The United Nations Human Development Report for 1997 noted that the nett wealth of ten billionaires amounted to "1.5 times the combined national income of the 48 least developed countries".[13] The heads of many large corporations are so wealthy that increasing their wealth by any amount could not conceivably enhance their material comforts. Nor could reducing their wealth by as much as 99% conceivably lessen those comforts. Yet the majority of such people—or at least those who get the most publicity—are evidently hell bent on seeking dominance of their industry and, in the process, making more money. There is no concept of 'enough'. The usual rationalisation of this attitude is 'doing the best by' or 'serving the interests of' the shareholders. An alternative explanation is that it is simply greed, in which case it would fit into the mould of a survival mechanism carried over from earlier evolutionary times when a surplus of food or other necessities was relatively infrequent. It also qualifies as one of many examples of 'Dominant Male Syndrome'— which, in these days of initials and acronyms, really deserves an abbreviation—DMS.

Another example of DMS is to be found in dictators who, through incompetence or corruption, or both, have presided over the disintegration of the nation for which they were responsible. Despite overwhelming evidence of that disintegration, it is so often the case that the said leader hangs on until he is assassinated or otherwise forcibly removed. In such cases the instinct to be dominant would seem to override all other considerations.

It is more difficult to generalise about politicians in a democracy. Nevertheless I find it hard to avoid the conclusion that a majority of politicians, at least in today's major political parties, are motivated primarily by a desire for power for its own sake and/or to implement their particular ideology, and only secondarily to do something genuinely for the good of their country.

In Chapter 12 I commented on the human propensity for 'reductionism' at the expense of developing an understanding of complex systems. What we are best at in this context is taking a complex system to pieces, both physically and conceptually, and determining how individual pieces work. We are not nearly so good at understanding the functioning of the whole system, at least in terms of the quantitative details of its dynamics and its regulation. Indeed I think I would argue that, the more we know of the details of a complex system, the more trouble we have in appreciating its overall functioning.

Until the late 18th century scurvy was a major hazard confronting anyone taking a long sea voyage. It was eventually recognised empirically that the disease could be avoided by eating sauerkraut, raw potatoes or drinking lime juice. In order to take those precautions and avoid scurvy it was not necessary to know anything about ascorbic acid ('Vitamin C'), its metabolic role or its so-called 'anti-oxidant' properties. With its identification, purification and commercialisation it was also implicated (controversially) in boosting the immune system but, when consumed in large doses, found to produce problems of its own, including encouraging the formation of oxalate kidney stones in some people and possibly a reversal of those 'anti-oxidant' characteristics.[14]

Again, it is not necessary to know anything about the function of different regions of the brain, about synapses, neurotransmitters or anything else at that level to appreciate that a human being, a domesticated goat, cow, dog,

kangaroo, parrot, lion, rhinoceros, gorilla that is treated kindly is likely to behave differently from one that is not. (I am not quite so confident about croc-
odiles.)

The arrival of hunter-gatherers in a new environment changed the ecology
of that environment, but not necessarily to a greater extent than would the
advent of another active species. In order to survive they would need to understand, or at least respond to, their habitat in terms of the habits of their
prey, the plants that were edible, the broad effects of other animals within the
area on their prey and on their edible plants—and effects of seasons. That is
also generally true of other species of wild animal—however one might like to
interpret 'understand'. In general, swidden farmers changed their environment to a greater extent than did hunter-gatherers and would therefore require an additional level of understanding. But that understanding need be
still only at an empirical level—and obviously it was soon acquired.

Fixed-field farming made a greater impact than either of the other two
'life-styles' and required still more understanding of what was going on.
Whereas the swiddener need only recognise empirically that an area of
ground needed time (c 30-40 years) to recover its fertility, the fixed-field
farmer had to appreciate that, if he wanted to survive, he had to take steps to
prevent erosion after tillage and to manure his soil. He did not have to know
the details—that the crops required nitrogen, phosphorus, potassium, trace
elements etc—only that he had to put back something derived from the crops
he had grown, or some equivalent. Such understanding was indeed acquired—but not by everyone. Some farms and some communities did fail.
Recognition that plants needed a range of specific elements for their growth,
the identification of non-biological sources of those elements and an underlying, if unrecognised, assumption that those sources were inexhaustible, led to
some far-reaching changes in the management of farms. The ecology of individual farms was simplified but the dimensions of the larger system, of which
those farms were part, were extended. The dynamics of that larger system
changed and the primary function of farms evolved, at least in the eyes of the
industry, from producing food to producing money.

By the standards of all other species, human engineering is staggering in its
skill and sophistication. But, here again, there are major discrepancies between the understanding needed in engineering design and construction, and
that needed to recognise the wider impacts of engineering projects. An engineering achievement with its own set of unintended consequences is the motor car—a remarkable machine that is almost indispensable for those of us
who live in rural districts of industrialised countries. It might well be, however, that this machine has caused more environmental and social damage in

cities that any other invention intended primarily for human convenience. It is a—if not the—major source of air pollution in most large cities and, of course, an important contributor to the wider problem of greenhouse gases. Since the first fatality in 1898, motor vehicles have been directly responsible for some 30 million human deaths. They currently kill about 500,000 and injure 15 million people annually in road accidents[15]—to say nothing of wildlife.

The growth of car numbers raises its own questions about carrying capacity. Throughout most of the industrial world the standard approach to traffic congestion has been to build more roads and freeways. Inevitably the new roads encouraged more cars and, equally inevitably, the new roads became clogged. Thus another positive feedback system was established. The increase in the number of cars and in the time required for journeys on saturated roads continues to exacerbate urban air pollution. The process has included the atrophy of public transport in most major cities. It is estimated that traffic congestion costs the European Union £100 billion annually—four times the amount spent on public transport.[16] If these possibilities were foreseen by those taking the decisions to build new roads, let alone by those who invented and made the first cars, they did precious little about it.

There is a growing awareness, perhaps especially in Europe and in San Francisco, of a need to place a limit on the construction of major roads and highways. I do not know if this perception arose in the conceptual context of dealing with a vicious circle or, in simpler terms, merely with a logistical problem of transport. But whatever the perspective, the functional similarities between the car-road interaction and that of the human population and its food supply should be apparent.

Failure to recognise even the most obvious basic characteristics of dynamic systems has led to some deliberate decisions that are best classified as insane. Perhaps the starkest and most demented of such decisions in the 20th century was Mao Zedong's 'Great Leap Forward'. By boosting steel production this was intended to transform China into an industrial nation. Raw material for steel furnaces included bits of scrap metal scavenged from the streets by, among others, six-year old children. Forests were cleared to provide fuel for the furnaces—and peasant farmers were directed to work in steel mills. Thirty million people died in the resulting famine[17]—a number matching that scored by the motor car, but achieved much more rapidly.

The 'Great Leap Forward' reflected an obsession or some kind of irrationality. It was also an extreme case of a disturbing phenomenon—the assessment or evaluation of a complex system by a single criterion. This is a practice that seems to have become more deeply entrenched as the number of 'experts' in various disciplines has increased. It is a practice that asks for trouble when applied to virtually any type of dynamic system and is particularly dangerous when the system is one that is managed.

The human body is undoubtedly the most comprehensively studied example of any multicellular physiological system—probably of any complex dynamic system. Any one of a number of measures *e.g.* blood pressure, pulse rate, body temperature, lung capacity, can indicate some disturbance or malfunction of the system, but rarely can a single measure identify the cause of such malfunction. That requires more information based on additional criteria. Conversely, a single test might give a normal result when an individual is supporting, say, an early malignant tumour. Similarly, a significant change in the population or biomass of a species of plant or animal in a forest can indicate that something is going on, but of itself, it is rarely enough to identify just what that might be. Yet there is a prevailing ideology that the assessment of managed systems, be they ecological, social or economic, can be done—indeed, should be done—by an economic measure.

I commented in an aside in Chapter 1 that the only thing I have in common with economists is that I don't understand economics. The dominant religion in the industrialised countries, especially in the 'West', seems to be the contemporary economic theory that, at least in Australia, is blessed by the designation 'economic rationalism'. I classify it as a religion because it is a system of beliefs that exists with remarkably little supporting evidence, that spawns an organisational structure with its own clergy, which knows that it is right and that those who think otherwise are heathens, heretics, or fools.[18] Like other religions, economics has its prophets. In the West they would seem to be either Adam Smith or John Maynard Keynes. For a while Karl Marx had a substantial following, then went out of fashion following the collapse of communism. There are signs that he might be about to make some sort of comeback.[19] No doubt there are others.

Economics has been called 'the dismal science'—but that is flattery. Science evolves through testing its hypotheses and discarding those that do not pass. From my perspective, contemporary economics 'advances' by citing its prophets, by basing decisions on models, regardless of their relevance or accuracy,[20] and ultimately by judging the success of everything by a single monetary measure.

More specifically, the conventional measure of the functioning of a nation's economy—indeed of a nation's well-being—is the so-called Gross Domestic Product (GDP). Recently this device has come in for increasing criticism because of its misrepresentation of the general 'health' of a society. Other, more comprehensive, measures have been proposed.[21] But, as Eckersley points out, although there is growing disquiet among ecologists, academic economists and others about the use of a single measure for such a purpose, little of this disquiet has filtered through to politicians or business leaders.[22]

The practice of assessing 'things' or processes by economic criteria has apparently permeated all segments of society, at least in the industrial world. Thus a report prepared for the British Department of the Environment includes tables entitled: "Impact of air pollution on property values"; "The impact of aircraft noise on house prices"; "The impact of traffic noise on house prices".[23] Tickell,[15] in his account of the havoc caused by the motor car, notes that the International Red Cross assessed the economic cost of motor accidents at "$53 billion a year in developing countries—equivalent to the entire international aid they receive". Another report,[24] suggesting that the measures needed to prevent global warming might not be 'cost-effective', assessed the money value of lives of people in the industrial world as up to fifteen times greater than that of their counterparts in the poorer nations.[25] This is reminiscent of the comment by L. H. Summers about the merits of dumping toxic waste in countries with low wages (Chapter 13). In March 1996 (and probably on many other occasions) the value of industrial shares on the New York stock market (the 'DOW') fell in response to news that unemployment in the USA had decreased.

There have also been attempts to estimate the money value of 'service' provided by ecosystems. In one case it was acknowledged that most of these services lie outside 'the market' and were therefore difficult to estimate, but were nevertheless considered to equal or exceed the total 'GNP' for the planet.[26] The total economic value of the biosphere was estimated to lie within the range of $16–54 x 10^{12}. Global 'GNP' was put at $18 x 10^{12}.

These estimates can be interpreted in several ways—not for their economic validity, but as indicators of attitudes or patterns of thinking. It might be that economists have realised that our habitat is important somehow or other and have tried to assess that importance in the only way they know. Alternatively, to pursue the religious analogy, conservationists have either been converted by the missionaries or else have adopted the strategy of 'if you can't beat 'em, join 'em'. In the latter case it would be a calculated decision to try this approach to bringing ecological and environmental dangers to the attention of the priesthood. Similar comments can be made about attempts to defend 'biodiversity' on the basis of the commercial dividends that might stem from a possible future discovery of a pharmaceutical agent within the ocean or a forest (Chapter 12).

Whatever the logic behind it, the assignment of money values to the natural habitat is a tacit acceptance of the dominant theology that nothing has any value by any other measure. It is also open to dispute about the price assigned to a forest, a river, to topsoil or the atmosphere. A lowering of the 'official' valuation could well justify the continuation or initiation of degradation.

Quite obviously the original significance and value of money lay in its enabling the exchange of goods between communities to adopt a more complex, and hence more versatile, pattern than was possible under a simple system of barter. This duly led to extended trade networks which provided communities with access to a wider range of goods and types of food than would otherwise have been possible. But the growing complexity of those networks, and everything that went with that complexity, eventually changed money from being a very convenient means to an end, to an end in itself. This is so patently obvious that it must have been said many times. But it leads to another generalisation—a rephrasing of one of the basic arguments that I have already advanced. Apart from the effects of population pressure, the dangerous mismanagement of the global ecosystem can be attributed above all to the assessment of function and performance by economic criteria. Money might not be the root of all evil but it certainly provides a very effective growth medium, not only for that particular commodity, but also for blindness.

The religion of economics has had two major sects, one of which has been (temporarily?) beaten into submission by its dominant rival. The current orthodoxy is that the movement of money should be left to the 'free market'. The justification for this seems not so much to be that 'the market knows best'—although that might have been the case in the 19th century—but rather a kind of economic Darwinism, the survival of the fittest.

In other guises and with other terminology, this is undoubtedly the older of the two sects, having existed in its essentials since the earliest civilisations. It seeks to employ workers at the lowest possible wages and will enslave them if circumstances allow. Since the Industrial Revolution it has sought to replace employees with machinery to the greatest extent possible. After the effective collapse of its rival sect, communism, it has become more arrogant and, among other things, declared that publicly-owned enterprises supplying services to the community at large should be 'privatised'—and thus be made more 'efficient'. In general these corporations then proceed to 'downsize' which, in English, means getting rid of a substantial number of their employees while simultaneously raising the remuneration of their senior executives to extraordinary levels. One of the common consequences of reducing staff, and simultaneously seeking greater profits, is that quite often insufficient attention is given to the maintenance of equipment.

I do not pretend to have reviewed this topic but my impression is that, when public utilities in an industrial nation have been sold to private enterprise, service to the public has deteriorated more often than it has improved.

A well-publicised example of the latter was a lengthy series of power failures throughout much of Auckland early in 1998 and in California in 2001. At about the same time and under broadly similar circumstances, there were power failures in southern Queensland. Britain has had its share of problems with water supply systems since they were sold off as, indeed, have other countries.[27] For several years the Australian state of Victoria had an 'efficient' system of funding its public hospitals. The system was known as 'Case Mix'; it was reviewed in late 1997 or early 1998. The review was very critical and, upon its release early in May 1998, the instigator of Case Mix, an academic economist, was interviewed briefly for an ABC Radio news item. He attacked the reviewers for consulting hospital staff rather than inspecting 'the books'.

In recent years the industrial nations have become wealthier as judged by the types of economic measures referred to above. But this increased wealth has mostly been accompanied by an increasing disparity between executives and employees and, more generally, between rich and poor. For example, in 1980 the average income of company executives in the USA was 42 times that of their employees. By 1999 the corresponding ratio[28] had risen to 419:1.

There is another anomaly arising from the contemporary orthodoxy. It is a paradox that not only blatantly contradicts anything that might pass for logic, but also seriously exacerbates human impact on the biosphere.

We are exploiting or managing the global ecosystem at many levels. Apart from soil degradation and other matters discussed in earlier chapters, one of the results of human activities is a rate of extinction of other species estimated at some 100 times the 'background' rate.[29] There are now arguments that, because of the extent and magnitude of the human impact, conservation areas need to be managed rather than left to their own devices if endangered species are to have a reasonable chance of survival and, hence, a safe level of biological heterogeneity is to be maintained.[30] Opinions differ about the total area of such conservation zones, but there is general agreement that 10-12% of 'each nation or ecosystem', which is more than generally available at present and for which there are proposals, is not enough.[31] The barriers even to this objective are population pressure and economics. The paradox lies in the recognition that essentially the whole planet is now managed, that it must be managed, that the ultimate criteria of management are financial—but the economy itself is sacrosanct and must be left to its own devices—to 'market forces'.

Not only must the economy be left to market forces, but it must continue to grow. Its growth is usually assessed by one of the measures (GNP, GDP) mentioned above. The supporting argument most frequently advanced, at least for public consumption, is that growth is necessary to provide jobs. This last statement is made as often as not in a socio-economic environment of 'downsizing'. The dogma has not gone unchallenged.[32]

In my naïveté, the concept of economic growth has the following connotations and implications. If the human population is to grow and maintain its current standard of living, the economy must also grow. If the Third World is to improve its standard of living, its economy must grow. Either of these processes would increase the rates at which resources are used and wastes produced. They would thus have finite life expectancies, the actual duration of which would be susceptible to some influence by technological innovation. It is scarcely conceivable, however, that such innovation could change the life expectancy of the whole process from finite to infinite.

There are arguments that the economy can grow in ways that do not consume physical or biological resources, and doubtless that is true up to a point. But I cannot imagine that economic growth under those conditions could make more than a very minor contribution to the maintenance of standards in a growing population or to improving them in one that is static. The dogma apparently does not recognise any concept of a steady state. In that respect it has much in common with the thinking of those business tycoons who, as I said earlier, show no sign of having any concept of 'enough'. What all this amounts to is that, if we are serious about sustainability, we cannot be serious about unlimited economic growth—and vice versa.

Together the doctrines of growth and the free market have spawned, or at least encouraged, 'free' international trade and the processes that have come to be known collectively as 'globalisation'.[33] Free trade and globalisation have some dimensions of direct relevance to environmental problems and, of course, to my arguments in Chapter 14. The real, as distinct from the stated objective of free trade would seem to be fairly obvious—to benefit large corporations based predominantly in the USA and, perhaps to a lesser extent, in Europe. The benefits flow from unrestricted access to foreign markets, from establishing bases in countries with valuable natural resources, where the cost of labour is cheapest and, in some cases, where there are no legal barriers to child labour. Another 'benefit' is that industrialised countries can bribe others to become dumping grounds for toxic or radioactive wastes.[34] One immediate consequence of all this is that control of a nation's economy—and everything that flows from that—is transferred in large measure from a national government to transnational industries. Free trade has its critics.[35]

Whether this is seen as a good or bad thing is likely to be influenced by perceptions of the relative competence, conscientiousness and concern for the real welfare of a nation—or the planet as a whole—on the part of government on the one hand and multinational business on the other. There are some who argue that, by and large, leaders of business are likely to be more intelligent than politicians and therefore will probably make a better fist of running things. A related line of argument might be that, while it is true that the primary motivation of business is to make money, the primary motivation of politicians, or at least those who are members of major political parties, is to get and retain power.

Morality does not figure very prominently in international trade, although there are occasional exceptions.[36] We have seen some examples of that truism in earlier chapters. Another fairly stark reminder is the dominance of arms over every other trade 'good' (as economists call such things). The modern explosion (!) of the arms trade began early in the 1960s and is currently dominated by the USA and Britain. Some direct consequences of this trade include deaths in war with estimates ranging from 50 million since the end of World War II, to 75 million over the period 1960-1995.[37] They also include government deception and hypocrisy in foreign policy such as that of the British government in its dealings with Malaysia over funding the Pergau Dam (Chapter 9) and with Indonesia in relation to its invasion of East Timor.[38]

Free trade has the capacity to undermine, if not overrule, national environmental policy. Early in 1998 the Canadian government was sued for $251 million by the Ethyl Corporation, a US company, for banning the manufacture of a toxic petrol additive within Canada. At about the same time two local authorities in Mexico were sued for refusing permission for US companies to establish dumps for toxic waste. Both cases were brought under the aegis of the North American Free Trade Agreement (NAFTA).[39] There can be little serious doubt that American corporations are the major beneficiaries of NAFTA or that their benefits are achieved at the expense, in many senses of that word, of Canada and Mexico.

More recently another prospective international trade treaty has been going through a secret gestation under the auspices of the OECD. I refer to the Multinational Agreement on Investment or, inevitably, the 'MAI'. To the extent that I understand it, this treaty was intended to give large corporations based in the OECD countries essentially unfettered access to the resources of other countries. It would have been illegal under the treaty for a government to discriminate in any way in favour of a domestic industry at the expense of a foreign competitor. Any national environmental safeguards or policies would have been at risk had they interfered with the commercial interests of a foreign investor. Apparently the intention was for the OECD nations to reach agreement and then, having done so, to offer other countries the opportunity of signing without the option of introducing any changes. The Agreement would have had the capacity to undermine or override the authority of national governments including, of course, those that were elected democratically. It has been referred to in at least one headline as 'The New World Government'.

The Agreement was subjected to objections from a number of participating countries, perhaps most vigorously from France which, among other things, became concerned about loss of its cultural identity. In addition, once news of the negotiations leaked past official barriers, a global campaign conducted through the Internet by individuals and non-government organisations evidently had a considerable effect. Reports subsequently described the MAI in terms ranging from 'on hold' to 'dead in the water'.[40]

There is another type of objection to globalisation. It is a theoretical concern, but no less serious for that. Globalisation in all its senses inevitably leads to uniformity—to cultural uniformity and, of immediate relevance to this part of the discussion, to economic uniformity. Or, to phrase that another way—to the loss of cultural and economic diversity. In terms of the functioning of complex systems, this is analogous to the loss of biodiversity—of genetic heterogeneity—in an ecosystem. The world is losing its social, cultural and economic resilience. Like an enormous field of wheat, the entire planet is becoming increasingly vulnerable to the same social or economic diseases, whatever they might be. And like that field of wheat, the global socio-economic system will need to be very comprehensively managed if it is not to succumb to the assorted slings and arrows to which it is exposed and against which it has lost its natural immunity. Or, to put this another way, the monoculture and globalisation are each dynamic equivalents of putting all one's eggs in the same basket.

Managing this homogeneous system will include the economic equivalent of applying pesticides—and have its own range of unintended, undesirable consequences. The actions of the IMF would seem to fall within that category. I do not have the impression that living conditions within countries that have obeyed IMF directives have improved because of that response. And again we encounter the stupefying inconsistency of a system which sets out to produce global homogeneity, which therefore must be managed but which asserts that the ultimate god is the free market which must be left entirely to its own devices. I argued in Chapter 14 that the global ecosystem is in serious danger of 'collapsing' to the extent of being unable to support more than a small proportion of the present human population—if that. It would seem that there are good reasons for suspecting that a collapse of the global socio-economic system might be more imminent. There are others far more familiar than I with the world of finance and economics who share such an opinion.[41]

Perhaps it is pertinent at this point to raise another theoretical question arising from the general theme of globalisation, uniformity and world government. Is there a level of population above which a society cannot be governed or managed successfully by a single administration, no matter how competent? I use the word 'successfully' to imply a general level of decency, equity, freedom from exploitation and not simply success by economic criteria. My own opinion is a very definite 'yes', but at this stage it remains just that—an opinion that I am not yet able to argue rigorously. My intuition also tells me that the limit would vary with the physical environment, with population density and with the general level of living standards. It also tells me that the current global population is too big to be governable by any kind of single world government, if that should be the effective outcome of the present trends towards globalisation.

The preceding sections, which might well read as yet another diversion from my main theme, have sought to identify some attitudes, preconceptions, ideologies that are likely to engender at least suspicion, if not outright antagonism to my central arguments. And, superimposed on those ideologies, is human difficulty in understanding and dealing with the dynamics of complex systems. Indeed, if I were to make one simple generalisation it might be that, once Man adopted agriculture, the most fundamental threat to the long-term survival of our species lay in that deficiency—in our collective inability to meet the requirements of the 'Seventh Law'. A derivative of that blind spot is an apparently widespread inability to recognise that a dynamic system that apparently works well over a significant period of time might not, for any of a number of reasons, be able to do so 'for ever'.

This type of faith or scepticism—it can take either form—shows itself in many ways. It is reflected in the ridicule that has been heaped on Thomas Malthus over the years (Chapter 10), in refusal to acknowledge that some natural resources are effectively finite and therefore exhaustible (*e.g.* Chapter 11), in some of the assumptions underlying the push to continue raising crop yields—for more 'Green Revolutions', in the underlying faith in continuing technological salvation—and so on.

Can we overcome the problem? Trying to answer that question requires either another book or a brief excursion into the odd sweeping generalisation. Necessity—to say nothing of preference—dictates the latter.

One option is to step back and look at our environment and its management in a much broader perspective than is currently common practice—or, as a former Australian Prime Minister was wont to say, we should consider the 'big picture'. Another way of expressing the same thing might be that we should bring more commonsense to the empirical assessment of the world around us. Among other things, that would mean changing the criteria by which the many systems on which we depend are assessed. Perhaps the most serious difficulty confronting that proposal lies in the basic proposition that one man's commonsense is another man's lunacy.

The other type of option is to resort to highly complex mathematical modelling. This can suffer from the length of time needed to obtain the information needed for a comprehensive model and from inevitable disagreement on the weight given to its countless components. Moreover, a model is a model is a model but, regardless of its validity or lack of it, it is likely to be believed if it looks sufficiently complex and suits the objectives or preconceptions of the observer. Of course it can be tested over time—but so can the consequences of the empirical approach. What this amounts to would seem to be a case for both approaches.

But a widespread basic failure to understand just what is going on is only part of the barrier confronting objective environmentalists who see real danger ahead and seek to persuade their fellow man—and woman—to take some steps to avoid that danger, or at least lessen its impact. There is no shortage of environmental critics who, out of conviction or cynical vested interest—or both—set out deliberately to discredit those advocating environmental caution.

In the anti-environmental books that have come to my notice, a good deal of attention is usually given to those issues that I have classified in Groups 1-3 (Chapter 12)—in other words, to problems that are likely to respond to healthy economic conditions and technological development. Generalisations made in this context, however, can be as startling as their counterparts from 'deep ecologists'. I have already referred to some of Beckerman's pronouncements in this regard (Chapters 11 and 12). Others tend to concentrate on polemics and on criticism of the more extreme Green arguments. They are not without their own idiosyncrasies.[42]

Many, perhaps most, industries have the potential to affect adversely public health, social well-being or the biosphere at large. The adverse effects might be indirect—through the unintended consequences of the 'First Law', or they might be direct when the physical activities or the material product of an industry are actually or potentially dangerous. Most industrial nations have regulations or laws intended to limit the damage that can be caused by industrial activity but, astonishing though it might seem, industries occasionally find ways to circumvent the restrictions imposed by those laws or regulations. We saw in Chapter 12, for example, that bans on the use of refractory insecticides in the industrialised nations led to the dumping of these substances in agrarian countries. We also saw the effectiveness of industrial lobbying of the US Congress to postpone banning the fumigant, methyl bro-mide, a substance implicated in the breakdown of stratospheric ozone.

Perhaps the tobacco industry has been the most unscrupulous in suppress-ing and distorting information about its activities and the dangers of its prod-uct. And, like the pesticide manufacturers, it makes up for lost sales in the 'West' by increasing its propaganda and sales in the Third World.[43] A phar-maceutical company in the USA was ordered to pay $100 million compen-sation after it suppressed research results showing that one of its products, 'Synthroid', was no more effective than a cheaper equivalent from another source. It was estimated that patients using Synthroid had, *in toto*, paid over $2 billion more than necessary because of this suppression of information.[44] When moves to phase out leaded petrol in Australia began, the lead industry argued and lobbied in a number of ways against this policy. One of its argu-ments was that benzene, a component of unleaded petrol, was more danger-ous than lead. It also organised its own surveys purporting to contradict oth-ers that had shown that children living in areas of high traffic density had ele-vated concentrations of lead in their blood.[45]

Nor are government departments blameless in this respect. For example, California's Environmental Protection Agency had intended to destroy re-cords of dissenting scientific opinions about the potential effects on public health of industrial emissions and the discharge of wastes. The agency was forced by the government to change this policy after protests by scientists and environmental lobby groups.[46] The British Ministry of Agriculture, Fisheries and Food (MAFF) has been less than frank in releasing information about the development of 'mad cow disease' and its implications for human Creutzfeld-Jacob disease.[47] These are just a few examples. For a very comprehensive and objective account of the various methods used by industries to counter evi-dence of their environmental impacts, see Beder (1997).

Of course it is to be expected that an industry will oppose moves that threaten its profits. But when the practices of that industry seriously under-mine public health, as is the case with tobacco, what do the staff and employ-ees of the industry really think about that? Do they genuinely disbelieve the evidence, notwithstanding their attempts at concealment? Do they believe the evidence but cynically decide they don't care? How many of them smoke? How many of them accept the evidence and the criticisms but are afraid to say so for fear of losing their jobs? Or is this another example of 'tribalism' in which members of an organisation automatically or subconsciously identify with and defend the objectives of that organisation?

When it comes to environmental challenges some of the questions need to be rephrased a little differently. It is probably true that the environmental threat currently receiving the most publicity and discussion is the 'green-house effect' with its implications for the global climate. Recognition that there is a problem led to the Kyoto conference in 1997 with various commit-ments to emission targets in the 21st century.[48] But there is organised denial of the phenomenon.[49] Do those responsible for the denial genuinely believe what they are saying and totally discount all the evidence? Are they seizing upon the inevitable quantitative uncertainties in that evidence and exploiting it beyond reason? Or do they believe it but consider that the impact of the changes in climate will not reach disastrous proportions within their own life-times? In other words, are they opting for something along the lines of 'a short life and a gay one' (using 'gay' in its old-fashioned sense)?

And, closer to home, what about the responses to my arguments? Repre-sentatives of industries whose profits depend in large measure on a growing population will probably point to the median or minimal forecasts by the UN and say there is nothing to worry about. They are also likely to say that the sooner the living standards of the Third World are brought up to those of the industrialised nations, the sooner the population will level off. In that they could well be right if the goal were achievable—which almost certainly it is not. And, of course, the Catholic Church and others with a broadly similar philosophy in such matters will argue that, inasmuch as we are the chosen species, nothing too nasty is likely to happen to us and therefore we should keep on multiplying. I assume that their policies on birth control and related matters had their origins in a perceived need to keep their numbers high enough to ward off threats from rival sects.

Representatives of agribusiness will presumably dismiss out of hand my ar-guments for a redistribution of the population between town and country with their attendant changes in farm management. Will they consider the ev-idence but say that I have taken no, or insufficient account of economic argu-ments? Will some acknowledge validity but conclude that the serious conse-quences are unlikely to be felt within their own lifetimes? Will some of them say, in effect, that my arguments amount to something along the lines of 'where every prospect pleases and only Man is vile'—or 'all the world is mad except thee and me'?

A former Australian Prime Minister (not the one who spoke of the 'Big Picture') was known, among other things, for reminding us that life was not meant to be easy. He was right, of course. It would be foolish indeed to pre-tend that the ecological situation is anything but dangerously unstable. The problems we have generated are vast and, although conceptually simple, they are immensely difficult to deal with in practice. The most fundamental and most daunting challenge probably lies in coming to terms with ourselves. Perhaps we should begin by acknowledging that we are quite a strange species of animal or, more specifically, of chimpanzee[50] that is showing precious little evidence of adapting to its ecological environment. We might also pause to consider whether *sapiens* is the most appropriate specific name for us. Clever, yes; smart, yes;—but wise? And, as someone or other once said—it's later than you think.

Notes

— Estimates of the farm 'surplus' for 8th-15th centuries AD —

The following values for populations, population densities and the sizes of towns are taken from McEvedy and Jones (1978) and McEvedy (1992). By early in the 8th century the population of an area embracing Europe, the Aegean, the eastern Mediterranean, Arabia, Persia, western Russia, Egypt and the Berber region of north Africa amounted to about 55 million. Population densities of 10.km^{-2} or more were reached in no more than about 5% of the total designated area. Over the period of the 6th-15th century, it is unlikely that the population of any city, with the possible exception of 15th century Cairo, exceeded 125,000. McEvedy lists one early 8th century city, Constantinople, with a population greater than 50,000. Of the other cities on or near the early 8th century major trade routes, four had populations of 23,000-49,000 and eleven of 15,000-22,000.

If we take the sizes of these various towns as the median of the range in each case, then the sum of their populations amounts to less than 1% of the total for the entire region under consideration. All of these towns must have had the ecological essentials of a city rather than an agrarian village to the extent that their food supplies were grown by others and transported to the town. They were not likely to be the only towns with those characteristics, however. I can but guess at the number of people in centres of less than 15,000 but which can be classified as 'urban' by the simple criterion of their inhabitants' subsisting on food grown by others. Such a guess is that the entire urban population, as thus defined, represented 1-2% of the total, that at least 98-99% of the total were farmers who, overall, produced a surplus of 2% or less.

Six centuries later the population of the whole region was about 109 million, of which some 79 million were residents of Europe. There were now nine cities with populations in the range 50,000-125,000, twenty with 23,000-49,000 and forty four with 15,000-22,000. Again taking the median of each range, the urban population thus defined accounted for about 2% of the total population. A similar urban proportion lived in Europe 'proper'. With the previous guess about small towns, 96-98% of the population was rural and, overall, farms produced a surplus of 2-4%, about double that of six centuries earlier. It was still overwhelmingly an agrarian civilisation. These are the numbers that applied immediately before the 14th century European pandemic of bubonic plague, the 'Black Death'. By late in the 15th century, when the pandemic had run its course, Europe had returned to its pre-plague level of about 80 million (see Fig. 5.1). The urban population and, presumably the farm surplus, had increased to something like 3-6% of the total by the same criteria and with the same assumptions as previously.

— A Simplified Classification of Types of Pathogen —

Simply and broadly, infectious diseases can be classified into several groups according to their means of transmission and what, if anything, they do between times when they are not actually infecting anyone. In this classification, which is abbreviated, greatly simplified, has diffuse boundaries between groups and is somewhat idiosyncratic, I have not attempted to distinguish systematically between the type of infectious agent (bacterium, virus, protozoon etc) although in some cases, specific pathogens are identified. My purpose in the listing is simply to give a feeling for the classes of disease that Man let himself in for when he took up farming and, in particular, moved into town.

1 Zoonoses. These are diseases whose agents normally parasitise non-human vertebrates and are transmitted to Man either through chance contact (as with rabies) or because the pathogen or parasite has blundered by killing its normal host and has to find somewhere else to go (as with bubonic plague). The agents of such diseases can lie within any of the categories mentioned above, with the qualification that fungal diseases are rare. The infectious agents can be transmitted by direct contact such as biting (*e.g.* rabies), by handling living or dead animals (*e.g.* toxoplasmosis, leptospirosis, anthrax), or by a vector, usually an arthropod. The vector can be an intermediate host in a complex life cycle (*e.g.* malaria), or it can be little more than a mechanical means of transmission—an animated hypodermic syringe (*e.g.* various viral diseases).

Eating animals, wild or domesticated, can also lead to a range of infections and infestations, especially if the animal is not (adequately) cooked.

2 'Modified' zoonoses. There are also diseases that are sustained and transmitted (with or without a vector) within a human population, but can also cross back and forth between human and other vertebrate communities. Malaria and yellow fever are two such examples. The boundary between this group and Group 3 is more blurred than most.

3 Specific human pathogens. These are diseases, commonly viral (*e.g.* measles, smallpox, influenza, hepatitis, AIDS) or bacterial (*e.g.* tuberculosis, syphilis) that exist predominantly if not exclusively within the human population. They may have their counterparts in other species, and in some cases probably evolved from them, but those counterparts are usually caused by a different strain of the organism or virus. They are transmitted between individuals through the air, through direct physical contact, or through inert material such as eating utensils, clothing or bedding.

4 Adventitious or opportunistic pathogens. There are some organisms, mainly bacteria and fungi, whose normal habitat is soil or decaying tissues of plants or animals. They are thus normally saprophytic but can become pathogenic if they gain suitable access to animal, including human, tissue. Access is commonly obtained through a wound or through the lungs. Infections caused in this way are often serious and, if untreated, can be fatal. Notable examples are *Clostridium tetani, C. perfringens*, and *Aspergillus fumigatus*. The first two are anaerobic soil bacteria that can cause respectively tetanus and gas gangrene; the third is a fungus whose normal habitat is decaying plant material, but which can cause the potentially fatal aspergillosis, or 'farmer's lung', if the spores are inhaled. *Clostridium botulinum* should also be mentioned. This bacterium is not pathogenic in the normal sense but it produces a very potent lethal toxin that, although a protein, is absorbed from the digestive tract. The toxin can also gain access through wounds. In the past the common source of the toxin for Man lay in handling, and perhaps eating, the (inadequately cooked) intestines of animals or the flesh of animals that had been dead for long enough to allow access and growth of the bacterium. To-day the most common source of botulism, at least in the industrialised world, is improperly processed canned foods. None of the diseases in this category is normally 'contagious', that is to say directly transmissible between individuals.

5 Gastro-intestinal infections and infestations. There is a wide range of gastro-intestinal diseases that can be caused by bacteria (*e.g.* typhoid fever, cholera, numerous types of dysentery), by viruses (various forms of dysentery) and

Feed or Feedback

by protozoa (*e.g.* amoebic dysentery). The diseases can range in severity from mildly inconvenient to fatal, and can include chronic debilitation—especially with protozoal infestations. In addition there are various metazoal infestations of the digestive tract, one of the commonest and best known being that of the roundworm, *Ascaris lumbricoides*. All of these diseases share a common basic mode of transmission in that organisms, virus particles or eggs are voided in faeces and subsequently ingested by another individual in food or water. Metazoal infestations such as that of *A. lumbricoides*, have a complex life cycle that is, however, restricted to the one host. *Ascaris* eggs are ingested, the larvae penetrate the gut wall, are transported in the blood stream to the lungs where they enjoy temporary residence after which they migrate to the oesophagus from which they are swallowed.

Another gastro-intestinal parasite that deserves specific mention because of the impact it has had on human health is the human hookworm, *Ancylostoma duodenale*. Like that of *Ascaris*, the first stage in the transmission of this parasite from one host to another is the discharge of its eggs in the host's faeces. Unlike *Ascaris* eggs, however, hookworm eggs hatch outside the body. The larvae gain access to the next host through the skin, they are carried in the bloodstream to the lungs and thence, like *Ascaris*, pass to the oesophagus and are swallowed. Inasmuch as they gain entry through the skin they are different from the 'standard' gastro-intestinal parasites and pathogens whose transmission route is anal → oral. Another important distinction is that, whereas *Ascaris* feeds on intestinal contents, the hookworm lacerates the gut lining and feeds on blood. For that reason, apart from any other, it is more dangerous than *Ascaris*.

— The Doubling Time of a Population —

The expression, 69.3/P, used to calculate the doubling time of a population where P is the annual percentage increase, is derived from the so-called exponential growth equation,

$$N_1 e^{kt} = N_2 \qquad (1)$$

where 'N_1' is the number (the population) at the beginning of the time interval, 't' (expressed in years in this case): 'N_2' is the population at the end of that interval: 'k' is a rate constant, in this case equal to 0.0P, the annual proportional increase of the population expressed as a decimal fraction: 'e' is the base of natural logarithms. Its value (to seven places) is 2.7182818... (A mathematics text should be consulted for its derivation and wider significance.)

Since we are concerned with the doubling of a population, N_2 has twice the value of N_1, and Equation (1) therefore simplifies to

$$e^{kt} = 2 \qquad (2)$$

Taking natural logarithms of both sides we get

$$kt = \ln 2 \qquad (3)$$

The value of $\ln 2$ is 0.693, so Equation (3) becomes

$$t_d = 0.693/k \qquad (4)$$

and we can now use 't_d' to denote doubling time. Since we want to express 'k' as a percentage, rather than the decimal fraction required by the equation, we multiply the numerator on the right of the equation by 100. When that is done Equation 4 becomes

$$t_d = 69.3/P \qquad (5).$$

— Calculation of the time taken for the mass of a population to equal the mass of the earth —

The equation for this calculation is one used routinely in determining growth rates, doubling times or the number of generations of microbial populations. The basic term is N_2/N_1, where N_2 denotes the mass or population of one item (in this case the larger, the earth) and N_1 is the smaller (in this case either the bacterial cell or the human population). Since we are concerned with a population that is growing exponentially, we need to know how many times N_1 must double in order to equal N_2 or, in other words, the number of generations. From this point everything is simplified by using logarithms, and the basic equation is

$$G = (\log N_2 - \log N_1)/\log 2 \qquad (1)$$

where 'G' is the number of 'generations' or doublings of (the mass of) the population needed for N_1 to equal N_2. That is converted to time by multiplying it by t_d (0.5 h for the bacterium, 40 y for the human population). The overall equation can be written as

$$T = t_d(\log N_2 - \log N_1)/\log 2 \qquad (2)$$

where 'T' is the time needed for N_1 to equal N_2.

— A Nitrogen 'Balance Sheet' —

In 1987 Australian wheat received 119,000 tonnes of fertiliser nitrogen— equivalent to 32% of nitrogenous fertiliser applied in Australia in that year. This was the highest proportion applied to any crop or pasture (Bellingham, 1989 cited by McLaughlin *et al.*, 1992). In that year the average wheat yield was 1360 kg.ha^{-1} and the average protein content of the grain was about 9%. (The yield is as cited by National Farmers Federation (1991/92); see also Table A4. The protein content (8.8%) is from the Australian Bureau of Statistics (1996b).) The grain therefore removed 122 kg protein, or 19.5 kg N.ha^{-1} from wheat fields (ignoring the application of seed). The area under wheat was 9,063,000 ha (National Farmers Federation, 1991/92). The grain therefore removed about 177 kt N.

1980 sources (Cheney *et al.*, 1980 cited by Galbally *et al.*, 1992). attributed 15.1 Mt.y^{-1} of combustion to cereal stubble. In 1989, wheat accounted for 65% of the total cereal harvest (Australian Bureau of Statistics, 1992b). Under such conditions wheat would account for about 5.5% of the total biomass burnt, equivalent to some 66 kt N. Harvesting the grain and burning stubble therefore removed about 243 kt N. So, even without leaching and erosion, it would seem that, in 1987, harvesting wheat removed from Australian soil about 124,000 tonnes more nitrogen than was supplied by fertiliser. Where did the balance come from?

Numbers cited in Chapter 11 attributed 1150 kt N to atmospheric deposition over the whole continent. The area under wheat is equivalent to about 1.2% of the total continental area. A rough calculation of likely distribution according to rainfall patterns suggests that about 1.5% of atmospheric deposition falls on wheat land, equivalent to about 17 kt N. The other potentially important source of nitrogen is biological fixation. A range of 0.1-25 kg N.ha^{-1}.y^{-1} has been suggested for the global range of rates of non-symbiotic fixation (Chapter 11). A very rough guess that Australian wheat fields perform at about a third of the highest rate, gives 8 kg.ha^{-1}.y^{-1}.

Leguminous crops in Australia have been estimated to fix nitrogen at around 200 kt N.y^{-1} (McLaughlin *et al.*, 1992). Wheat is not a legume and systematic crop rotation is not a common practice among Australian wheat growers, although it is done by a small but growing group of 'organic' farmers. The area under leguminous crops at the time referred to by these calculations amounted to about 17% of the area under wheat. With the assumption that, at any one time, about 5% of the wheat land was having a 'spell' with legumes, then the fixation rate cited above would lead to symbiotic fixation in wheat land of some 60 kt N.y^{-1}.

Finally, there is the efficiency of the use of fertiliser nitrogen. I shall be slightly more optimistic than the 30% cited in Chapter 11 for irrigated Californian crops, and take 40% efficiency for the uptake of fertiliser nitrogen by wheat. When all that is done, and still ignoring leaching and erosion—some of which are already included in the 'efficiency' value, we arrive at a nett annual loss of 46,000 tonnes N from Australian wheat plots. This is equivalent to a rate of about 5 kg N.ha^{-1}.y^{-1}. If all the nitrogen were bound, this would equate to a loss of organic matter of about 100 kg.ha^{-1}.y^{-1}—less than the range given in Chapter 10 (*c* 400-1200 kg.ha^{-1}.y^{-1}) for rates of loss of organic matter. But given the uncertainties in my calculations and the evidence from a number of sources cited in Chapter 10 that, with cultivation, loss of organic matter is commonly exponential rather than arithmetic, the apparent discrepancy is of little significance.

Rather than construct a table, I have gone through this discursive exercise in order to emphasise the types of uncertainty that are inherent in such calculations. Tables can have an air of authority that, in this case, would be unwarranted. Moreover, this system is both relatively simple and quite well documented. An analogous assessment for all foodstuffs on a global basis is a very daunting business.

At least as far as wheat is concerned, the system is apparently not far out of balance. Given the assumptions I have made, the final nett rate could have gone either way—slightly positive or, as at present, slightly negative. It should be remembered, however, that there is no factor for leaching (which can include leaching to groundwater) or erosion. If included, these processes would increase the present negative balance.

To take the raw numbers: fertiliser supplied 119 kt of nitrogen, the grain removed 177 kt. If the system is indeed roughly in balance, then this suggests that about two thirds of the wheat crop under this kind of régime is dependent on fertiliser nitrogen.

— Water Pollution —

Broadly speaking, pollutants can be classified as those that are frankly toxic to all or most forms of life (*e.g.* heavy metals, various synthetic organic compounds including a range of pesticides etc) and those that are intrinsically harmless but, in sufficient concentration, change aquatic ecology to the point where the environment is lethal to many types of organism.

The adjective, oligotrophic, and its counterpart eutrophic, are most often associated with lakes but they can be used legitimately with any body of water. Oligotrophic water is normally considered to be 'healthy'. It is distinguished by a low concentration of suspended and dissolved nutrient substances, a high concentration of dissolved oxygen, low microbial populations but high fish populations. When applied to lakes it has the connotation that the lake is geologically young. Eutrophic water is pretty much the opposite. It is commonly turbid, has relatively high total nutrient concentrations (in solution, in suspension and in the microbial biomass), low concentrations of dissolved oxygen and low fish populations. In advanced stages of eutrophication the water smells foul. In some circumstances it is frankly toxic.

The overriding physical and chemical (as distinct from the biological) determinants of these conditions are water temperature, the depth of water and its movement (currents and vertical mixing), the concentration and chemical nature of suspended particles and of solutes. Temperature is important not only for its effect on biological activity but also for its effect on the solubility of gases, in particular, oxygen. Unlike most solids, gases are more soluble at low than at high temperatures. For example, at 'normal' atmospheric pressure at sea level, the solubility of oxygen in water is 4.9% (m/v) at 0°C, 3.2% at 25°C and 2.5% at 50°C (Weast, 1982).

The majority of aerobic aquatic organisms must obtain their oxygen directly from solution. If the water temperature rises, total metabolic activity and hence the rate of oxygen consumption will increase, but oxygen solubility will decrease. Organisms requiring oxygen will thus be at a disadvantage on two counts. Oxygen dissolves in water from the atmosphere and is released into the water from aquatic photosynthesis. Both of these processes lead to the highest concentrations of oxygen near the surface for the obvious reasons that that is where atmospheric oxygen is dissolving and because there is a limit to the depth of penetration by light. The euphotic zone, in which photosynthesis is possible, extends to a depth of about 100 m in clear water. Dissolved oxygen diffuses slowly so, if the water is stagnant, in most cases metabolic activity would completely exhaust oxygen in deep waters. Just how long that would take would depend on the numbers of animals and microorganisms and on the water temperature. With some mixing, however, and no more than moderate total metabolic activity (reflected in the measure known as BOD—biological or biochemical oxygen demand), oxygen does reach the deeper layers. Indeed, even at the bottom of the deepest ocean trenches, 10,000 m and more, there is accumulated organic matter, the waters are well oxygenated and there is a remarkable variety of highly specialised deep-sea organisms. That this can happen is due in part to the very low metabolic rates of organisms at those temperatures (around 3°C) and enormous hydrostatic pressures. In some specific zones bacterial populations are also supported by volcanic activity through hydrothermal vents that supply inorganic nutrients and raise temperature locally. If the body of water is shallow, there will also be photosynthetic activity by plants growing on the bottom sediments. Clearly the quantitative details of these various processes will be very different in a fast flowing river, a deep lake or the open ocean.

At first sight there might be an apparent paradox in the statement that, in oligotrophic waters, nutrient concentrations are low but fish populations are high. There are important differences, however, between terrestrial and aquatic environments in the dynamics of the systems, the processes that are rate-limiting and, in particular, the roles of oxygen. Other conditions being favourable (adequate water supplies, no lethal plagues etc), total population or biomass of terrestrial animals is limited ultimately by the biomass and growth rates of edible plants. It is not limited by oxygen concentration and there is no serious direct competition for nutrients between, say, higher animals and soil microflora. Indeed, a high microbial biomass is normally indicative of productive soil. Fish are poikilothermic—their body temperature is essentially that of the surrounding water. In all but very exceptional circumstances, therefore, their body temperatures—and hence their metabolic rates (per unit of mass)—are lower than those of homeotherms. For a given body mass, a fish normally needs less food per unit of time than does a bird or mammal. In oligotrophic waters there might not be much for the primary grazers to eat, but they have time to find it.

What they do have to worry about, as I have said, is oxygen, and their biggest potential competitors here are microorganisms, especially bacteria. In oligotrophic waters bacterial numbers are low (around $500.ml^{-1}$ or less for heterotrophs: bacterial suspensions become just visibly turbid in a small flask at densities in the order of $100,000.ml^{-1}$). That can change dramatically if nutrient material enters the waterway in significant quantities. The details and extent of the impact will vary widely according to the chemical composition of the pollutant, the proportions of the nutrients it carries that are soluble and particulate, its turbidity, its pH, its flow rate and temperature—and, of course, the volume, flow rate, temperature and depth of the receiving body of water. If the pollutant consists predominantly of organic nutrients such as discharged in sewage, effluent from a fruit cannery or an abattoir, the essentials of its impact will be roughly as follow.

The organisms that will respond fastest to the discharge are heterotrophic bacteria. Depending on their type they can have generation (doubling) times of about 5 hours at 10°C, 3.5-5 h at 15° and 1.5-2.5 h at 20°. No fish population, indeed no animal population, can even remotely approach such growth rates.

If there is enough nutrient in the discharge, the aerobic bacteria—both strict and facultative—will grow to a level at which they consume oxygen faster than it can be replaced; the water then becomes anoxic. At that stage only the anaerobes, strict and facultative, will be able to multiply. Unless an enclosed body of water or a sluggish river becomes completely polluted and putrid, there will be concentration gradients of nutrients and dissolved oxygen between the core of the discharge and its edges. The types of microorganisms that predominate will change along the gradients—in simple terms with aerobes at the edges, anaerobes in the middle and facultative organisms in the intermediate zones. None of this is intended to deny a role of other microorganisms such as protozoa and fungi, but their metabolic and growth rates under circumstances such as I have outlined will be overshadowed by those of the bacteria.

Sessile animals in the anoxic zone will be killed as will many in the transition region. This will happen because of asphyxia whether or not the microbial population is producing toxins. What happens to the fish will depend also on the physical dimensions of the process and whether or not the fish can escape from the anoxic water. If they can they will probably survive unless they succumb to toxaemia. If they cannot escape they too will die from asphyxia.

A common situation producing a discharge of inorganic nutrients arises when excess fertiliser is leached from farmland into a body of water. (It also happens when sewage treated at least to the secondary stage (see below) is discharged but, in that situation, there is also a significant output of organic material.) Nitrogen can also enter waterways and soil from the atmosphere, usually through the medium of rain. When this process is severe enough to be called pollution, the source of the nitrogen is usually combustion—predominantly industrial or from motor transport in cities. There can be significant transfer of nitrogen through the atmosphere from the urine of farm animals, particularly when they are confined at high densities as in feedlots. (See Amann and Klaassen (1995) and *Ambio* **23** No. 6 (1994), entire issue.) In relatively densely populated regions such as much of Europe and parts of Scandinavia, eutrophication by nitrogen of some soils is sometimes as serious as aquatic eutrophication. Moore (1995) cites rates of N deposition over much of

Europe in the mid 1970s of around 2-6 kg $N.ha^{-1}.y^{-1}$ whereas recent rates for some Dutch woodlands amounted to at least 60 kg $N.ha^{-1}.y^{-1}$.

The organisms responsive to inorganic pollution are initially the auto-trophs—those that derive their nutrient elements from inorganic sources. A critical determinant of which organisms predominate, however, is the source of their energy. There are bacteria, the 'chemolithotrophs', that derive their energy from inorganic reactions, such as the oxidation of ammonia (NH_3 or NH_4^+) by oxygen to nitrite (NO_2^-) or nitrate (NO_3^-), or the reduction by mo-lecular hydrogen of nitrate to molecular nitrogen or to ammonia, or the re-duction of sulphate (SO_4^{2-}) to sulphide (S^{2-}), The physico-chemical condi-tions needed for these reactions and the accompanying assimilation of nitro-gen and carbon are so exacting, however, particularly in respect of oxida-tion/reduction potential (Eh), that they are unlikely to be dominant in the type of situation outlined.

Of far greater quantitative importance under these circumstances are the photolithotrophs, organisms that photosynthesise and thus have access to plenty of energy, at least near the surface of the water during the day. Photo-lithotrophic bacteria are restricted, at least initially, because they are strictly anaerobic. The important organisms are the algae and the cyanobacteria ('blue-green algae'). The cyanobacteria are procaryotes but, like the eucaryotic algae, they produce oxygen as a by-product of photosynthesis activities. Eu-caryotic algae can grow well enough in water containing adequate concentra-tions of the essential nutrients that, in their case, includes combined nitrogen. The cyanobacteria, on the other hand, have a competitive advantage because a number of their species can fix elementary nitrogen. Some of them produce heat-stable toxins. Three genera commonly responsible for toxin production are *Anabaena*, *Microcystis* and *Oscillatoria*. An outline of the chemistry and biochemistry of some of the toxins can be found in Carmichael (1994).

When inorganic pollutants accumulate in a body of water, the biological sequence of events is roughly as follows. If the discharge includes nitrogenous compounds, there will be a fairly rapid response of algae and cyanobacteria in the surface layers of the water. In the deeper layers there might be a slower re-sponse of some chemolithotrophic bacteria. At night the photosynthetic or-ganisms will resort to heterotrophic metabolism using stored carbohydrates as an energy source and they will consume oxygen. As their numbers increase, some species might form mats that will be lifted to the surface of the water by the oxygen bubbles released during photosynthesis. These mats will block both light penetration and oxygen diffusion into the body of the water. Moreover, heterotrophic bacteria will now begin to multiply on soluble me-tabolites that are excreted or diffuse from the algal cells, as well as on dead cells, and will consume any dissolved oxygen. If the concentration of pollutants is high enough, this process will lead to a full-blown eutrophic body of water, stinking and toxic, distinguished by its pea-green colour and sometimes by an algal mat on the surface.

This outline is, of course, a simplification. The difference between the impact of predominantly organic or inorganic pollutants is rarely as definitive as described. For example, the autolysis of heterotrophs can provide inorganic nutrients at concentrations sufficient to support an algal bloom. There are also aerobic photoheterotrophic bacteria that make a significant contribution to the cycling of carbon in marine environments (Kolber *et al.*, 2001)

In an oligotrophic body of water none of the elements such as calcium, iron, magnesium, potassium, or the essential trace elements is likely to be present at concentrations so low that, if all physical conditions were favourable, that element would limit the initial growth rate of the microbial population, although any, perhaps especially iron [for example, Behrenfeld *et al.* (1996), Coale *et al.* (1996), Cooper *et al.* (1996) LaRoche *et al.* (1996) and Pakulski *et al.* (1996)], might limit later growth rates and ultimately the total biomass. I should emphasise here that, in a natural ecosystem, the concentration of each element in solution reflects a dynamic, not a static situation. In other words, if their concentrations are more or less constant, that constancy reflects a steady state inasmuch as all are in a continuous state of turnover through the existing biota. Some solutes in a steady state concentration can also be effectively at equilibrium with the surrounding land—the bottom and shores of the waterway.

In most natural situations of this type, early growth rates are more likely to be limited by the major nutrient elements, nitrogen, phosphorus or carbon/energy. Nitrogen and phosphorus commonly occur in oligotrophic rivers at concentrations of around 0.2 and 0.02 mg.l^{-1} respectively (*e.g.* Kennish, 1989; NSW Water Resources Council, 1994a). The limiting element might not necessarily be the same at different temperatures, levels of pH or light intensity. In sea water, the situation is somewhat different because the phosphate ion can form any of several insoluble salts or complexes. Thus, in oligotrophic sea water, the concentration of dissolved (inorganic) phosphorus is normally so low that it does not rate a mention in concentration tables such as those in Kennish (1989). At first sight this might seem sufficient to prevent eutrophication of sea water since, no matter what was discharged into the sea, phosphorus would always be limiting. Of course it is not as simple as that. If, for example, phosphorus were discharged as an ester, the fate of the phosphorus would be determined in large measure by the relative rates at which that ester was assimilated directly by microorganisms, or hydrolysed followed by competition between biological assimilation and chemical precipitation.

Thus the discharge into a waterway of material that contains relatively high concentrations of these elements in states that are readily assimilable usually produces a rapid response in the microbial population. When there is ample phosphorus but little or no bound nitrogen or organic carbon in the discharge, the organisms best placed to take advantage of the situation are the cyanobacteria.

— Sewage Treatment —

The treatment of sewage commonly involves the following major steps:
1 Filtration through a coarse screen to remove large items.
2 An oxidation stage.
3 Passage through a wide shallow settling pond, an environment that is also conducive to supplementary oxidation and uptake of inorganic nutrients by algae. If the algae are removed from the effluent, this is an important (tertiary) stage in sewage treatment. When circumstances allow, the final effluent might be allowed to filter through a bed of sand before eventual discharge.

The critical part is the oxidation stage in which organic matter, including cells of the strict anaerobes, is metabolised by aerobes. In essence this involves oxidation of part essentially to CO_2 and water, and assimilation of part into the biota which then become a separable sludge. This is achieved in the main by either of two methods that differ widely in their engineering details but are broadly similar in the microbial ecology of the processes they support.

The essence of the process is to provide a gelatinous matrix with a very large surface area onto which the organic matter, both particulate and soluble, is adsorbed. In the presence of an ample supply of oxygen the adsorbed material is either oxidised or incorporated into the cells of the complex microbial community that forms the matrix and mediates the oxidation. A key organism is *Zoogloea*, sometimes called the 'sewage fungus', but in fact a bacterium. (The term 'zoogloea' is sometimes used also for a gelatinous mixed microbial colony without necessarily implicating the genus *Zoogloea*.)

The older of the two principle types of equipment used for this purpose is the 'trickling filter' in which a concrete cylinder is filled with stones of shapes that ensure they cannot pack tightly together. The sewage is sprayed onto the surface of the stones and trickles down over them, eventually to a drain at the bottom. When the filter is fully operational, the stones are covered with the zoogloeal slime which is thin and has a very large surface area. Oxygen is readily accessible.

The second method is the so-called 'activated sludge' process. In this the sewage is fed into a tank—usually rectangular—in which it is agitated and aerated very vigorously. The solid particles of the sewage—the sludge—rapidly accumulate a zoogloeal slime similar to that of the trickling filter and present a very large surface area for adsorption and oxidation. When that point is reached the sludge is said to be 'activated'.

Both types of treatment are supplemented by other stages such as settling ponds, clarifiers and, in the case of the activated sludge works, mechanisms for returning some sludge to be used as a 'nucleus' for incoming sewage. Thereafter there might be a tertiary stage involving a retention pond with a residence time of some weeks, a period that can vary widely with circumstances. In some circumstances a physico-chemical technique, such as membrane filtration, might be used to reduce the concentration of nutrient elements to a lower level than is otherwise normally obtainable. There the effluent, which might or might not be chlorinated, is discharged somewhere—mostly into water, sometimes onto sand or soil whence it filters into water, in a few places onto land where it is used for irrigation.

The total phosphorus content of effluent treated according to these general principles is commonly within a range of about 5-9 mg.l^{-1}. In other words, 'standard' tertiary treatment removes less than 40% of the phosphorus —more than 60% is discharged in the effluent. The city of Melbourne has two major treatment facilities. One is a farm at Werribee where some 4500 ha of pasture is irrigated by raw sewage at a rate of around 180,000-190,000 Ml annually. The irrigating liquid is ultimately discharged into Port Phillip Bay. About 38% of the total phosphorus is removed in the course of irrigation. The other facility is a conventional activated sludge works handling some 140,000 Ml.y^{-1} and discharging into the sea. It also removes 38% of the total phosphorus, most of which is incorporated into the sludge. (D. Gregory, Melbourne Water, 1992. Personal communication). The total phosphorus content of a representative type of dried sludge is around 2.8% (Kyle and McClintock, 1995).

'Advanced' treatment, as used in the Australian Capital Territory, normally reduces the total phosphorus concentration in the effluent to less than 1 mg.l^{-1}. This treatment includes batteries of activated sludge tanks in series and the stimulation of microbial activity by supplementing the sewage with iron salts.

The nitrogen content of sewage effluent is a much more variable commodity to the extent that there can be wide variations in the relative proportions of nitrate and ammonium ions according to the oxidation/reduction history of the sewage; analyses for the effluents from some small treatment works put nitrate + ammonium nitrogen in the range of 2.5-3 mg N/l.

— Tables —

TABLE AI *Global estimates of annual production in various ecosystems*

Ecosystem	Primary Production (g.m⁻².y⁻¹)		Total Mass (dry) (kg.m⁻²)	
	Mean	*Range*	*Mean*	*Range*
Forest				
Closed	1600	1000-2500	35	6-60
Open	1300	600-2000	35	6-60
Coniferous	1300	600-2000	35	6-60
Grassland (temperate)	600	200-1500	1.6	0.2-5
Crops	650	100-3500	1	0.4-12
Rangeland (semi-arid)	90	10-250	0.7	0.1-4

From Milthorpe (1982)

TABLE A2   *Densities of selected elements in Eucalyptus and Pinus forests*

Forest Type	Density of Element*			
	C	N	P	S
Eucalyptus				
(38-*c* 60y)	106-293	248-426	12-38	26-54
(roots)		64	9	–
Pinus				
P. radiata (22-29 y)	143-192	286-453	33-67	–
(unfertilised)				
P. pinaster (14 y)				
Unfertilised	101	219	16	35
Fertilised (P)	123	279	21	46

* The dimensions are kg.ha^{-1} except for carbon which is tonnes.ha^{-1}. Except for the row labelled '(roots)', all densities refer to the content of the element in the biomass above the ground. From Hingston and Raison, (1982) and Milthorpe (1982).

TABLE A3 *Mean Content of Elements in the Biomass (above ground) for*
Various Types of Forest

Forest Type	Element and Content (%)		
	N	P	S
Eucalyptus			
Canopy	0.106-0.129	0.005-0.009	0.010-0.021
Understorey			
N-fixing spp	0.86-1.06	0.017-0.028	0.066-0.120
Non-fixing spp	0.37-0.58	0.016-0.029	0.079-0.136
Pinus			
P. pinaster	0.233	0.017	0.036
Northern Hemisphere and Tropical (mean values)			
Boreal	0.347	0.039	0.045
Temperate Coniferous	0.228	0.016	–
Temperate Broadleaf	0.320	0.022	0.022
Tropical	1.127	0.064	–

Condensed from Hingston and Raison (1982).

TABLE A4 *Wheat yields from selected regions and periods*

Region (Period)	*Yield*[1] *(kg.ha⁻¹)*	*Source*
Southern Mesopotamia (2400 BC)		
Wheat + barley	1955	Jacobsen & Adams (1958)
Wheat	325	(Chapter 4)
England (manorial system) 1283-1349		
Gross	515 ± 200	Pretty (1990)
Nett[2]	385	
W. Europe (1850-1905)		
Winter einkorn	835 ± 135	Gregg (1988)
Spring einkorn	645 ± 75	
Winter emmer-spelt	1045 ± 190	
Spring emmer-spelt	755 ± 100	
N.W. Europe (excluding France)		
1850	990-1200	Grigg (1982)
1969-1971	4150-4620	
Austria, France, Italy		
1850	670-770	
1969-1971	2390-3650	
Spain, Greece, Russia		
1850	450-460	
1969-1971	1270-1850	
Norway		
1850	570	
1969-1971	3130	
USA		
1870	780	
1960	1610	
1982	2760	Pierce (1990)

TABLE A4 *(Ctd)*

Region (Period)	Yield [1] (kg.ha⁻¹)	Source
Sweden		
1870	1395	Grigg (1982)
1960	3220	
Nepal (experimental plots)		
1989-90		
High fertiliser	3330 ± 365	Sharma (1992)
Low fertiliser	1720 ± 240	
Australia		
1870	750	Grigg (1982)
1950/51-1959/60	1135 ± 180	Natl Farmers Fedn
1980/81-1990/91	1385 ± 280	(1992)
World Average		
1950/51-1959/60	1095 ± 90	
1980/81-1989/90	2150 ± 165	

1. Expressed to the nearest 5 kg.ha⁻¹ ± the standard deviation where available.
2. The nett yield is that remaining after subtracting seed for the next season's planting.

TABLE A5 *Soil Losses of Nitrogen and Phosphorus from Wheat in Europe in the 8th and 19th Centuries*

Date (approx)	Population (millions)	Wheat Consumed (Mt)	Farm Surplus (%)	'Efficiency' of recovery[1] (%) City N	City P	Farm N	Farm P
700	55	16.5	2	10	70	40	95
1850	265	79.5	67	10	70	40	95

	Proportion of 'Night Soil' Recovered from City[2] (%)	Loss of Element — Proportion of Total (%) N	P	Annually from Soil[3] (kg.ha⁻¹) N	P
700	10	61	7	3.2	0.1
	50	61	7	3.2	0.1
1850	10	86	64	15.5	2.9
	50	84	46	15.0	2.1

(1) 'Efficiency' refers to the proportion of the nutrient element, nitrogen or phosphorus, that survives leaching, storage, transport etc and is returned to arable soil in a form available to plants. (2) For the purpose of these calculations it has been assumed that all human excrement produced on farms was returned to the soil. (3) This allows for the element applied to the soil in seed but it does not allow for adventitious fertilisation from natural sources (see text).

Supplementary notes for Table A5

The losses of nitrogen and phosphorus from farm soil in one year were calculated from either of the following equations:

$$T = (M-A) \{E(I-S) + ESR\} \qquad (1)$$
$$T = (I-A/M) \{(M-C) + REC\} \qquad (2)$$

where:

T = the total mass (tonnes) of the element returned to the soil:

M = the total mass (tonnes) of the element consumed in the year's crop (grain only):

A = the amount of the element assimilated by the population. (Assimilation occurs only in individuals who are growing—see Chapter 2. The term 'A' is thus a function of the number of children and pregnant women in the population. It is used because the dead are rarely buried in farmland but its numerical impact on the results in Table A5 is negligible in the present context):

E = the 'efficiency' of recovery by the crop of the element from human excrement on the farm (see also footnote to the table):

e = the corresponding 'efficiency' for recycling city nightsoil. The difference between 'E' and 'e' is a function of the time the stuff is left lying around before collection and redistribution to farm soil. During that time there will be losses of both elements, especially nitrogen.

r = the proportion of city nightsoil collected and returned to farmland (expressed as a percentage in Table A5, as a decimal fraction in the equations). No rural equivalent of 'r' is included since the proportion of rural faeces and urine returned to the land is taken as 1. This is optimistic and it takes no account of distribution on the farm.

s = the farm surplus expressed as a decimal fraction; it is based on the distribution of population between town and country (Chapter 5), but the two thirds surplus listed for the mid 19th century might be distorted slightly by international trade: Despite the application of the equation to one year we are concerned with an effect that will exert itself over centuries during which the contents of rural cesspits and human dung heaps will be dispersed by a range of natural agencies. Food scraps are omitted from the terminology and calculations since the food under consideration is grain from which the amount of scraps would be negligible. See also Fig. 14.1 for a more comprehensive assessment of the the the implications of recycling urban sewage.

The factors and assumptions not shown in Table A5 and on which the calculations were based include: the annual *per capita* consumption of cereal grains was 300 kg; this was treated entirely as wheat; grain (wheat) contributed half the mass of the average diet (see Slicher van Bath, 1963): the average nitrogen and phosphorus content of wheat grain are taken as 1.5% and 0.38% respectively, although it is likely that the nitrogen content was slightly higher in the 19th century (based on Donald, 1964/5; Watt and Merrill, 1963): the nett annual yield of wheat in the 8th century was 350 and in the 19th century, 1200 kg.ha^{-1}. The amount of each element assimilated (A) was derived from calculations not shown because of the insignificance of the final value. 'Efficiency' is based on material cited by Loomis (1978) and Smil (1993). Finally, this table, like tables generally, probably has an air of authority. It depends on so many assumptions, however, that it should be seen as no more than a reasonable approximation of what it purports to illustrate.

TABLE A6 *Patterns of freshwater consumption for various regions in 1987*

Region	Annual Consumption					
	Total (km³)[a]	Fraction of Resource[b] (%)	Per hd (kl)	Purpose (%)		
				Ag.[c]	Dom.[c]	Ind.[c]
World	3240	8	644	69	8	23
Africa	144	3	245	88	7	5
Asia	1531	15	519	86	6	8
North & Central America	697	10	1861	49	9	42
USA	467	19	1868	42	13	45
South America	133	1	478	59	18	23
Europe	359	15	713	33	13	54
USSR (former)	358	8	1280	65	7	27
Australia	14.6	4	898	70	12	9

(a) This is the amount of water actually used. It can include both surface and ground-water for all regions but Australia for which the quantities refer specifically to surface water (see below). (b) This denotes the fraction represented by (a) of the total amount of freshwater annually available. (c) The abbreviations denote Agriculture, Domestic Use and Industry respectively. All material, except that for Australia, was derived from World Resources Institute (1994). That publication had an entry for Oceania that was seriously distorted by anomalous 1975 data for Australia. The Australian material in this table was derived from the Australian Water Resources Council as cited by the Australian Bureau of Statistics (1992a). Global totals as used by the World Resources Institute have been left unchanged. Groundwater is discussed elsewhere. Figures for 1999 indicate that globally, irrigation consumed some 2500 km³ and accounted for 65% of total water consumption by the human population (Gleick, 2000).

TABLE A7 *Some Australian rates of soil erosion caused by water*

Region	Erosion Rate $(t.ha^{-1}.y^{-1})$	Comment	Reference
Australia (general)	0.27-0.45	continental mean	(a)
	from <1	pasture	
	to 382	sugarcane	
	0.015-6.34	many circumstances	
	70	northern rangelands	(b)
	150	eastern tropical cropping	
	209	parts of arid zone	
NSW	31-87	various: bare fallow	(c)
	2.40 ± 3.25*	cropping plots	
	0.24 ± 0.40*	pasture	
	350	storms on cropland	
Queensland	from 2.1	no tillage	
	to 61.0	bare fallow before sowing	
	from 1.11	wheat, no tillage	
	to 9.92	sunflower, normal tillage	
		(1982-1986)	
(Same fields)	from 0.00	wheat, no tillage	
	to 16.13	sorghum, normal tillage	
		(1985-1986)	
Western Australia	140-686	potato	
		(slope 7.8-20 '%')	

* Mean ± standard deviation. (a) Various sources cited by McLaughlin *et al.* (1992): (b) various sources cited by Wasson *et al.* (1996): (c) Various sources cited by Edwards (1991).

TABLE A8 *Some 'national' rates of soil erosion caused by water*

Region	Erosion Rate $(t.ha^{-1}.y^{-1})$	Comment	Reference
USA	30		(a)
	13		(b)
	5-170	cultivated	(c)
	0.03-3	'natural'	
China	< 2	'natural'	(c)
	150-200	cultivated	
	280-360	bare soil	
India	0.5-1	'natural'	(c)
	0.3-20	cultivated	
	10-20	bare soil	
UK	0.1-0.5	'natural'	
	0.1-3	cultivated	(c)
	10-45	bare soil	
Nigeria	0.5-1.0	'natural'	(c)
	0.1-35	cultivated	
	3-150	bare soil	
Ivory Coast	0.03-0.2	'natural'	(c, d)
	0.1-90	cultivated	
	10-750	bare soil	(c)
	108-170		(d)
Tanzania	0	forest or thicket	(d)
	7	cultivated	
	146	bare soil	

(a) Pimentel (1976) cited by Goudie (1986): (b) Crosson (1995): (c) various sources, 1948-1981, cited by Morgan and Davidson (1986): (d) various sources, 1972-1977, cited by Goudie (1986).

TABLE A9 *Annual discharge rates of sediment, nitrogen and phosphorus into the sea from the east coast of Queensland*

| | | Type of Catchment | | |
Pristine	Grazing	Cropping	Urban	Total
		Area (km²)		
53145	381606	13844	1920	450515
		Percentage of total		
12	85	3	0.4	
		Sediment (kt)		
863	11686	1614	77	14240
		Percentage of total		
6	82	11	0.5	
		'Average' Specific Discharge Rate (kg.ha⁻¹ catchment)		
162	306	1166	401	316
		Nitrogen (t)		
4192	56627	11169	120	72108
		Percentage of total		
6	79	15	0.2	
		'Average' Specific Discharge Rate (kg.ha⁻¹ catchment)		
0.8	1	8	0.6	2
		Phosphorus (t)		
601	8092	1676	136	10505
		Percentage of total		
6	77	16	1	
		'Average' Specific Discharge Rate (kg.ha⁻¹ catchment)		
0.1	0.2	1.2	0.7	0.2

Summarised and condensed from Moss *et al.* (1993). The 'average specific discharge rates' for each component were calculated from the sums of loads and areas listed for each catchment by Moss *et al.* See also Table A10.

TABLE A10 *Some Specific Rates of Sediment Flux from Rivers on the East Coast of Queensland*

Land Classification	Sediment Flux (t.km^{-2}.y^{-1})		
	Minimum	Maximum	Median
Pristine	3	68	18
Grazing	13	270	71
Cropping	32	676	178
Overall	13	244	60

Summarised from Moss *et al.* (1993). This table gives a statistical summary of specific export rates of sediment from each catchment as listed by Moss *et al.* Compare with Table A9 in which specific rates were calculated by a different method and are somewhat different from those listed here.

TABLE AII *Some values for the content of organic matter in soil from various regions*

Region	Organic Matter (%)	Comment	Reference
Uruguay (prairie)	3-5		(a)
Former USSR			
'Dark Grey Forest'	5.2-10.5		
chernozems	>4		
Sonoran Desert	0.8		
Hawaii	3.8		
Puerto Rico			
humid	3.5		
(semi ?) arid	1.8		
Uttar Pradesh			
Tarai Region	2-3.5	1971 estimates	
Kenya			
Highlands	>4	1956	
wheat cultivation	2.5-4	estimates	
West Africa			
savanna	1.4		
North East Brazil			
shifting agriculture	1.3-2.3		(b)
virgin soil	3.2		
Australia			
desert loams[1]	*c* 1		(c)
alpine humus[1]	*c* 50		

(1) These two numbers denote the approximate range for all Australian soils, but see also Table A12. (a) Various sources cited by Pitty (1979). (b) Tiessen *et al.* (1992). (c) Charman and Roper (1991).

TABLE AI2 *Some values for the content of organic matter in soil from various districts in Australia*

District	Organic matter (%)	Comment	Reference
Five states	1.6-4.6	cereal cultivation	(a)
Northern NSW		Top 5 cm	
virgin[1]	2.4-4.6	(median 3.9)	
cultivated[1] 1-20 y	1.6-4.1	(median 2.1)	
9 Victorian districts[2]	0.6-10.5	(median 2.2)	(b)
Eastern Australia krasnozems[3]			
virgin soil	*c* 11.53	Top 15 cm	(c)
South Australian acid soils			
five sites	1.9-8.1		(d)
6 Southern Queensland sites		Top 10 cm	
virgin soils in 'cereal belt'	1.5-4.2	(median 2.95)	(e)

(1) The same soil type in each of seven comparisons. (2) Rainfall 280-760 mm; median 584 mm. (3) The range for krasnozems throughout eastern Australia is 3.8-25.7%; the range for Queensland is 1.9-10.5 %. (a) Charman and Roper (1991). (b) Leeper and Uren (1993). (c) Oades (1995). (d) Amato and Ladd (1994). (e) Dalal and Mayer (1986b). I have multiplied any original results quoted as '% carbon' by 1.9 (see text, Chapter 10).

TABLE AI3 *Some rates of deposition of nitrogen from the atmosphere*

Region	Deposition Rate $(kg\ N.ha^{-1}.y^{-1})$	Comment	Reference
USA	< 5	desert	(a)
	> 30	near feedlots, Midwest.	
	13-30	Wisconsin 'average' c 20	
	21-83	uptake of NH_3 by soil near New Jersey (1964 report)	
England			
(Rothamstead)	4.4	1889-1903	
	5.4	1960-1964	
	8.6	1969-1970	
N.E. America	5-25	(mid 1990s)	(b)
Northern Europe	5-60	(mid 1990s)	
Europe	2-6	(mid 1970s)	(c)
Netherlands (some parts)	$\geq$ 60	(1995)	
England (Cumbria)	30	(1995)	
N.W. Scotland	c 6	constant for past 40 y	
Australia	0.3-3	'average' 1.5 (1981)	(d)
Global	7-9	(1975)	

(a) Various sources cited by Legg and Meisinger (1982). (b) Various souces cited by Wedin and Tilman (1996). (c) Various sources cited by Moore (1995). (d) Lamb (1981) and Burns and Hardy (1975) cited by McLaughlin *et al.* (1992).

TABLE A14 *'Production' and export of phosphorus in Australian primary produce*

Item	Phosphorus		
	Content of item (%)	'Produced' in item (tonnes)	Exported in item (tonnes)
Live animals	1.00	'2125'	2125
Meat	0.16	4803	1777
Dairy (800 mg.l^{-1}) + eggs	0.15	5205	563
Animal feeds	0.30	6372	2349
Fruit + veg.	0.04	1684	307
Misc. foods	(0.5?)	?	368
Cereals	0.35	76682	54162
Sugar cane	0.04	9698	
Tea	0.04	0.3	0.1
Coffee	0.30	3	3
Nuts	0.39	152	36
Oilseeds	0.30	753	635
Tobacco	0.12	14	2
Total		107491	62327

Sources: Australian Bureau of Statistics (1991a and b; 1993a). Phosphorus contents of the various commodities were obtained for the most part from Watt and Merrill (1963) and Nutrition Search (1979). The phosphorus content of whole animals was taken as 1%, the value commonly used for large animals. Bones contain about 80% of the phosphorus of an animal. Despite that, bones have not been listed as such either in this table or in Table 11.1 because most bone material is reused either as fertiliser or as 'bone meal'. With the exception of sugar cane, items listed as produced but not as exported are presumed not to have been exported. Sugar is exported but it contains no phosphorus.

TABLE AI5 *Population densities relative to farmland and proportion of population classified as 'urban' in selected regions and countries*

Region	Population Density (ha⁻¹)		
	Cropland	*Cropland + Pasture*	*Urban (%)*
'World'	4.0	1.2	45.2
Africa	4.1	0.7	34.7
Asia	7.4	2.8	34.0
Nth + Central America	1.5	0.7	74.0
Sth America	2.8	0.5	78.0
Europe	3.7	2.3	75.0
(former) USSR	1.3	0.5	68.1
Australia	0.4	0.04	85.2
Bangladesh	14.1	13.2	19.5
Canada	0.6	0.4	78.1
China	12.8	2.5	30.3
Egypt	22.7	22.7	44.8
India	5.5	5.1	26.8
Japan	27.4	24.2	77.9
North Korea	12.0	11.7	61.3
South Korea	21.4	20.5	77.6
USA	1.4	0.6	76.2

Population densities were calculated from 1991 statistics for cropland, 1989-91 statistics for areas of 'permanent pasture' and 1995 estimates of populations. Australian population densities are derived from Australian Bureau of Statistics (1992a). All 'Urban Proportions' are estimates for 1995 (World Resources Institute, (WRI) 1994). The definition of 'urban' is uncertain and varies from one country to another. Countries listed include only those with 'significant' agriculture, defined arbitrarily in this case as having more than a million hectares of cropland. Because 1995 population estimates are used, cropland densities are slightly higher than those listed (as reciprocals) by WRI.

Notes

Notes with Chapter 1

1 It is sometimes wrongly assumed that photosynthetic eubacteria ('ordinary' or 'proper' bacteria), which almost certainly preceded the cyanobacteria, were responsible for oxygen evolution (see, for example, Calder, 1984). If the physiology and biochemistry of those organisms were similar to those of their modern counterparts, then they had no part in it. Unlike cyanobacteria, eucaryotic algae and higher plants, photosynthetic bacteria do not split water with the liberation of oxygen. Instead they use either hydrogen sulphide or some organic compounds as an 'electron donor' and release either elementary sulphur or an oxidised form of the organic compound. Cyanobacteria, or organisms very like them, have been found in Early Archean basalt of north-west Western Australia and dated to at least 3465 million BP (Schopf, 1993) but it is unlikely that the atmosphere contained significant concentrations of oxygen until about 2300 million BP (see Kasting, 1993). At the time I write, the oldest known macroalga seems to be *Grypania spiralis*; it was discovered in Michigan and is about 2100 million years old (Han and Runnegar, 1992). Even though it was almost certainly photosynthetic it needed some atmospheric oxygen (to sustain its dark metabolism), but not at present concentrations. Oxygen began to approach present atmospheric levels about 2000 million BP (Kasting, 1993).

2 Some computer models use mathematical procedures that are functional homologues of pathways found in ecosystems. For that reason, they can help our general understanding of how complex biological systems work. But unless sufficient quantitative information can be obtained to incorporate into these models, and unless the models are sufficiently comprehensive to represent all the critical components of a system, their predictive value is likely to be restricted to broad generalisations that can often be matched by intelligent guesswork.

3 *E.g.* Flood (1980).

4 Gladkih *et al.* (1984).

5 For example, Bellwood *et al.* (1992), Diamond (1997), Flood (1990), Grigg (1974), MacNeish (1991).

6 MacNeish (1991).

7 Cohen (1977).

8 Even-handedness requires another dialogue at this point, but a thorough search of the records has revealed only one comment, uttered when a foraging band realised that it had been beaten by another to an expected harvest of wild grain.

"Bugger! We'd better start growing our own."

9 See also MacNeish (1991); Cowan and Watson (1992).

10 Frisch (1988).

11 Harrison *et al.* (1977).

12 Cohen (1989).

13 Wahlquist (1982).

14 Flood (1989). The significance of middens can be overestimated. For example, the energy available from a red deer carcase is roughly equivalent to 52000 oysters, 157000 cockles or 31000 limpets (Bailey, 1978, cited by Barker, 1985). Bailey and Barker, however, left out the 'roughly' and equated the deer carcase with 52267 oysters! Nevertheless, some middens have been carefully evaluated. A group of about 500 middens near Weipa on the Cape York peninsula, Queensland, contains shells from some 9 billion cockles. Accumulation began about 1200 BP, producing on average about 27 tonnes flesh annually, enough to feed 18 people a year (Flood, above).

15 See, for example, Ager (1992).

16 Positive feedback loops can occur in many types of situation. For example, they be found in virtually all forms of addiction. Under these circumstances the feedback loop engenders progressively increasing consumption of the addictive substance which, if not interrupted in some way, will probably lead to the death of the addict.

Vicious circles can also work in a 'negative' direction as, for example, when an animal or a community does not have enough energy to let it obtain the additional energy that it needs for survival. That, of course, is a technical description of starving to death; the general type of process is also commonly known as a 'downward spiral'.

17 The doubling time (in years) of a population that is growing exponentially can be calculated by dividing 69.3 by the percentage annual growth rate. See Appendix for the derivation of this relation.

18 This principle has its theoretical basis in thermodynamics. The critical property is entropy which is effectively a measure of disorder. A dilute, or widely dispersed resource has a high entropy and needs a correspondingly high expenditure of free energy to concentrate it, that is to lower the entropy to a point where the material, whatever it might be, can be used. There are obvious industrial/economic examples of the principle. The equation underlying the generalisation is: $\Delta G = \Delta H - T\Delta S$ where G is the Gibbs free energy, H the enthalpy, loosely defined as the total heat, T is temperature and S the entropy. The symbol Δ, denotes the change in the property which it precedes. A textbook on thermodynamics should be consulted for the derivation of this equation. See also Note 16.

19 The inhabitants of a site in Syria, settled about 11000 BP and known as Tell Abu Hureyra, subsisted for more than a millennium on cultivated plants and hunted animals, principally gazelles. In its first phase, which lasted for about 1000 years, the population was some 200-300. It was then apparently vacated for about five centuries and resettled in the Neolithic period, around 9500 BP with a population in the range of 2000-3000 subsisting as before except that about 20% of the animals consumed now included sheep and goats, which might have been herded. About 1000 years after the Neolithic resettlement of the site, gazelle numbers were reduced to a very low level and the inhabitants resorted seriously to animal husbandry (Legge and Rowley-Conwy, 1987). Cohen (1997) also makes the point that, in general, there was a delay of at least 2000 y between the first use of wild cereals and the first use of cereal 'domesticates'.

20 Boyden (1987).

21 See, for example, Grigg (1982).

22 Handwerker (1983) cited by Bogucki (1988).

23 It is possible that the proportion of the nitrogen isotope, ^{15}N, in tissues of infant bodies and in skeletal remains can indicate whether or not a child has been weaned. Preliminary evidence suggests that the advent of maize cultivation in the American Great Plains did not affect the time to weaning (see, for example, Ross, 1992).

24 In the European area, average Palaeolithic heights were 170-174 cm for males and 156-157 cm for females. By the Neolithic, these had declined to 164-167 and 153-154 cm respectively. In the Mediterranean region the corresponding heights were 177 and 166 cm for the Palaeolithic, dropping to 169 and 156 cm in the Neolithic (see Cohen, 1989), At the beginning of the 19th century, the average height of English farm workers was 155 cm for males and 145 cm for females (Whitlock, 1990). There is other interesting information to be drawn from skeletal remains. For example, some archaeologists are willing to infer an agrarian existence merely from a high incidence of dental caries even without corroborating evidence from food refuse (Cohen, 1989).

25 Bogucki (1988); Cavalli-Sforza *et al.* (1993). See also Diamond (1997).

Notes with Chapter 2

1 Any ecology text will (should) give an account of succession. It is discussed in some detail, for example, by Odum (1971) who bases his description on succession in northern hemisphere ecosystems.

2 A simple physical illustration of the difference between an equilibrium state and a steady state can be given by two boats, one of which has a hole in the hull but is kept afloat by the use of a pump; the other's hull is intact. Both boats retain similar positions in the water but one has to expend energy to do so and is in a steady state. The other is completely at rest, that is to say, at equilibrium.

3 The Second Law of Thermodynamics can be expressed in several ways. One of the more commonly used definitions was formulated by the physicist, R. Clausius, in the middle of the 19th century. In essence it says: '*It is impossible to construct a machine which is able to convey heat by a cyclical process from one reservoir (at a lower temperature) to another at a higher temperature unless work is done on the machine by some outside agency.*' A major thermodynamic stumbling block in the way of reversibility lies in the property of entropy (see Chapter 1, Note 18).

4 Despite the irreversibility of complex systems such as these, it is quite possible for some individual reactions within a system at steady state to be freely reversible—and hence at equilibrium. For example, if the reaction:

$$\rightarrow A + B \rightleftharpoons C + D \rightarrow$$

occurs within a complex system such as a metabolic pathway, it is quite possible for the system as a whole to determine that the concentrations of the four reactants, A, B, C and D, are such that the reaction is genuinely at equilibrium and is freely reversible, even though the larger pathway within which it is embedded is not.

A rough hydraulic analogy can be seen in a river, part of which is dammed. The river in its entirety flows into the dam, over the spillway and beyond. Overall the flow is irreversible; that is to say, the water cannot be returned to a point upstream without the application of energy. Within the dam, however, the flow rate is so slow that, for most practical purposes, the dammed water is at rest, or, in other words, at equilibrium. The bigger the dam in relation to the total volume of water in the river, the closer will that approximation be to an accurate description of the state of the water.

5 The relative nature of the constancy or stability needs to be emphasised. The system is forever in a state of change but, during the early stages of development, the direction and rate of change are dominated by the processes of succession. In a mature ecosystem, the biomass, quantitative species composition, total metabolic activity etc will sometimes oscillate with a fair degree of regularity, sometimes change chaotically for intrinsic as well as seasonal and climatic reasons. Mathematical models can provide a general theoretical basis for the oscillations (and other changes) but, because of the complexity of mature ecosystems, it is probably never possible to predict accurately the precise form of the various changes that they undergo. In general terms, however, the causes of oscillations (and other changes) may be said to include the different kinetic characteristics, specifically the different 'rate constants' of the innumerable processes or reactions that constitute an ecosystem—or any other complex dynamic system, including physiological and socio-economic systems. See also Keller *et al.* (1996).

6 This is a reason why you are wasting your money if you have your body frozen after death in the hope of some future resurrection. The genetic material, DNA, is subject to damage from background and cosmic radiation, but if you are frozen the normal repair mechanisms do not function. Give your money to a good charity instead.

7 Recognition of this is apt to cause dismay among conservationists and glee in the forest industries, the latter using it to justify even such extreme practices as clear felling on the grounds that subsequent regrowth will consume more carbon dioxide and produce more oxygen than the old forest. The situation is not nearly so simple, however. The total balance will depend in the first place on what happens to the trees that were destroyed. If they are used in such a way that a significant proportion of the wood or foliage is rapidly degraded as by burning, by natural decomposition, or even by paper manufacture, the amount of carbon dioxide released will exceed anything that young trees can assimilate for many years. Even though a young plantation or new natural regrowth will have a high P/R quotient, the total nett rate of photosynthesis is low simply because the plants are small and the total plant biomass is low.

8 See also Note 5.

9 *E.g.* Bogucki (1988).

10 *E.g.* Cohen (1977).

11 *E.g.* Flood (1989).

12 Mellars and Reinhardt (1978) cited by Bogucki (1988).

13 Especially in arid zones, Aboriginal Australians used fire to produce a complex mosaic of burnt and unburnt patches. Among other things, the mosaic was an important factor in the survival of medium-sized mammals in periods of drought. The burnt patches themselves were simplified by the burning and underwent a 'normal' process of succession during regeneration. The effect over the wider area, however, was a pattern that was more complex than would have been the case had the burning not taken place (Morton, 1990). See also Pyne (1991) and Whelan (1995).

14 *E.g.* Creagh (1992).

15 A useful mnemonic, for those who have encountered the subject or a university department so named is 'Oh Christ! Not History (and) Philosophy (of) Science!'

16 The general equation depicting the exchange of oxygen and carbon dioxide with carbohydrate is:

$$[CH_2O] + O_2 \rightleftharpoons CO_2 + H_2O$$

The 'forward' reaction denotes respiration and is accompanied by the release of energy. The 'reverse' reaction denotes photosynthesis and requires energy—supplied in this case by light. The formula $[CH_2O]$ represents carbohydrate without specifying its type or the number of carbon atoms it contains.

17 Denitrification is represented by the following general equation:

$$2NO_3^- + 12H \rightarrow N_2 + 6H_2O + 2e$$

where the 12 hydrogen atoms are derived from the energy metabolism of the bacterium and the 2e on the right side of the equation represents the two excess electrons (negative charges) of the nitrate ions on the left side. Nitrogen fixation is represented by the following general equation:

$$N_2 + 6H \rightarrow 2NH_3$$

As with denitrification, the hydrogen atoms are derived from energy metabolism. The ammonia formed in this reaction reacts with an organic acid, α-ketoglutaric acid, to form the amino acid, glutamic acid and its amine, glutamine, from which the nitrogen can then be passed on to other compounds.

18 The amount of sulphur in the atmosphere is estimated to be about 0.036×10^8 tonnes (see Blair, 1980). This amounts to about 7×10^{-10} of the mass of the atmosphere. Although biological activity can release volatile sulphur compounds [predominantly but not wholly as dimethyl sulphide, $(CH_3)_2S$, carbonyl sulphide (COS) and hydrogen sulphide (H_2S)], relatively quickly from soils and aquatic sediments, it is not conceivable that sulphur could be recovered biologically from such a high dilution at a rate that would have any significant effect on the overall cycling rate of sulphur. This is likely to be true notwithstanding the higher concentration of these gases at low altitudes. But that is not the whole story. Hydrogen sulphide is oxidised in the atmosphere to sulphur dioxide (SO_2) which, in turn, is converted through several steps to sulphuric acid (H_2SO_4). This is not a gas but it can remain suspended as an aerosol until it eventually settles or is brought to earth in rain. In addition, however, there is a very substantial man-made emission of sulphur dioxide from the combustion of fossil fuels, mainly coal. For example, in 1989 the total emission of sulphur dioxide in Central Europe, the then USSR, Western Europe and the USA amounted to 56 million tonnes—of which the USA was responsible for 37% (World Resources Institute, 1992). The residence time in the atmosphere of sulphur dioxide (and sulphuric acid) is relatively short—usually about a day—but it is sensitive to the height of the smoke stack from which it was emitted, wind velocity and rain. Because of the short residence time, acid rain, which is toxic to plants and aquatic animals, is largely confined to the industrial regions of the Northern Hemisphere and to inoffensive countries that are unlucky enough to lie down wind of them. Blair (1980) has commented that the total sulphur dioxide emissions "in all of man's recorded history" are less than that of a single major volcanic eruption. A large proportion of volcanic sulphur, however, is carried into the stratosphere and distributed widely over the planet before it eventually settles over the course of some years.

19 Phosphorus pentoxide (P_2O_5) is carried in particulate form in ash from forest and other fires. If the particles are fine enough and the convection from the fire powerful enough, they can be carried over considerable distances before settling. But settle they will, as likely as not into the sea. An estimate of phosphorus loss caused by low intensity fires is about 300 mg $P.m^{-2}$ from a fuel bed of 1.7 $kg.m^{-2}$ (Milthorpe, 1982).

Two volatile phosphorus compounds are produced microbiologically by the reduction of phosphate (PO_4^{3-}) under anaerobic conditions in swamps and marshes. They are phosphane (PH_3) and diphosphane (P_2H_4). Diphosphane is spontaneously inflammable and is probably the source of the ignition of methane to produce the 'will o' the wisp' phenomenon. Combustion of phosphane and of diphosphane produces P_2O_5, which being particulate, will soon settle out in the absence of powerful convection currents. Small quantities of phosphane are also formed in the digestive tract of animals and expelled in flatus. It continues to be produced in small amounts in dung. In the absence of ignition, spontaneous or otherwise, the phosphane has the capacity to be carried considerable distances in the atmosphere but will progressively oxidise to P_2O_5 and eventually settle. See also Chapter 11.

20 The relevant metal elements also lack volatile metabolites and, at least in principle, are used irreversibly in agriculture. Calcium, magnesium and potassium, the three that are used most prolifically in fertilisers, are more abundant than phosphorus and are unlikely to be exhausted before that element.

21 Odum (1971).

22 *E.g.* Olsen and Dean (1965).

23 Pierce (1990); Ensminger (1991) cited by Rifkin (1992).

24 There seems to be no generally accepted "…ivore" term for coprophagic animals—which is perhaps an oversight on someone's part.

25 Silvester and Musgrave, (1991).

Notes with Chapter 3

1 The evidence in these matters is equivocal. For example Grigg (1974) comments "... it can be argued that herding animals everywhere seem to have been domesticated by sedentary agriculturalists, whilst animals living in areas occupied by hunters and gatherers were not domesticated; that the distribution of pastoralism is peripheral to sedentary agriculture, whilst nowhere have there been pastoral nomads who were without knowledge of agriculture." But in the next sentence he adds, "It must be noted however that the earliest evidence of domesticated sheep and goats in South West Asia occurs in sites lacking evidence of domesticated plants." Elsewhere he notes that "The first archaeological evidence of domesticated cattle is from Greece and is dated at 8500 BC or (sic) 6500 BC." Animal domestication preceded plant cultivation in Egypt by perhaps a thousand years (see Stanley and Warne, 1993); animals and plants were domesticated at about the same time in the 'Fertile Crescent' (see Harlan, 1971). See also Maisels (1990).

2 I have omitted from this list the unused portions of plants left in the field when a crop was harvested. Such plant residues would benefit the soil in a number of ways, including offering protection against erosion, but unless the plants were legumes, their contribution to nutrient flow could be seen as reducing the total removed rather than supplementing the total returned. This is a fine and somewhat pedantic distinction, of course; it is made in the interests of simplicity rather than for any other reason.

3 Barker (1985).

4 E.g. Bogucki and Grygiel (1983); Wells (1983); Price and Petersen (1987).

5 Arrhenius (1938).

6 Provan (1973).

7 Infant mortality of sedentary agricultural communities was evidently higher than that of nomadic hunter-gatherer societies, at least partly because of poor sanitation (see Chapter 1).

8 Dawson (1881).

9 Grigg (1974).

10 For example, Barker (1985), Bogucki (1988), Gras (1925), Grigg (1974, 1982)—and many others.

11 The details, needless to say, are more complex and, among other things, will depend on the nutrient status of the soil of each area and the response of the crops on the one hand and the pasture on the other to changes in that status. If, however, the effects of the transfer of manure are assessed by direct estimation of soil nutrients, then the consequences will be as described in the text.

12 Rollefson and Köhler, summarised by Anon (1990).

13 Jacobs (1969).

14 For example, Boyden (1987), Cohen (1977).

15 Saunders (1992) states that Roosevelt has exploded "the myth that tropical rainforests never gave rise to true civilization". He is referring to a complex culture in lowland Amazonia.

16 Average present-day annual rainfalls in regions of early civilisations are: South West Asia, less than 300 mm, but the Taurus and Zagros mountain ranges, that feed the 'Fertile Crescent', have higher falls, exceeding 1000 mm in places; Egypt, 250 mm; The Indus Valley (from Harappa to Mohenjo-Daro), less than 300 mm in the north to *c* 750 mm in the south; China (Yellow River), *c* 650-900 mm; Peru (Chimú and Inca), less than 250 mm (the Chimú constructed canals to bring water more than 70 km through mountainous country to their farms (Ortloff, 1988); Central America, from *c* 800 - more than 1750 mm over the range occupied by the Maya. Notwithstanding the official rainfall of the areas, parts of the Yucutan peninsula occupied by the Maya are described as having "... scant rainfall and scrub vegetation" (Hammond, 1986).

17 *E.g.* by Knapp (1988).

18 Needless to say there are computer models for assessing the rates of accumulation of surplus given various assumptions of yields and frequencies of crop failures (see, for example, Gregg, 1988).

19 One of the less plausible but more entertaining explanations of the 'failure' of aboriginal Australians to adopt agriculture was suggested by Randhawa (1980). The 'explanation' was that Australia lacked native mammals that could be used as draught animals. Given that the stump-jump plough was invented in Australia—albeit after the arrival of Europeans—it is perhaps a pity that more imagination was not applied to the potential of kangaroos as draught animals.

Notes with Chapter 4

1 Boyden *et al.* (1981).

2 *E.g.* Gimpel (1992).

3 24,000,000 Africans are estimated to have been transported to the Americas by the British; less than half of them survived the voyage across the Atlantic (Walvin, 1992).

4 See Finley (1980) for more detail and comment on the various aspects of slavery referred to in the preceding paragraphs.

5 The trade in slaves between sub-Saharan Africa and the Mediterranean began in Ancient Egypt. Slaves were supplied from markets in Djibouti to the Arab emirates in the 1950s and slavery of Africans by Africans reportedly persisted in a number of sub-Saharan countries at least until the late 1970s (Hughes 1993).

6 It is estimated that there were some 55 million children in enforced labour in India in the early 1990s (*Background Briefing*, Australian Broadcasting Commission, 4th January, 1994, first broadcast April, 1993; this is also the source of the statement about child labour in Britain and the USA).

7 The use by the Japanese of World War 2 prisoners for construction work had an ideological motive probably no less important than the economic one.

8 Mumford (1961).

9 R. Lanciani, cited by Mumford (1961).

10 Rome was devastated by plagues in 23 BC and 65, 79 and 162 AD.

11 See Moss (1972) and Mumford (1961).

12 There is a close parallel here with the practice of medicine when medication produces side effects that might be serious enough to prompt supplementary medication to counteract those side-effects—and so on. The medical profession, however, is probably more aware of the phenomenon of side effects that most other professions dealing with complex dynamic systems.

13 Some modern buildings can hold populations much greater than most ancient and many modern cities. For example, the World Trade Center in New York, damaged by a bomb in February 1993 and destroyed by aircraft in September 2001, held up to 100,000 people, the population of ancient Athens.

14 For some other descriptions and definitions see, for example, Johnson and Lewis (1995); McTainsh and Boughton (1993).

15 For the sake of simplicity I include salination with the accumulation of 'toxic substances'. Sodium chloride, the commonest of the salts implicated in salination, is not specifically toxic in any usual biochemical sense as, for example, are salts of lead or cadmium. Indeed, at low concentration, both sodium and chloride ions are essential for all forms of life and, at relatively high concentrations, are essential for some specific forms. The toxicity of high salt concentrations is partly a result of lowering the water availability—the water 'potential'—and partly a direct physico-chemical effect of the ions and undissociated salts on both cell and organismic physiology (see Brown, 1990).

16 Loomis (1978). The nitrogen removed in the crop was taken as 20 kg (this is equivalent to a protein content of the grain of about 12.5%) and the replacement from natural sources as 14-26 kg comprising (in kg $N.ha^{-1}.y^{-1}$): dust, rain and bird droppings, 8-12; seed grain, 4; leguminous weeds, 2-10. In addition Loomis allowed a minimum of 5 kg nitrogen from farmyard manure.

17 It can be an oversimplification to ascribe a rate- or growth-limiting role to a single element. For many practical purposes, however, it is reasonable to do so provided that basic physical requirements such as adequate light, water, temperature and pH are met.

18 Something of this sort may well have been implicated in the results of a series of measurements of yield in manured and unmanured experimental plots at Woburn, England over the five decades from 1877-1927. Barker (1985, citing Rowley-Conwy, 1981) illustrated the results graphically to make the point that the manured fields sustained average yields of barley and wheat about twice those that were not manured. The yields of both types of grain under both sets of conditions progressively declined, however, over the fifty years. For example, using numbers interpolated from Barker's graphs, the yield of unmanured barley dropped at an average rate of about 25 $kg.ha^{-1}.y^{-1}$ over the course of the experiment (the correlation coefficient of the linear regression to which the interpolated points were fitted was -0.76). The mean yield and standard deviation of unmanured barley over the period was about 840 ± 470 $kg.ha^{-1}.y^{-1}$. The corresponding values for unmanured wheat were; rate of drop in yield, about 12 $kg.ha^{-1}.y^{-1}$ (correlation coefficient, -0.56); mean yield and standard deviation over the five decades, approximately 730 ± 300 $kg.ha^{-1}.y^{-1}$. As Barker noted, the manured yields were about twice those of the unmanured crops but the rate of decline was about the same. The very high standard deviations result predominantly from annual fluctuations rather than the progressive decline in yield.

19 Knapp (1988).

20 Jacobsen and Adams (1958).

21 The density of bulk wheat, and of other cereals of similar grain size is about 0.77 kg.l⁻¹.

22 Jacobsen and Adams (1958).

23 Stevens (1981).

24 National Farmers Federation (1992). See also Table A4.

25 See, for example, Hillel (1992), Jacobsen and Adams (1958), and Knapp (1988).

26 Herodotus, *The History. Book 2*, cited by Stanley and Warne (1993a).

27 Stanley and Warne (1993b).

28 Hillel (1992).

29 One mm of soil is roughly equivalent to 1.3 tonnes.ha⁻¹ (see Stocking, 1987). Given the uncertainties inherent in measurements of this kind, it is difficult to know if the apparent difference in rates of silt deposition between the Nile and the Mesopotamian rivers is significant. Modern flow rates are instructive, however. Over the period 1912-1962 the Nile flowed at an average rate of 2650 m³.s⁻¹, peaking in September at an average rate of 8180 m³.s⁻¹ (measured at Aswan). These measurements were taken after the construction of the Aswan Low Dam in 1902 and encompassed its modification in 1912 and in 1934. The completion of the High Dam in 1964 effectively ironed out the spring peak. The Tigris, over the period 1931-1966 averaged 1320 m³.s⁻¹, peaking in April at 3210 m³.s⁻¹ (measured at Fatha, Iraq). The Euphrates, over the period 1932-1966, averaged 906 m³.s⁻¹, peaking in May at 2380 m³.s⁻¹ (measured at Hit, Iraq) (Beaumont, 1981). Thus the average annual flow rate of the Nile was about 19% greater than that of the other two rivers combined. The peak flow rate was about 46% greater than the combined peak flows of the two Mesopotamian rivers.

Notes with Chapter 5

1 For example Boserup (1965).

2 Grigg (1974); Spencer and Thomas (1978) cited by Goudie (1986).

3 Grigg (1974).

4 Grigg (1982).

5 Bray and Trump (1982); Spencer and Thomas (Note 2).

6 Weiss *et al.* (1993).

7 Randhawa (1980).

8 See, for example, Grigg (1982).

9 For example, Barker (1985), Gras (1925), Grigg (1974, 1982), Slicher van Bath (1963), Whitlock (1990)—and many others.

10 There could be exceptions if a second crop were susceptible to or, for any other reason, encouraged the growth of pathogens that attacked the first crop. Cooter (1978) has suggested that bare fallow periods of 2-3 years might be needed to remove pathogens (with a reasonable degree of confidence).

11 The extent to which soil or its constituent nutrients are lost under these conditions is, of course, heavily dependent on the type of soil, topography and rainfall. Pierce (1990, Fig. 7.5) cites an instructive comparison from Hawaii. On flat ground bare fallow lost ten times as much total nutrient as a ploughed cowpea—maize rotation. On a slope of 7%, the bare fallow lost eleven times as much. There was severe soil erosion throughout much of Western Europe in the late Middle Ages and again from the second half of the 17th to late in the 18th century, possibly in response to the widespread use of the bare fallow. At the end of the 18th century fodder crops replaced the bare fallow and there was subsequently much less erosion. There were probably also other factors involved in the widespread soil degradation of the 17th century—including another pandemic of bubonic plague, a high incidence of infectious diseases generally (see Fig. 5.1) and a protracted economic depression with a profound effect on the price of farm produce. (Slicher van Bath, 1963).

12 Cooter (1978); Pretty (1990).

13 Barker (1985). Over the period 1500-1820, European yield/seed quotients, averaged for wheat, barley, oats and rye, changed in the following way: England, France and the Low Countries, a progressive increase from 7.4 to 11.1; France, Spain and Italy, fluctuations within the range 6.2-7.0; Germany, Switzerland and Scandinavia, an overall increase from 4.0 to 5.4; Eastern Europe and Russia (1500-1799), fluctuations within the range 3.9-4.7 (Grigg, 1982). Slicher van Bath (1963) lists yield/seed quotients for wheat, barley, oats, rye and peas for various (predominantly western) European countries from mid 16th—early 19th century. The average quotients for wheat for that period ranged from 2.5 to about 28. The highest quotients were recorded for the Netherlands in the late 18th century.

14 Pretty (1990).

15 Compare with Table A4.

16 Average annual milk production in modern India is about 250 l/cow; in the USA it is about 6750 l/cow (Pierce, 1990). A good yield from a grass-fed cow in Australia is of the order of 4000 $l.y^{-1}$.

17 Slicher van Bath (1963) has this comment about feudal farm productivity:
  "The support of even a few people required a great deal of land. Thus the property of the chapter of St Symphorien at Autun contained about 100 farms, yet their produce was barely enough for the sustenance of 15 canons and a few servants."

18 Denitrification in the strict biochemical sense of reducing nitrate or nitrite to elementary nitrogen as described in Chapter 2 is not likely to be very important here. Neither nitrate nor nitrite occurs to any significant extent in either faeces or urine. If the dung heap also contained urine, the most important chemical reaction leading to loss of nitrogen is likely to be the hydrolysis of urea to ammonia with the subsequent loss of some of the ammonia by evaporation and/or leaching. The equation is:

$$(NH_2)_2CO + H_2O \rightarrow 2NH_3 + CO_2$$

19 Definition of 'feudal' and 'feudalism' has engendered a good deal of academic debate among historians. A succinct description of the system has been given by Dodgshon (1987):

"For the great mass of society, feudalism was ... about labour services, feudal dues, and servitude. At this level, or at the level of the ordinary peasant, it was a system of exploitation and appropriation.

"At the root of feudalism as a system of exploitation and expropriation was the dependence of peasants on lord. These ties of dependence ranged from outright slavery, through varying degrees of serfdom and unfreedom, to being one simply of jurisdiction by a lord over a peasant."

Slicher van Bath (1963) comments that, among other things, a feudal system functioned as 'a natural economy', dependent on payment in kind. He also points out, not surprisingly, that the structural and organisational arrangements of feudal estates varied throughout the different regions of Europe.

20 See, for example, Gras (1925).

21 Ralph Whitlock (1990) describes some calculations made in 1826 by "that old political firebrand William Cobbett" in the following way:

"He then turns to the needs of an average family of agricultural labourers—'a family of five persons; a man, wife and three children, one child big enough to work, one big enough to eat heartily, and one a baby'. Such a family, he asserts, would want five pounds of bread, a pound of mutton, two pounds of bacon and, on average, a gallon and a half of beer per day. (He takes no account of green vegetables which, he says, '...human creatures ought never to use as sustenance!'). He regards the quantities he allows as by no means excessive: '...just food and drink enough to keep working people in working condition'. The cost at prices then current would amount to £62.6s.8d. per year. Yet at the agricultural wages then prevalent (and he allows 9 shillings per week as against the 7 shillings more usual in Wiltshire), the annual income of the family would be only £23.8s. The lot of the agricultural worker he therefore describes as 'skin, bone and nakedness' ...

"Summing up, he points out that the farms of Milton were producing enough food (on his allowance) for 2,510 persons or 7,500 if they were to be fed at the same rate as the half starved labourers who produced it."

I have referred to the average height of English farm workers from this period in Chapter 1. For several centuries earlier, however, English farm workers had eaten relatively well (again, see Whitlock).

22 See Barker (1985).

23 Langdon (1982).

24 Slicher van Bath (1963).

25 Unless stated otherwise I have taken population sizes, here and elsewhere, from McEvedy and Jones (1978) and, specifically for mediæval Europe, from McEvedy (1992). McEvedy and Jones have outlined the uncertainties in their numbers; those uncertainties should be taken as applying equally to populations quoted herein. See also Fig. 5.1

26 But see Langdon (1982).

27 Ladurie (1978) has provided a remarkable account of life and work in the Pyrenean village of Montaillou over the period 1294-1324.

28 See also Wilkinson's (1973) discussion and interpretation of economic 'progress'—while bearing in mind that his term, if not his concept of 'ecological equilibrium', is misleading (see Chapter 2).

29 Whitlock (1990).

30 This aspect of English rural history has been discussed at length. See, for example, Newby (1987), Whitlock (1990).

31 The first English patent for a threshing machine was granted in 1788, the year of colonisation of NSW. The machines were in common use by 1820 (Whitlock, 1990) and became the focus of sabotage by rural workers who saw them as yet another threat to their livelihood. It would nevertheless be misleading to imply that the mechanisation of agriculture was necessarily the dominant factor in migration to towns. Indirect effects of other areas of mechanisation were sometimes no less important. For example, as Wilkinson (1973) has reminded us, the increasing mechanisation of transport from the late 18th century, even at the level of making bicycles, progressively reduced the need for horses and thus enabled pasture that had previously fuelled transport to be converted to cropland. In addition, of course, there were economic, social and psychological factors that profoundly influenced migration from the country to towns. But notwithstanding all of that, the movement of people to the cities would not have been possible had it not been underpinned by an agricultural surplus.

32 Newby (1987). Coward (1988) comments that in 1790 rural labourers in Britain "outnumbered urban workers by two to one: by 1831 that ratio had been almost reversed". See also Note 34.

33 Mitchell and Deane (1962) cited by Wilkinson (1973).

34 Reay (1990). Slicher van Bath (1963) cites the following numbers for the proportion of 19th century western European populations actively working in agriculture, forestry and fishing: England: 1831, 28%; 1871, 21%: France: 1827, 63%; 1866, 43%: Germany: 1882, 42%: Italy: 1871, 51%; 1881, 46%.

35 Singer (1993) has summarised the manner in which these two religious teachers have influenced modern rationalisation and justification of greed.

Notes with Chapter 6

1 Unless, of course, he is constipated. Even so that condition will merely delay, not eliminate, the full expression of the problem.

2 In general, pathogens are more likely to be virulent if they can satisfy either of two sets of requirements. The first is that they are normally transmitted to a new host before the first one has recovered or died. The second is that they have an effective survival mechanism outside the body of the (primary) host. This can be achieved through an intermediate host/vector, through the ability to multiply in some other medium such as water, or simply through some form of dormancy such as resistant spores, eggs, or durable viral particles. (See also Ewald, 1993).

3 Cohen (1989).

4 The same argument applies to other species, of course. In particular, herbivores that maintain(ed) very large herds, often densely packed, as well as bats and various species of marine birds that nest in dense colonies are likely to experience serious epizootics from time to time. Goldsmid (1988) has summarised the infectious diseases to which aboriginal Australians might have been exposed before European occupation.

5 Human occupation of caves in the Levant and elsewhere during the Pleistocene was accompanied by rodents (see, for example, Bunney, 1994). During such interludes the human inhabitants could have been exposed to rodent-borne zoonoses, including bubonic plague. Movement away from the cave should normally have hastened the end of an outbreak.

6 The surviving populations would have had enhanced immunity to the organisms encountered under these conditions. Enhanced immunity and its corollary, the ability to carry virulent organisms and viruses asymptomatically in a travelling community, was at least as important as firearms in enabling Europeans to overcome aboriginal populations in Australia and especially in America. The devastating susceptibility of indigenous Americans to European diseases brought by Columbus and later by Cortés has been attributed to the lack of close contact with domestic animals of the types prevalent in Europe (see, for example, Meltzer 1992). Indeed, Meltzer makes the additional point that, although the Aztec capital, Tenochtitlán, had a population in 1492 estimated at about 200,000, it had no 'crowd diseases'. I cannot believe that and, in any case, recent genetic evidence has demonstrated beyond reasonable doubt the indigenous existence of tuberculosis in Peru (Holmes, 1994a). Moreover, an absence of indigenous 'crowd diseases' is not a necessary condition for susceptibility to imported diseases; all that is required, assuming there is no innate immunity, is a lack of previous contact with those specific diseases. Unless there is some kind of cross immunity, contact with a different range of diseases is largely irrelevant. In addition, however, it seems that indigenous Americans were genetically 'unusually homogeneous' (Black, 1992). The effect of this would be to ensure that patterns of innate susceptibility to infection were very widespread throughout the population.

The only redress that the American Indians are thought to have had was to send syphilis to Europe through the agency of the Spanish sailors. Archaeological evidence of possible syphilitic bone lesions found in England and dated earlier than 1492 has now thrown some doubt on this assumption (Culotta, 1993: Zimmer, 2001).

7 Retrospective diagnosis of infectious diseases is not without its uncertainties—and its critics. For example, Twigg (1985), cited by Dixon (1994) has suggested that the Black Death was not bubonic plague at all, but anthrax.

8 McEvedy (1992).

9 Fig. 5.1 does not show the full impact of this because of the 100-year spacing of the points on that part of the curve. Neither does this average toll for Europe convey the whole story, of course. The death rate for the British Isles was closer to 50% according to some estimates (see Keys, 1993). In Italian towns, between 25 and 40% of the population died within a few months of the arrival of the disease (Cipolla, 1992).

10 The notion of 'contagious' diseases, that is diseases that could be spread by direct contact with someone who was sick or with objects he had touched, is very old. The ancient Hebrews saw sickness as a punishment from heaven but nevertheless saw fit to isolate lepers, discard 'unclean' items or materials and to avoid certain types of food such as pork and shellfish. The transmissible nature of infectious diseases was recognised by writers such as Titus Lucretius Carus in the first century BC and by Giovanni Boccaccio in the 14th century AD.

Fracastorius di Verona, who is credited with founding the discipline of epidemiology in the 16th century, attributed the transmission of infectious disease to 'seeds' (seminaria), probably the first recorded suggestion of a particulate agent as the cause. Nevertheless, he accepted the conventional wisdom that the seminaria were provided in the first place by supernatural forces. Despite Fracastorius's sound reasoning and the impressive amount of evidence he accumulated, the Miasma Theory continued to hold sway largely because of the recognition that disease could also be transmitted through the air, a process that at the time could not be reconciled with particulate infectious agents.

There were many milestones along the way to the proposal and ultimate acceptance of the 'germ theory' of disease; most provided only indirect evidence whose real significance was not recognised until much later. Some of the more important such milestones were the first microscopic sighting of microorganisms in the 17th century by the Dutchman, Antony van Leeuwenhoek, the direct transmission of infectious material in the 18th century by the surgeon, John Hunter, who inoculated himself with exudate from a patient with gonorrhea but unfortunately gave himself syphilis at the same time, the process of vaccination refined by Edward Jenner in the late 18th century from the widespread and much older technique of variolation, the first direct demonstration (by Agostino Bassi in 1836) of the transmission of a pathogen, in this case a fungus that infected silkworms, the ending of a localised cholera epidemic when John Snow closed a contaminated public water pump in Broad Street, London, in 1854, the introduction of antiseptic surgery by Joseph Lister in the 1860s and ultimately the definitive work of Louis Pasteur and Robert Koch in the late 19th century.

11 See Curson and McCracken (undated) for the citations.

12 Cipolla (1992).

13 This and the preceding quotation are cited by Cipolla (1992).

14 Spender (1991). Where I have inserted an asterisk in the quotation Spender had a footnote that read "This could be a reference to Thomas Povey's establishment; he had a bathroom at the top of his house and presumably had piped some form of running water."

15 Post (1985).

16 'Flux' was the comprehensive term for dysentery in all its forms.

17 The concept of infection was not in any way incompatible with the Miasma Theory. The infectious agent could stick to skin, clothing etc and, since it was volatile, could be transmitted in breath—no doubt especially when the breath smelt bad.

18 See, for example, Derry and Williams (1960).

19 For example, Dyos and Wolff (1973).

20 Stephens (1849).

21 The problem of rented houses was not confined to Adelaide, needless to say. Wohl (1973) has pointed out that England legislated in 1872 for local governments to appoint medical officers. The local governments (the 'vestries') consisted largely of men determined, in Wohl's words, "to obey cherished precepts of low rates (council taxes) and laissez faire". Many of the vestrymen were the owners of slum properties. Wohl quotes one medical officer who, on his appointment, was greeted by the vestry chairman with the words "Now Doctor, I wish you to understand that the less you do the better we shall like you".

22 From Wrigley and Schofield (1981) cited by Jackson (1988).

23 The Aboriginal population was by then decreasing at what might well have been a comparable rate, that is to say with a half life in the order of 7-8 years. A doubling time of 7.6 years in a population that reached sexual maturity around 14 years of age and had a high death rate can be achieved, of course, only with significant supplementary immigration.

24 Melbourne was settled in 1834. By 1890 it had a population of 491,000, making it perhaps the 22nd largest city in the world (see Chapter 7). In 1891 the official population of Sydney and its suburbs was 378,872 (Coward, 1988). It was inevitable that a growth rate as high as Melbourne's over those first 56 years would produce severe growing pains in its various supporting services. Its sewerage system began effective operation in 1897. In 1890 the death rate was about 20/1000, which was higher than Sydney's for that specific year but right on the regression line of Sydney's rates (Fig. 6.3).

25 Cited by Coward (1988).

26 Until indicated otherwise, the material under discussion is taken from Coward (1988).

27 The terms 'cesspit' and 'cesspool' seem to be largely interchangeable. There would seem to be some merit in reserving '... pit' for when the hole is first dug and '... pool' for when it has been in use for a while.

28 The possibility of being admitted to hospital was an excellent reason for trying to stay healthy. Although probably no worse than British hospitals of the time, Sydney's 'Rum Hospital' fell somewhat short of an ideal place for treating the sick. The hospital was commissioned by Governor Macquarie to meet the needs of the rapidly growing population. Because the British government would not supply funds for the purpose, Macquarie turned to private enterprise. He granted to the hospital's building contractors a monopoly on importing rum. The contractors included the Surgeon-in-Charge, D'Arcy Wentworth who, together with his two fellow contractors, became wealthy from importing rum at 3 shillings a cask and selling it at 40 shillings. Evans (1983) describes the hospital's hygiene thus:

"The hospital was complete by 1816 and patients were moved from the dirty and overcrowded old George Street North Hospital. However even in the new Rum Hospital, the nurses were drunken and debauched convicts. ... The convict nurses stole the patients' food and anything else they could find, and the patients in turn stole from each other. The corridors were full of patients suffering from dysentery, so weakened by the disease that they could only crawl to the outside privies on their hands and knees. At night they were locked up in the wards by the nurses and the sick were left totally unattended by the staff, with no access to the outside privies. Since there were no catering facilities, the evening meal was cooked by the patients themselves in the ward, unhygienically surrounded by the reeking buckets that served as toilets at night. Hygiene was virtually non-existent and the kitchen for the staff doubled also as the mortuary."

29 Calculated from the NSW Registrar-General's vital statistics as cited by Coward (1988).

30 Australian Bureau of Statistics, monthly summary, December, 1993.

31 For example by Goldsmid (1988), Watt (1989), Curson and McCracken (undated), Coward (1988), Watt (1989), Attwood, Forster and Gandevia (1984), Attwood and Kenny (1987).

32 Calculated from statistics cited by Coward (1988). 'Typhoid' and 'diarrhoea' are put in quotes because of uncertainties in the diagnosis of typhoid fever at the time and the range of organisms that can cause the symptoms of diarrhoea.

33 The rate of decline in infant deaths was also apparently linear with a correlation coefficient of -0.95 for infant deaths as a percentage of the total. Calculated from Coward (1988).

34 One of the many consequences of sedentism combined with a restricted water supply and the wearing of clothes was lice infestation. Lice, which have the capacity to transmit some very dangerous diseases (typhus is one of the most notorious), have been a serious human problem throughout history. A convict transported to Australia in the Second Fleet was estimated by the ship's surgeon to have 10,000 lice "upon his body and bed" (see Goldsmid, 1988). Not only did an adequate water supply make comprehensive washing possible but it also changed expectations and standards. Probably the next most important technological development in overcoming lice infestations was the vacuum cleaner.

35 Coward (1988) has discussed the political and social hurdles that had to be jumped in order to dispose of human waste in ways less dangerous than allowing it to accumulate in the backyard cesspit, or throwing it into the street.

36 Scott (1952).

37 The distribution of population between town and country implies an average farm surplus of slightly over 20% (see Chapter 5). I have recalculated the amounts of faeces from Scott's primary figures. They are slightly different from the masses obtained by converting his totals to metric units.

38 A survey conducted by Scott and his colleagues on 81 farms in West Shantung led to the classification of 34 farms and their immediate surroundings as 'Light' and 47 as

'Heavy' according to the level of infestation with ascaris eggs. The percentage of infested samples from different locations, for the 'Light' and 'Heavy' groups respectively, were: Pit Latrine, 85 and 94; Courtyard, 60 and 62; Living Room, 34 and 47; Street, 44 and 47. Scott also tabulated the number of eggs found in the various samples but added that the sampling method used detected only about 1% of the eggs actually present.

Notes with Chapter 7

1 In some cases European farmers achieved such a transfer manually by cutting turf and applying it to another field (see Slicher van Bath, 1963).

2 Wool removes nitrogen and small amounts of sulphur but no other crucial elements in significant quantities.

3 Slicher van Bath (1963) should be consulted for quantitative information about rates of application of manure at different times on various farms. To cite the information here would require more qualification than is warranted. See also Chapter 6 for Cipolla's comments on sanitary standards in northern Italy and elsewhere in Europe in the 17th century.

4 The statistical distribution of population between town and country depends, of course, on definitions, in particular the smallest size of a settlement classified as a town. Lampard (1973) adopted 20,000 as the minimum and went on to make the following observation: "Of fifteen countries, outside England and Wales, with more than 16.9 percent of their population urbanized, all, with the exceptions of the Netherlands and Scotland, had achieved that level since mid-century. By 1891, Scotland with 42.4 and Australia (six colonies) with 40.9 percent of their populations urbanized were the only countries to have surpassed the level of England and Wales in 1851-35 percent." The size of cities also depends on where their boundaries are drawn, a task that is not always as self-evident for modern conurbations as it was for old walled cities. But see also Lampard's comments in Note 13.

The 2:1 urban:rural distribution that I am using is loose in that it refers broadly to farmers and 'others'.

5 Melbourne provides a well documented example of this process. The city was founded in 1834 and grew very rapidly. It went through the inevitable phase of cesspits and these were duly superseded by removable cans. Dingle and Rasmussen (1991) describe how that system was affected by growth of the city:

"Contractors were allowed only to empty pans under the cover of darkness. In many suburbs they could not begin before 11 p.m. and had to be finished by 3 a.m. ... In earlier years the metropolis had been dotted with manure depots to which the nightmen took their cargoes. By the 1880s loads were taken instead to the fringes of the metropolis, to be purchased by market gardeners and used as fertilizer. ... About 150 loads were collected nightly, each load averaging 2.5 tons. Much of it was carted 10 or 15 miles by men who were at work from sixteen to eighteen hours a day.

"By 1890 this system was in danger of collapsing. Suburban expansion pushed out the frontiers of settlement; consequently nightmen from central areas had further to travel to find market gardeners willing to accept their cargoes. No longer could they complete their work in darkness. ... Nightmen sometimes tipped their loads illegally on vacant allotments, into the Yarra, on roadsides, or anywhere they could do so and remain undetected: 'If the night is wet and they cannot get to their destination on time, they open the cart on the way and let the stuff out on the road.'"

6 In the early 20th century, larger Chinese cities were divided by a 'line' of demarcation into two zones. Residents of the outer zone either had their nightsoil etc removed free of charge, or they were actually paid for it. Those inside the line, being further from the city's boundary, had to pay for removal of their excreta. (Mark Elvin, personal communication).

7 Because there are natural processes by which elements are added to soil (which usually means transferred from somewhere else; see Chapter 2), there is a 'critical' yield of any crop below which it can be harvested indefinitely without apparent soil impoverishment. Most investigation and speculation about this phenomenon has focused on nitrogen. Loomis (1978), for example, has suggested a range of rates for the various avenues of nitrogen replenishment in mediæval English soils. But replenishment rates of any element, and hence critical yields, must necessarily vary widely from one location to another for reasons of topography, climate, the types and densities of plants and animals in the area, the proximity and types of industrial activity and, of course, the underlying chemistry, physics and ecology of the soil. Because of the various states of nitrogen in an ecosystem (NO_3^-, NO_2^-, N_2, NH_3, NH_4^+ and organically bound), because of the volatility of N_2 and NH_3, the capacity of the ions to be adsorbed onto particles of clay etc, the sensitivity of that adsorption to pH and the sensitivity of the whole network of reactions to temperature, water content and movement, soil porosity, oxidation/reduction potential, soil biota etc, the rates of loss or replenishment can vary greatly from one set of circumstances to another. In some well known experiments undertaken at the Rothamsted Research Station in England from the middle of the 19th century to the middle of the 20th, unmanured wheat plots containing about 3 t.ha^{-1} total nitrogen were allowed to revert to 'wilderness'. After 81 years these plots accumulated as much total nitrogen (about 7 t.ha^{-1}) as wheat plots receiving farmyard manure (35 t.ha^{-1}.y^{-1}). This was equivalent to an average rate of 49 kg.N.ha^{-1}.y^{-1}. The accumulation, however, was heavily dependent on the herbaceous cover that grew in the 'wilderness' and, of course, it took place in soils that were not deficient in other essential elements. The rate of replenishment does not and cannot apply to wheat fields (see also Legg and Meisinger, 1982).

8 Despite which, about half of present day English and Welsh sewage sludge is used on farmland (Kinnersley, 1994).

9 The phosphorus content of 'scoured' cattle bone is about 12.5% (Slansky 1986).

10 Basic slag contains about 8% P.

11 Van Slyke (1988). About 58% of China's surface is above 1000 m, about 33% is above 2000 m (Smil, 1984).

12 Smil's (1993) Fig. 2.6 includes a plot of 'cultivated' land *per capita*. The *per capita* area for the end of the 19th century is equivalent to a population density of about 5.9.ha^{-1} (cultivated). The same figure also indicates that, at the end of the 19th century, there was a total of about 780,000 km^2 cultivated land which, with a population of 420 million (Smil's Fig.1.2), gives a population density of about 5.4.ha^{-1} (cultivated). Given the uncertainties in the primary data and in interpolating from steep curves, that is reasonable agreement. But Smil's Fig. 4.7 includes a plot of 'arable' land from 1850-1990. The area corresponding to 1890 amounts to about 1.05 million hectares which gives a population density of 400.ha^{-1}! Either the figure does not refer to the whole of China, a possibility not evident from the text or legend, or the units on the relevant axis should be km^2, not ha. In that case the population density would be 4.ha^{-1} (arable).

13 Lampard (1973) lists the world's 30 'largest agglomerations' for 1890, 1920 and 1960. The list for 1890 ranged from London at 4,212,000 to Baltimore at 434,000. Melbourne, then a mere fifty six years of age, came in at 22nd with a population of 491,000. (Melbourne also had an unenviable reputation for typhoid fever—see Note 5.) No Chinese cities were listed, apparently because of uncertainty about their actual sizes, although Lampard says that "Peking was almost certainly" over 500,000. Skinner (1977a), however, estimated that by the 1840s "China's four largest cities —Peking, Soochow, Canton and the Wuhan conurbation—ranged in size from 850,000 down to 575,000" and that, between 1895 and 1911, Peking, Shanghai and Wuhan all exceeded one million.

Nevertheless, the contrast between China and Australia in the degree of urbanisation is stark. In the 1840s the major geographical regions of China ranged, according to Skinner (1977b) from about 4%-8% urbanisation. Lampard commented about the general situation in 1890: "Apart from Australia, in so many respects a *rara avis*, no country yet had more than 30% of its population urbanized" (but see also Note 4). Even with its inconsistencies, there is probably no other simple statistic that says so much about the gulf between the farming practices of these two countries.

14 King (1911). In about 1908, Dr F. H. King, a former Chief of the Division of Soil Management of the US Department of Agriculture, visited China, Japan and Korea to study their farming methods. He published his findings in a book that has recently been reprinted. He viewed the scene through rose-coloured spectacles but, because of the time at which his observations were made and the factual nature of many of them, they are of interest.

15 Walker (1988).

16 Hancock (1972).

17 Mark Elvin (personal communication). See also Elvin and Liu (1998).

18 See Smil (1984, 1993) and, for insight into some of the consequences of the 'Great Leap Forward', Jung Chang (1991).

19 Scott (1952).

20 Unless indicated otherwise, the factual material in this section is from King (Note 14).

21 From the quantities used, 40 ounces (about 1130g)/adult/day, he was evidently referring to the total of faeces + urine.

22 High density agrarian societies commonly face chronic shortages of fuels for cooking and domestic heating.

23 The area and its integrated system of farming are described in detail by Ruddle and Zhong (1988). See also Korn (1996).

24 For example, Stavrianos (1988).

25 Iinuma (1973).

26 See Bellwood (1991) for an outline of the significance of the Austronesian languages.

27 Clarke (1971).

28 From a nutritional standpoint a simple comparison with the yields of other types of crop, for example wheat, can be deceptive. The water content of the tubers is about 73%; a yield of 18 tonnes is therefore equivalent to about 4.85 tonnes dry mass. The water content of wheat is about 13%. The protein content of the fresh tubers is low (about 1.8%) and the 18 tonnes would provide about 325 kg protein. A similar amount of protein could be obtained from about 2.7 tonnes of wheat. Moreover, this simple comparison takes no account of the type of protein, that is to say its amino acid composition and hence its nutritional value. Clarke refers to evidence of mild protein deficiency in the community.

29 Clarke considered diseases of the respiratory tract to be among the commonest causes of death, a situation he attributed to "the smoky atmosphere of the low-roofed houses". Given the ease of growth of wood-rotting fungi in the houses, despite the smoke, lung infections by the spores of moulds such as *Aspergillus fumigatus* might also be suspected. This suspicion is encouraged by the relatively high incidence of tinea as a general skin infection rather than its restriction to areas where two surfaces of skin are close or in contact, as is its usual role in drier climates. Scabies was common, there were some parasitic infestations, he was not aware of tuberculosis, venereal diseases or yaws. Some (19.5%) of the Bomagai-Angoiang and a neighbouring clan had been found in 1963 to be carrying the malaria parasite.

30 Various equations have been used to calculate 'carrying capacity' in this type of situation. Taking the maximal amount of 'cultivable' land (under cultivation + accessible) as 587 ha, Clarke used two such equations to calculate carrying capacity and some other dynamics of the Bomagai-Angoiang's farming system. Both equations gave the minimal cycling time for a garden as 39.85 years (1.25 years' cropping and 38.6 years' fallow). One gave a sustainable population of 158, the other 198; the actual population at the time was 154. These three numbers respectively are equivalent to populations densities of 27, 34 and 26.km^{-2} of 'cultivable' land.

31 Clarke commented that, as assessed by its biodiversity, the Ndwimba basin was ecologically healthy and that the system was effectively in a steady state (although he said "equilibrium"—see Chapter 2). He contrasted the complexity of the forest, most of which was secondary growth, with the extensive grasslands elsewhere in the highlands where populations had exceeded the carrying capacity of a swiddening system and degraded it. The growth of the population in those districts apparently followed the banning of tribal warfare by the colonial administration as well as access to the medical facilities provided by the patrol officers. (See also Note 30).

32 Gary Larson, the American cartoonist, has produced a cartoon depicting some Pilgrims, their wives, and Indians seated together at a sumptuous meal. The table is waited on by two Indians and a very large female settler. One of the Pilgrims is holding up his right hand, with a dark ellipsoidal knob on the end of each extended finger. The legend reads, "Thomas Sullivan, a blacksmith who attended the original Thanksgiving dinner, is generally credited as being the first person to stick olives on all his fingers". Presumably the olives were imported.

33 Gras (1925).

34 Bolton (1989).

35 See, for example, Evans (1983); Jones and Raby (1989).

36 The First Fleet disembarked 1438 individuals (excluding children) of whom 733 were convicts, 543 male and 190 female (Watt, 1989).

37 Bolton (1989); Hancock (1972).

38 I have omitted reference to Tasmania because I am trying simply to identify some of the more significant ways in which agriculture caused rapid soil degradation on the mainland where, by and large, it is worse than in Tasmania.

39 For example, Fry (1994).

40 There is no shortage of detailed accounts of Australia's environmental history. See, for example Adamson and Fox (1982), Ashton and Blackmore (1987), Chartres *et al.* (1992), Donald (1964/5), Dovers (1994), Fry (1994), Hancock (1972), Hardy and Frost (1989), Lines (1991)—and many others. Australian agricultural and environmental history has benefitted more than any other major agricultural nation, with the possible exception of New Zealand, from photography. Professional photographers appeared on the Australian scene in the 1840s, less than sixty years after the founding of the colony. Thus, at the time of writing, there are photographs covering more than 70% of the period of European settlement. The availability of such photographs has been used to great advantage by Ashton and Blackmore.

41 Ashton and Blackmore (1987). I find it difficult to assess the accuracy of statements like this since early estimates of areas under forest could have been little better than rough guesses. There are also some conflicting claims. Rolls (1994) has argued, with more emphasis on poetic licence than on measurement, that there are more trees in New South Wales now that when it was first settled by Europeans. He says "The commonest remark of those practical men who first saw Australia was, 'You won't have to clear it to cultivate it'." If it comes to guessing the commonest remark of the early immigrants I think I can do better than that—but it might not be of immediate relevance. Rolls also argues on the basis of what settlers or explorers could or could not have done had the land been as densely wooded as it is now. He claims, for example, that Blaxland, Wentworth and Lawson could not have found their way over the Blue Mountains so soon "if the country had carried the present dense growth of tall eucalypts". My own experience of walking in the Blue Mountains makes it difficult for me to take such an assertion seriously. Apart from topography, the principal impediment to walking in the Blue Mountains is scrub or undergrowth, not trees. In any case, it is now evident that Aborigines, feral cattle and an escaped convict crossed the Blue Mountains before B. W. and L. (Cunningham, 1996).

While it is beyond doubt that, at the time of European occupation, there were extensive grasslands in New South Wales, and that some are now wooded, it is also beyond doubt that there has been extensive nett deforestation since that time. See also Note 42. An objective assessment of the treatment of Australian forests is to be found in Dargavel (1995). He states that, as a direct result of two centuries of European settlement, "By the 1980s, about 26 million hectares had been deforested, leaving 43 million hectares as the forest estate which has now been supplemented by 1 million hectares of forest plantations." Deforestation is continuing.

42 Walker (1978) shows a photograph of a (presumably) late 19th century wooden bridge across a steep gully. The bridge is supported by an enormous stack of horizontal logs that fill the gully. With some guesses about dimensions, I calculate that the bridge was supported by about 3700 m³ of wood. I cannot tell from the photograph whether the bridge is a footbridge or for vehicles; I assumed the latter and guessed its width to be 2.5 m. If it were, in fact, a footbridge, the calculated volume of wood should be reduced accordingly.

In Australia, as elsewhere, deforestation for both agriculture and the forestry industry have removed most of the old trees. Early photographs of the Australian timber industry, photographs of the kind that are found on the walls of small rural museums, university forestry departments etc, commonly depict timber cutters standing beside fallen trees with a diameter of at least 2.5 m at the base. The photograph of a car within the arch of a Californian redwood is well known. Less well known are comparable Australian photographs, not of cars under trees but of people living inside them. In 1908 a Victorian journal, the Weekly Times, published a photograph of a Gippsland family whose house had been burnt down; they had made a comfortable temporary home in such a eucalypt. The family comprised both parents, three children and a dog. The space within the tree accommodated "two large beds, tables, chairs and sundry other furniture" (Walker, 1978).

Trees of that size are many centuries old. When, as a part of a management policy, forests have been allowed to regrow after clearing, they do not and cannot contain such trees because there has not been nearly enough time for them to mature. Their aesthetic value is obvious. Their ecological importance includes the protection and nesting opportunities that their hollows provide for numerous species of birds and mammals. Younger trees do not offer these facilities and, as a result, many Australian native animal species face an uncertain future.

43 Governor Macquarie recommended that "... Small Families from the Middling Class of Free People ... should receive Fifty to One Hundred Acres of Land According to the number of his Family, all to be victualled at the Expense of the Government for Eighteen Months." Cited by Fry (1994).

44 See, for example, Patricia Clarke's (1986) biography of Mary Braidwood Mowle (née Wilson).

45 Hancock (1972).

46 See, for example, Hancock (1972); Dovers (1994).

47 Small mammals generally have higher specific metabolic rates, higher specific growth rates and, for the most part, shorter and more frequent breeding cycles than large mammals. (A 'specific' rate in this context means the rate per mass of organism.) A physiologist colleague of mine has commented wryly that if one's objective is merely to increase the annual yield of meat per area of land, it would be simpler and more effective to grow rabbits than to fiddle around with the genetic manipulation of cattle, sheep or pigs. The ability of rabbits to displace native animals of comparable size (small macropods, bilbies) can be attributed partly to litter size, breeding frequency and the relatively specialised diets of the native animals (but see also Morton, 1990 or, in an edited version, 1994).

48 I have seen in the late 1940s rabbit populations so dense that the ground in the middle distance appeared grey. Then, as one approached the earth moved, or rather seemed to—not in the sense that that expression is commonly used these days—but visually as the rabbits ran to their burrows. Their departure exposed bare soil. Populations of such density were on the verge of collapse. They did, in any case, collapse along with the populations of farm animals, in drought. But they recovered faster than the farm animals—fast enough to keep up with the grass when it began to grow again. The rabbit population was brought under some sort of control in the 1950s with the deliberate introduction, by a group of people headed by Frank Fenner, of the highly specific myxoma virus. Without myxomatosis most of the infested rangeland would by now have been degraded beyond any realistic hope of recovery. Where that happened neither the rabbits nor the larger herbivores would have survived. The disease saved the land from 'complete' degradation and, by so doing, saved the rabbit from itself. It was a form of culling.

There are two major limitations to the continuing effectiveness of myxomatosis. The original strain is transmitted by mosquitoes and thus is not transmitted easily, if at all, in arid regions. Later strains using fleas as vectors are undergoing trial. The second limitation is that an animal population that is not weakened by other challenges, such as starvation, will inevitably develop resistance to an infectious disease. Rabbit haemorraghic disease virus (rabbit 'calici' virus) has recently been introduced into Australia with a similar objective to the introduction of the myxoma virus. Unless supplementary measures are invoked, the results are likely to be similar—an initial sharp decline in rabbit numbers followed by a steady increase as resistance is developed. Rabbit populations have probably also been kept in check in recent years by an alarming increase in the numbers of feral cats and foxes, and it is possible that something like a balance has been reached, although it is a balance with all three species at destructive levels. Foxes were introduced to Australia for the same reason as rabbits— blood 'sport'. They have caused and are continuing to cause enormous damage to native animal populations, but they have undoubtedly had a significant effect on rabbits. I have the impression that the fox has suffered less at the hands of graziers than has the native dingo.

49 Evans (1975).

50 Jenkins's underlying humanity is illustrated by the following two passages:
"Our present and former legislators are to be blamed for handing over so much good land to the same people. When land is taken from one tenant and given to another, it is to be expected that the latter improves it, but the lands of Victoria have been diminished in value from the time they were so barbarously taken from the natives, and when the golden rules of humanity were so deliberately flouted" (January, 1873).

"I read that President Grant of America has said that he would exterminate the Indians for the sake of Civilization. O Lord deliver us if we must civilize the world through killing those whom God has stationed on the land" (March, 1873).

51 A similar attitude persisted well into the 20th century. It is reflected in a European tendency, especially when writing about agriculture, to refer to forest or woodland as 'waste'. Donald, an Australian Professor of Agriculture, commented in a lecture delivered in 1964 "... it (fertilised leguminous pasture) is not only raising the productivity of settled areas, but is enabling the conversion of *useless* scrub and poor forest into pasture land of satisfactory productivity" (my italics); (Donald, 1964/5). On the other hand, from almost the beginning of its colonisation, there were some individuals in New South Wales advocating various forms of environmental conservation (Bonyhady, 2000). The late 19th century saw the emergence in Australia of what Bonyhady has called "The first conservation group in the world".

52 Jenkins (Note 50) noted in 1874 that a farm on which he was working produced 32 bushels of wheat per acre (2145 kg.ha^{-1}) whereas some neighbouring farms were yielding only 3.5 bushels (235 kg.ha^{-1}).

53 Robinson (1976).

54 Johnson (1994).

55 Gras (1925).

Notes with Chapter 8

1 The 'unsaturated zone' to an average of about 1m below the surface which, of course, includes topsoil and subsoil, is estimated to contain about 150,000 km^3 water, or about 0.01% of the global total. But of that, about 40% is saline and about 5% is permanently frozen. 'Genuine' subterranean water accounts for slightly less than 0.6% of the earth's total.

2 Philip (1978); see also Shiklomanov (1993) cited by World Resources Institute (1994); de Villiers (1999).

3 Dury (1981). See also Table A9 and Chapter 10.

4 See Clapp (1994).

5 Rosen (1973).

6 Derry and Williams (1960).

7 Cited by Rosen (1973); italics in citation.

8   In the light of this sensitivity I continue to be impressed by the apparent ease with which such organisms manage to become established in human intestines in the first place—especially in modern environments with high standards of hygiene.

9   The major components of such transport chains are haemoproteins (cytochromes). They are components of membranes—the mitochondrial membranes in eucaryotes, the cell membrane in procaryotes and, in the case of photosynthetic electron transport, chloroplast membranes.

10   For example, Alexander and Stevens (1976); Imhoff *et al.* (1971); Smith (1998).

11   By way of comparison, laboratory liquid media for growing algae to high densities commonly contain phosphate at concentrations equivalent to 6-18 mg P.l^{-1}; a medium for growing the nitrogen-fixing bacterium, *Azotobacter*, is used with a phosphate concentration equivalent to 18 mg P.l^{-1}. More general bacteriological growth media may contain phosphorus at ten times that concentration.

12   Australian Bureau of Statistics (1992a), Table 4.2.2 and (1996a), Table 6.5.3.

13   Australian Bureau of Statistics (1992a).

14   The poet, Dorothea Mackellar described Australia as a country, among other things, "Of ragged mountain ranges, Of droughts and flooding rains." A friend of mine has described Australia's climate as a series of droughts punctuated by floods. Another friend has said that we don't have a climate—we have only weather. A more prosaic account of the climate can be found in Colls and Whitaker (1990).

15   A calculation based on the actual totals of the tabulated entries gives 73%. See above in text.

16   Australia: State of the Environment (1996); Australian State of the Environment Committee (2001). World Resources Institute (1998) rather curiously attributes only 33% of Australian water consumption to agriculture. See also Gleick (2000), Smith (1998).

17   The first scientific account of the potentially lethal effects of cyanobacterial toxins arose from an examination of eutrophication in the Murray estuary. The study was undertaken by George Francis of Adelaide and was published in *Nature* on 2nd May, 1878. Francis assigned the organism that he considered mainly responsible for the toxin to the genus *Nodularia*.

18   Mackay and Landsberg (1992).

19   Uncertainties in sampling methods together with wide variations in flow rates demand caution in interpreting published statistics on these river systems. The Australian Bureau of Statistics (1992a) gives (in Table 4.1.4) a volume of 24,300 Gl for the mean annual runoff of the Murray-Darling drainage division and a mean annual 'outflow', presumably measured at the mouth of the Murray, of 12,200 Gl. Table 4.2.2 gives the total mean annual water use for the Murray-Darling basin as 8,660 Gl of which 7650 Gl, or 88% of the total, goes to irrigation. The 7650 Gl used for irrigation in the Murray-Darling system accounts for 75% of the total volume of irrigating water used in Australia (10,200 Gl). Wahlquist (1995), citing the Murray-Darling Commission, quotes an 'average' annual volume of 10,774 Gl diverted for irrigation (from rivers within the basin) and an outflow of "4,867 gigalitres, one-third of the natural flow".

Mackay and Eastburn (1990) have this to say: "The average annual flow in the Murray is around 11,000 gigalitres, but this average figure is misleading. Murray flow varies from virtually nil during drought years up to 40,000 gigalitres during a wet year. So variable is its flow that, despite all of the water now stored and diverted for irrigation upstream, it is impossible to prove statistically that river flow below the Darling junction has decreased significantly over the 100 years for which reliable flow records have been kept."

Anderson (1995a), citing a recent government report, states that "... only 25 per cent of what flows into the (Murray-Darling) basin ever reaches the sea".

20 Australian Bureau of Statistics (1992a).

21 Wahlquist (1994); Beale (1995a). Here, as in the USA, water to irrigators is heavily subsidised. Wahlquist cites the charge to irrigators of water as $2.50.Ml^{-1} whereas the actual cost of supplying the water is estimated to be about $80.Ml^{-1}. The issue of new irrigation licences was stopped in 1982. The cost of a licence then was $160; in the late 1990s they traded for around $540,000. Irrigated cotton plantations in northwestern NSW are the province of big agribusiness as was much of the development of irrigated agriculture in the American West. See, for example, Reisner (1986); Simon (1994).

22 Beale (1995b). Somerville and Briscoe (2001). See also Wiles and Turner (1996).

23 Bek and Robinson (1991).

24 Bek and Robinson (1991), p22.

25 Although the phosphorus concentrations are disturbingly high for a rural river, they are low by the standards of some European rivers. According to Lean and Hinrichsen (1992), the 'average' (total?) phosphorus concentration in the Seine for the period 1986-88 was about 0.8 mg.l^{-1}. The corresponding figure for the Po (see below in the text) is about 0.25 mg.l^{-1}.

26 Sources of nutrient discharge to streams are commonly classified as 'point' (sewerage outfalls, irrigation drainage outfalls, urban stormwater, fish farms) or 'diffuse' (leaching from the land—from forests, from pasture and from cultivated land). Nutrient loads discharged into streams within the Murray-Darling Basin are of the following order (dry year—wet year; t.y^{-1}). Total Phosphorus: (a) point sources, 650-900; (b) diffuse sources, 250-4,300. Total Nitrogen: (a) point sources, 3,900-5,300; (b) diffuse sources, 1,600-28,000 (Murray-Darling Basin Commission, 1992).

A major potential source of pollution is the cattle feedlot. In 1990 there were more than 260 large feedlots in the Murray-Darling Basin with a total capacity of over 300,000 animals. Most (about 80%) were in Queensland and northern NSW. There are state guidelines for dealing with waste from these establishments, but guidelines are not always ideal and are not always followed. Feedlots and piggeries have considerable potential for polluting waterways. See Chapter 13.

27 World Resources Institute (1992). By 2001 The Netherlands was treating approximately 98% of its sewage, Denmark 90% and Turkey about 15% (OECD 2001). The proportions of primary, secondary and tertiary treatment vary significantly within the OECD.

28 World Resources Institute (1992); Pearce (1995a).

29 Morbilliviruses constitute a group of viruses that include the agents responsible for measles, canine distemper, rinderpest and an apparently new disease that killed 25 race horses and their trainer in Queensland in 1994. See Harvell *et al.* (1999) for a comprehensive summary of emerging diseases of marine organisms and likely contributions of human activities to the process.

30 Pearce (1995a). I am not in a position to check these numbers, but I would add the comment that it is unlikely that inorganic nitrogen is distributed exclusively between ammonia and nitrate without significant concentrations of at least one intermediate oxidation state, nitrite.

31 Including the British coast (*e.g.* Anon, 1993; Pearce, 1995b). As might be expected, the Black and Caspian Seas and the coastal waters off Hong Kong are very badly polluted. In the late 1970s Hong Kong was estimated to discharge "more than 6000 tonnes of human and industrial solids into the harbour each day" (Boyden *et al.*, 1981). A more recent estimate is "2 million tonnes of largely untreated waste" (including liquid) discharged daily (Hunt, 1993). Nor are they confined to the northern hemisphere. Since the late 19th century Sydney has discharged raw sewage into the sea with inevitable pollution of the nearby beaches. From the beginning of the 20th century there has been organised protest, mainly by surfers, against the practice (*e.g.* Coward, 1988). The official response has been to extend the outfalls further offshore and into deeper water—in other words move the problem further out of sight. There was protest against that plan on various environmental grounds, including the ecological effects of discharging sludge close to the bottom in water that was less turbulent than that closer inshore. Some of the early palpable effects of the new practice included diminished pollution on the beaches close to the outfall but additional or new pollution on some other more distant beaches. 'Various' (2001) deals comprehensively with past and present human impacts on the Baltic Sea. See also World Resources Institute (1998) for an assessment of the global dimensions of 'nutrient overload' and Tilman *et al.* (2001) for projections of its dimensions by 2050. Further aspects of the environmental dynamics of phosphorus and nitrogen are discussed in Chapter 11.

32 Moss *et al.* (1993). For a geographically broader treatment of this general topic, see 'Various' (1995a).

33 Burke (1994).

34 The correlation for nitrogen is very close (P < 0.00001 for log population *vs* log [NO_3^{-1}]) (Peierls *et al.*, 1991). The correlation between log 'soluble reactive phosphorus' discharged and log population density was also close (P < 0.0001) but was closer with log urban population density (Caraco, 1995).

35 Anderson (1994).

36 Other sources estimate the total amount of 'biologically available' (fixed) nitrogen produced annually to be some 210 Mt from 'human sources' and 140 Mt from 'natural sources' (Vitousek *et al.* 1997 as cited by World Resources Institute 1998). See also 'Various' (2002).

37 World Resources Institute (1994).

38 See, for example, Pearce (1992).

39 Cited by World Resources Institute (1994).

40 See Chamberlain (1992). The problem was compounded by the introduction of exotic Nile perch (Goldschmidt, 1996).

41 I say 'normally' because some arid countries, provided they are wealthy enough (*e.g.* Saudi Arabia), are desalinating seawater for domestic use. The desalination processes, which are very dependent on energy, are also effective in removing other contaminants.

Notes with Chapter 9

1 Water is consumed in a literal sense by plants during photosynthesis. The water molecule is split to produce molecular oxygen; the hydrogen atoms are used in the reduction of CO_2 to carbohydrate, ultimately to be transformed to virtually all plant and animal tissues and a goodly proportion of microbial biomass. On the other hand, respiration and the combustion of organic matter produce water. Since the pre-industrial era, the concentration of CO_2 in the atmosphere has increased by some 27% —from about 280 to 356 ppm (World Resources Institute, 1994; see also Monnin *et al.*, 2001). If we make the very simple assumption that this arose equally from combustion/respiration of carbohydrates and burning hydrocarbons then, over the same period, 'atmospheric' water would have increased by about 40%. Based on the amounts cited at the beginning of Chapter 8, this would be equivalent to an increase of 0.0003-0.0004% in the planet's water, or about 5000 km^3.

2 About 6% of the total runoff from the planet's river catchments is estimated to be lost through human activities. Most of the loss is associated with irrigation, but evaporation from reservoirs is also significant. (L'vovich *et al.* 1990, cited by Dynesius and Nilsson, 1994).

The effect of human activities on the level of the oceans is, of course, much more difficult to estimate. On the one hand, some studies imply that the sea level has been rising at a mean rate of about 1.7 mm.y^{-1} for the past 80 years and that the rise is accelerating (see Bilham and Barrientos, 1991 for the original citations). Sahaglan *et al.* (1994) argue that human activities, excluding those associated with global warming, are contributing to a rise of 0.54 mm.y^{-1}. On the other hand, Gornitz *et al.* (1994) estimate that changes in sea level over the previous 60 years included a nett human contribution responsible for an average *fall* of 1.63 mm.y^{-1}. Processes included in this assessment were groundwater 'mining', deforestation, runoff, loss of wetlands, dams and irrigation.

3 Moody *et al.* (1991) cited by Dynesius and Nilsson (1994).

4 Johnson and Lewis (1995); Pearce (1994a and b).

5 Pearce (1994a, 1995c). See also Stone (1999); Lindahl-Kessing (1998).

6 Anon (1995a).

7 In most situations, of course, rain also falls on flood plains where dams cannot be built. In special cases, such as the Nile, Tigris and Euphrates, the amount of runoff from the flood plains themselves is negligible and the theoretical maximum or 'exploitable yield' is, for all practical purposes, close to the actual maximum.

8 See Reisner (1986).

9 The calculations for these rates used demographic information listed in 'Economist' (1990). Gross rates of water consumption were obtained from World Resources Institute (1994).

10 Jacobs (1990), Eastburn (1990).

11 NSW Water Resources Commission (1983) cited by Australian Bureau of Statistics (1992a).

12 Porosity is the proportion of the total volume of rock occupied by 'holes' or spaces; it is usually expressed as a percentage. Permeability is a measure of the flow rate of water through the rock; it is commonly expressed as cubic metres per day per cross-sectional area of the rock ($m^3.d^{-1}.m^{-2}$) (Dury, 1981). Since there is no term for thickness in the expression, it follows that a thick section is less permeable than a thin section of the same rock. The flow rate is also necessarily a function of the pressure difference across the rock.

13 The terms 'saline', 'salinity' and 'salination' in this context are not restricted to sodium chloride; they encompass any salts. Inland rainwater has a total salt concentration of less than 20 mg.l^{-1}; coastal rainwater can reach 60 mg.l^{-1}. Water with (nontoxic) salt concentrations up to 500 mg.l^{-1} is normally acceptable for drinking, but the taste will vary with the type of salt; a popular brand of commercial mineral water has a total salt concentration of about 560 mg.l^{-1}. Water with more than 700 mg.l^{-1} total salts has a distinctly salty taste and, with more than 1.5 g.l^{-1} is unpalatable (NSW Water Resources Council, 1994b). Seawater has a total salt concentration of about 35 g.l^{-1}.

14 Australian Water Resources Council as cited by Australian Bureau of Statistics (1992a). The relevant information was tabulated (Table 4.1.9) in this reference. The table contains a number of arithmetical errors as a result of which all totals but one (volume of saline water) are wrong. The numbers I have used in the text were calculated directly from the primary listings.

15 NSW Water Resources Council (1994b).

16 See, for example, Bowler (1990).

17 Pearce (1995d). Pearce's article gave the relevant concentrations as mg.l^{-1} instead of µg.l^{-1}. This error has been acknowledged (Pearce, personal communication). See also Bagla and Kaiser (1996); Chowdhury (1999); Gleick (2000); Masibay (2000); Nickson *et al.* (1998).

18 Asswad (1995).

19 Reisner (1986).

20 The groundwaters of the Murray Basin have a number of important characteristics that affect their salinity patterns. Water close to the intake or recharge areas is essentially 'fresh'. As it moves into the basin, however, it becomes more salty both through prolonged contact with soluble subterranean salts and through evapotranspiration.

This is perhaps another example of natural bloodymindedness in that one of the most important natural mechanisms for controlling the level of the water table is also one of the most important mechanisms for raising the salt concentration of that same groundwater. For this reason the most saline groundwater is often that which is closest to the surface—as is generally the case in the Murray Basin.

There are several different aquifers under the Murray Basin with different salinity characteristics. They range from 'fresh' to 30 g.l⁻¹ for the Renmark and Murray Group Aquifers, to 20 g.l⁻¹ for the Shepparton Formation Aquifer, and up to 50 g.l⁻¹ for the Pliocene Sands Aquifer—which is closest to the surface in much of the Basin. There is some leakage between aquifers. (See Evans *et al.*, 1990).

21 A comprehensive account of salination in the Murray Basin has been given by Macumber (1990).

22 See Bowler (1990); Rutherford (1990).

23 This is the depth at which capillary action in the pores of the soil can raise the water level, against gravity, to the surface. It is commonly 1-2 metres below the surface.

24 Eastburn (1990). According to Anderson (1995a), about 11,000 Gl is diverted annually from rivers within the Murray-Darling Basin. For the period 1988-89–1992-93, the Australian Bureau of Statistics (1996b) gives an average annual "diversion for irrigation" within the Murray-Darling Basin of 10,232 Gl. See also Chapter 8.

25 I have taken 94% of the volumes tabulated by Jacobs (1990) in accordance with the proportion of the diverted water already cited. See also Eastburn (1990).

26 See Mackay (1990) for more details.

27 Macumber (1990).

28 Australian Bureau of Statistics (1992a).

29 John Powell, Murray-Darling Basin Commission (personal communication, November 1995). See also Goss (1995) and Australian State of the Environment Committee (2001).

30 On average, a rice crop consumes water equivalent to about 1500 mm rain. This is 2½ times the requirement of maize or soybeans, twice that of wheat and about 1.7 times that of sugar cane or cotton (Beirne and Pratley, 1988). In 1991, NSW had 85,000 ha under rice, much of it in the Murrumbidgee Irrigation Area. Rice represented 44% of all irrigated cereal cultivation in that state. There was no rice cultivation in Victoria but there were 4000 ha of the crop in Queensland (Australian Bureau of Statistics, 1992b). In 1999-2000 Australia had 170,000 ha allocated to rice, virtually all of it in NSW (Australian Bureau of Statistics, 2001).

In 1989, approximately 54,800 ha of Australian cropland were irrigated by trickle or microspray. This amounted to 3 or 4% of the total area of irrigated land. The uncertainty arises from a discrepancy in Table 4.2.7 of the Australian Bureau of Statistics (1992a). The discrepancy originated in the entry for the Northern Territory, but it is not apparent in which category the mistake occurred or whether it was primarily typographic or arithmetic.

Irrigation rates (Ml.ha⁻¹.y⁻¹) for specific products are approximately: Horticulture, 6.5; Dairy Products, perennial pasture 11, annual pasture, 3.5; Sugar Cane, 5.5; Cotton,

7; Rice, 15.5; Pasture, 3.5. Total annual water consumption by those activities amounted to 11,607 Gl (Australia State of the Environment, 1996). A later assessment (Australian State of the Environment Committee, 2001) gives a total annual Australian rate of irrigation for 1996-97 of 17,940 Gl. Another perspective is given by figures such as a requirement of 17,000 kg water to produce 1 kg cotton or 4,700 kg water to produce 1 kg rice (Kendall and Pimentel, 1994). See also Chapter 8.

31 With objectives such as this in mind, the 'National Landcare Program' was established in the 1980s after discussions of the Australian Federal Government with the National Farmers' Federation and the Australian Conservation Foundation.

32 Close (1990).

33 Woodford (1995).

34 Reisner (1986); Smil (1987). Reisner also lists losses in storage capacities for a number of American dams.

35 Le Guenno (1995).

36 For example, Hillel (1994), Lowi (1993). In the context of depleting groundwater supplies and a water table sinking at the rate of 3 m.y^{-1}, the following statement is attributed by J. C. Randal (*Guardian Weekly* 31st May, 1992) to Elias Salameh, founder of the University of Jordan's Water Research and Study Centre: "No matter what progress irrigated agriculture makes, Jordan's natural water at this pace will be exhausted in 2010. Jordan then will be totally dependent on rain water and will revert to desert. Its ruin will destabilize the entire region." See also de Villiers (1999). Gleick (2000) has summarised, for the period 1503-1999, cases of international or interregional conflict over access to water.

37 Shiklomanov (1993) cited by World Resources Institute (1994). A later projection put total consumption in 2000 at about 5170 km^3 with agriculture responsible for about 60% of that (Biswas, 1998). A high degree of uncertainty in this type of estimate should be self-evident. Biswas was honest enough to call his quantities 'guestimates'. Gleick (2001) indicates graphically a total water consumption in 2000 equivalent to 3953 km^3.

38 World Commission on Dams (2000).

39 For the period 1989-91, some 17% of the world's cropland was irrigated (World Resources Institute, 1994). Given the uncertainties inherent in statistics of this nature, it would seem that at least a billion people depend completely on irrigation for their survival. But that assumes that the world's food is distributed more or less equally which, of course, it is not. Two extreme cases are Egypt (population 58.5 million) and Pakistan (135 million). All of Egypt's and 80% of Pakistan's cropland is irrigated and is vulnerable to salination.

40 World Resources Institute (1994) state that the total volume of water used annually accounts for only about 8% of the annual runoff of the world's rivers. Using somewhat different criteria, Postel *et al.* (1996) estimated that about 18% of "accessible runoff" is used "directly for human purposes". 'Accessible runoff' is total runoff adjusted "for geographic and temporal inaccessibility to estimate the proportion that is realistically available for human use". Inclusion of some supplementary considerations in their calculations led them to conclude that Man is 'co-opting' about 23% of the total renewable freshwater supply. Tilman *et al.* (2001) state that "Humans currently appropriate ... about half of usable freshwaters." See also Vörösmarty *et al.* (2000).

41 O'Loughlin (1980).

42 Tisdall (1980).

43 In the late 1970s the Tasmanian Hydroelectric Commission sought to build a dam on the lower Gordon River (the 'Gordon below Franklin') in the south west of the island. The State Government initially rejected this proposal but approved the construction of a dam further upstream, above the confluence of the Gordon and Franklin rivers (the 'Gordon above Olga'). This alternative would have flooded less of a pristine wilderness area, remarkable for its beauty as well as its ecological significance. The Commission and some lobby groups exerted pressure on the Tasmanian Upper House to reject the Government's decision and to revert to the original proposal of damming the Gordon below the Franklin. The generating capacity of that scheme was rated by the Commission as 282 MW maximum and 180 MW average compared with an average output from the Gordon above Olga of 118 MW. The Commission's arguments, while acknowledging possible environmental constraints that might be imposed on them by government edict, were based virtually entirely on its estimates of Tasmania's future energy needs and economic considerations. Comparisons with thermal power stations were also economic, not environmental (*A Report by the Hydro-Electric Commission on the Effect on Power Development of a Decision not to Use the Hydropotential of the Franklin River.* October 1980). Some of its central calculations for these comparisons were found to be faulty.

44 Reisner (1986). See also Simon (1994).

45 World Resources Institute (1998), World Commission on Dams (2000).

46 The World Commission on Dams (2000) states that "in India and China together, large dams could have displaced between 26-58 million people between 1950 and 1980."

47 According to Friends of the River Narmada (www.narmada.org) the planned height is 136.5 m.

48 Ahmad (1999); Tickell (1992).

49 The Bank's reasons for withdrawing from the loan were not quite as simple as this account might suggest. An informative summary of events within the Bank at the time has been provided by George and Sabelli (1994).

50 Friends of the River Narmada (www.narmada.org).

51 Editors (1993), Kumar (1993), Miller and Kumar (1993), Sen (1995), Wilks and Hildyard (1994).

52 Harding (2002),

53 Kumar (1995a).

54 van Slyke (1988). Pearce (1995e) gives the annual discharge as 700 km³.

55 van Slyke (1988) has given an informative account of the Yangtze, its history, hydrology, geomorphology and recent sociology.

56 World Resources Institute (1994).

57 Chau (1995).

58 See, for example, Dai Qing (1994); Barber and Ryder (1993).

59 As cited by Alwen (1995).

60 Norway is distinguished by such an approach to power generation. Electricity is Norway's cheapest form of energy which it produces in excess and sells to Sweden. Yet one is not often conscious of large dams. Where possible, generators are installed even in small waterfalls and rapids.

61 Editors (1994).

62 For example, see Warwick's (1996) report on the implications of a proposed dam on the Epupa Falls in Namibia.

Notes with Chapter 10

1 Davis (1991). Dyson (1996), in a very pedantic book review, had this to say: "Almost two hundred years ago, in 1798, Robert Malthus published his famous *Essay on the Principle of Population as It Affects the Future Improvement of Society*, deducing dire predictions of future misery from two hypotheses that he called the geometrical increase of population and the arithmetical increase of subsistence. His predictions of inevitable poverty failed because his hypotheses were faulty. The increase of subsistence turned out to be much faster than arithmetical. Nevertheless, uncritical belief in Malthus's predictions helped to hold back political and social progress in Britain for a century."

The failure to take into account either actual growth rates of the human population or the wider implications of time scales is common in this type of argument. The point about time scales is that there seems to be a widespread inability to recognise or acknowledge that a dynamic system that apparently functions very well (by selected criteria) over a short period can actually be self-destructive over a long one. My defence of Malthus might be seen as an apparent contradiction of the significance I attached in Chapter 1 to agriculture's bringing about a positive feedback relation between population and the food supply. It is not. The topic is considered further in Chapters 14 and 15.

Cohen (1995a) has dealt objectively and very comprehensively with this and related topics. Among other things he has listed, in chronological order from 1975-1994, twenty six definitions of human 'carrying capacity'. See also Moffat (1996).

2 Livi-Bacci (1992). The global totals in Livi-Bacci's Table 5.1 from which this figure is derived are the sums of his 'Rich' and 'Poor' categories for all years except 1960 where there is an arithmetical error. For that year I have used the sum of his two subdivisions rather than his total. Because I have plotted the results logarithmically, a treatment that greatly reduces the impact of the error, I have not consulted the original material to ascertain in which group the actual mistake occurred. Again, given my use of logarithms, I have ignored minor discrepancies between Livi-Bacci's totals and global totals listed for 1950 and 1990 by the World Resources Institute (1994). The 1995 total is from the latter source.

3 See, for example, Charman and Murphy (1991), Leeper and Uren (1993), Pitty (1979)—to name but a few.

4 Several sources cited by Pimentel *et al.* (1995a).

5 Most erosion from North American farmland, for example, occurs with spring rains (Johnson and Lewis, 1995).

6 There is no shortage of evidence of accelerated erosion early in history. See, for example, Blaikie and Brookfield (1987, 1987a); Goudie (1986); Johnson and Lewis (1995); Runnels (1995), and Chapters 3-5.

7 Hamer (1982) cited by Seckler (1987). Kirkby (1980) says "In current soil conservation practice, it is normal to plan for acceptable rates of erosion ... equivalent to a loss of about 0.2-1 mm/year from the surface". Those rates are roughly equivalent to 2.8-14 $t.ha^{-1}.y^{-1}$. He goes on to comment "the balance between mechanical soil erosion and chemical soil formation needs to be examined more carefully". Edwards (Note 9, below) cites a 1939 reference (Bennett) for a formation rate of 1 $mm.y^{-1}$ (12-14 $t.ha^{-1}.y^{-1}$) for North American cultivated soils supplied with organic matter. He queries the validity of this figure which he says is the "basis for the frequently adopted figure of 11.2 t/ha/yr as a maximum allowable soil loss for deep soils in the USA". He questions whether "any soil loss is acceptable" and adds the comment that the "figures quoted" are probably an order of magnitude too high.

8 Goudie (1986).

9 These rates are derived from Edwards (1991). A simple statement such as this, however, cannot make adequate allowance for regional variations. For example, on the Southern Tablelands of NSW a 'duplex' soil has an estimated age of 500,000 y. At another location in the same general area, a similar type of soil is estimated to be no older than 20,000 y (R. J. Wasson, personal communication, August 1997).

In converting soil quantities from depth (mm) to mass (tonnes) I have used Edwards's (above) density (for Australian soils) of 1.4 $g.cm^{-3}$, so that 1 mm of soil is equivalent to 14 $t.ha^{-1}$. Some other authors use slightly lower densities, *e.g.* 1.28 $g.cm^{-3}$ for northern hemisphere and tropical soils (for example Seckler, 1987).

10 See, for example, Crosson (1995); Pimentel *et al.* (1995a and b); Seckler (1987); Stocking (1987).

11 Wasson *et al.* (1996).

12 Wasson *et al.* (1996) point out that 47% of the Australian land surface drains to the interior of the continent. Of that, 74%—or 35% of the total land surface—produces no measurable runoff. An interesting comparison here is with the USA where a recent estimate of erosion by water is '1 billion tons' (1 Gt) annually (Glanz, 1994). After effectively completing this manuscript I encountered the article by Trimble and Crosson (2000) which describes some of the difficulties and resulting misinformation associated with attempts to measure erosion rates in the USA.

13 Gully erosion moves predominantly subsoil and will not have such an immediate widespread impact on agricultural production as a similar amount of soil lost to sheet and rill erosion.

14 Australian Bureau of Statistics (1996b).

15 All people employed in agriculture (401,900), including service industries (Australian Bureau of Statistics 1996b).

16 World Resources Institute (1994). A later estimate allocates 1.51×10^9 ha to cultivation and 3×10^9 ha to pasture and rangeland (Scherr, 1999 as cited by Gleick, 2000).

17 Pimentel (1976, cited by Goudie, 1986) has estimated that water removes 4×10^9 tonnes and wind 10^9 tonnes soil annually from the contiguous states of the USA. According to statistics cited by the World Resources Institute (1992), wind erodes about half as much land globally as does water.

18 Rosewell *et al.* (1991).

19 Kirkby (1980).

20 Reisner (1986).

21 Sources of the quotations and the other information cited since Note 19 are Glantz (1994a); Coffey (1978) cited by Goudie (1986); Goudie (1986); Johnson and Lewis (1995); Reisner (1986).

Doyle (1996), citing material compiled by the US Natural Resources Conservation Service, states:

> "In 1992 wind and water caused tolerable levels of erosion on 68 percent of cropland, an improvement of 21 percent over 1982. Some of the improvements were the result of crop rotation and better tilling methods but more important have been the efforts of the Conservation Reserve Program in which the government pays farmers to remove environmentally sensitive cropland from use.
>
> "Nevertheless, some cropland in the eastern three fifths of the country was eroding excessively in 1992 ...".

22 Zonn *et al.* (1994).

23 Heathcote (1994).

24 McTainsh and Leys (1993).

25 Sneath (1998).

26 Tegen *et al.* (1996).

27 World Resources Institute (1994). Later figures indicate a global average annual rate of deforestation of 11.26×10^6 ha.y^{-1} for the period 1990-1995 (World Resources Institute, 1998). This is equivalent to an annual loss of 0.32% of the 1990 area.

28 See, for example, Daily (1995).

29 See Tables 19.3 and 19.4, World Resources Institute (WRI) (1992). There are differences in land classification between these tables and Table 17.1 in WRI (1994). In the former, $6,092 \times 10^6$ ha are classified as 'Permanent Agriculture and Stabilized Terrain' and $3,486 \times 10^6$ ha as 'Natural Area'. The degraded area is given as $1,964 \times 10^6$ ha. Thus the total vegetated area amounts to $11,542 \times 10^6$ ha, 17% of which is degraded. Table 17.1 in the 1994 edition, however, lists three types of vegetated land—'Cropland', 'Permanent Pasture' and 'Forest and Woodland'. The combined area of these groups amounts to $8,697 \times 10^6$ ha, of which the total degraded area from WRI (1992) is equivalent to 23%. The same degraded area is 15% of the earth's total land surface.

The assessment classifies degradation as: 'Light'—a small decline in agricultural productivity but retaining full potential for recovery: 'Moderate'—a considerable reduction in agricultural productivity, amenable to restoration only through considerable financial and technical investment: 'Strong' or 'Severe'—no existing agricultural capacity, amenable to restoration only with major international assistance: 'Extreme'—incapable of supporting agriculture and not reclaimable. The total area of land listed in the last three categories amounts to 12,155 x 10^6 km^2—14% of the combined area of crops, pasture and wooded land (WRI 1994) or 11% of the 'vegetated land' (WRI 1992).

Another assessment puts soil lost from crop production at about 10 million ha.y^{-1} from a global total of 1.5 x 10^9 ha arable land. Half of the loss is from "the 590 million hectares in arid and semi-arid regions, including the North American prairies and the Central Asian steppes" (report of United Nations Environmental Programme as cited by Pimm, 1997b).

30 For example, Leeper and Uren (1993).

31 The term, 'organic matter', is often used loosely to mean the non-living organic components of soil and exclude the organisms—or at least to give that impression. Organisms present in a soil sample are necessarily included in estimations of organic carbon (*e.g.* Rayment and Higginson, 1992). A further complication in estimating organic matter is that elementary carbon (charcoal) will also be included in any estimate based on the oxidation of carbon to carbon dioxide. This can be quite significant in Australian soils which, partly because of the frequency of bushfires, commonly contain substantial quantities of elementary carbon, sometimes amounting to as much as 30% of total soil carbon (Skjemstad *et al.* 1996).

32 See, for example, Chartres *et al.* (1992). Bernal *et al.* (1993) use a conversion factor of 1.72. Yet another rule of thumb is to determine the amount of bound nitrogen and multiply it by 20 (Leeper and Uren, 1993). All methods have their drawbacks.

33 The 2% organic carbon minimum was advanced by Greenland *et al.* (1975), cited by Chartres *et al.* (1992). Morgan and Davidson (1986) are my source of the 2% organic content minimum. Another classification (Elliott and Leys, 1991) grades 'erodibility' with organic matter thus: high organic matter (> 3%), low erodibility; moderate organic matter (2-3%), moderate erodibility; low (1-2%) to very low (< 1%) organic matter, high erodibility.

In examining literature in this general area I have occasionally wondered if some authors using the terms 'matter' or 'content' do, in fact, mean 'carbon'; the converse is less likely. But, given the nature of this book and the time that would be involved in systematically answering the question, I decided not to pursue it beyond bringing the possibility to the reader's attention.

34 Lucas *et al.* (1977), cited by Chartres *et al.* (1992). The second decimal place is probably questionable but there seems to be little doubt about the general validity of the statement. Nevertheless some types of evidence that might seem to support the broad principle can be open to other interpretations. For example, over the period 1967-91, 'average' Australian wheat yields (excluding those from Queensland because of lack of information) rose from about 1.1 to 1.5 t.ha^{-1}.y^{-1}. Over the same period, however, average protein content of the grain fell from about 11.5% to 10% (Australian Bureau of Statistics, 1996a).

The increase in yields can be attributed partly to the increase in application rates of inorganic fertilisers over that period. (But probably not entirely—water régimes and climatic variations might also be implicated. See, for example, Godden *et al.* 1998). The decrease in protein content of the grain is consistent with a decline in soil organic carbon. But the results can also be seen in a different light. On the figures quoted, the yield of protein over the period increased from 126.5 to 150 kg.ha^{-1}. An alternative interpretation of the drop in protein content of the grain could be that the rates, and ultimately the extent, of production of the major constituents of the seeds (polysaccharides etc and protein) responded differently to conditions that, overall, stimulated higher yields. In other words, the changing conditions stimulated polysaccharide synthesis to a greater extent than they stimulated protein synthesis.

It is virtually axiomatic that a change in the growth rate of any organism, for whatever reason, will be accompanied by different degrees of change in the rates of production of the various constituents of that organism. An analogous effect occurs in ecosystems in which a change in the rate of production of biomass can be expected to affect the quantitative species composition of the system. This type of response can easily be demonstrated experimentally by varying the flow rate of the growth medium when a number of different species of bacteria are growing simultaneously in continuous culture.

35 Bird *et al.* (1996). Conclusions about residence times of specific molecules, as distinct from generalisations about the amount of a class of substance, require the estimation of the proportions of different isotopes of the relevant element.

36 Some recent estimates (Dixon *et al.*, 1994) give the following overall densities (t.ha^{-1}) for carbon in vegetation (v) and soil (s) for forests in the high, mid and low latitude divisions of the earth. High:- v, 64; s, 343: Mid:- v, 57; s, 96: Low:- v, 121; s, 123. The authors also list values for geographical regions within the latitude belts. The overall global densities for carbon amount to 86 t.ha^{-1} (total 359 x 10^9 tonnes) in vegetation and 189 t.ha^{-1} (total 787 x 10^9 tonnes) in soil. Earlier analyses had indicated an approximately equal distribution between the two components. Another estimate, using slightly different criteria, assigns 1220 x 10^9 tonnes of organic carbon to the top metre of the planet's soils—about 50% more than in the 'standing biomass' (Sombroek *et al.*, 1993). The authors comment that this estimate does not include plant roots and soil fauna.

37 For example, over the decade from 1976-86, the organic carbon content of three types of soil in Western Australia declined faster with conventional cultivation (CC) than with direct drilling (DD). The following numbers denote organic carbon content (%) in each soil type at the beginning and end of the decade. Earthy Sand, 1976, 0.79; 1986, CC 0.51, DD 0.64: Sandy Loam, 1976, 1.55; 1986, CC 1.30, DD, 1.53: Sandy Clay Loam, 1976, 1.03; 1986, CC 0.85, DD 0.92. Estimates for intervening years are also available. Hamblin and Kyneur (1993) cited by Australian Bureau of Statistics (1996b).

38 Charman and Roper (1991).

39 Oades (1995).

40 Hunt (1980) cited by Charman and Roper (1991).

41 Dalal and Mayer (1986a).

42 Dalal and Mayer (1986b).

43 Spain (1983) cited by Chartres *et al.* (1992).

44 Smil (1993).

Notes with Chapter 11

1 Nutman (1971), Evans and Barber (1977); both cited by Prasad (1986).

2 Various sources cited by McLaughlin *et al.*(1992).

3 Respectively—Hutchinson (1944) cited by Alexander (1971), Winteringham (1980) cited by Prasad (1986), and others cited by Freiberg *et al.* (1997).

4 Winteringham (1980) cited by Prasad (1986). In this case, as well as the reduction of NO_3^- to N_2 as discussed in Chapter 2, denitrification includes production of the intermediate, nitrous oxide (N_2O).

5 In both natural and managed situations, conditions arise that, when averaged over the seasons, do indeed approximate a steady state. An oft quoted experiment from Rothamsted, England, illustrates this in an interesting way. The experiment was basically a comparison of two types of wheat field, one manured, the other not. It extended from the first half of the 19th century to the second half of the 20th. The unmanured plots reached a steady state level of about 3 t $N.ha^{-1}$ in the top 23 cm of soil (the wheat was harvested normally). The manured plots received farmyard manure at a rate of 35 $t.ha^{-1}.y^{-1}$; over the course of about 50 years they approached a steady state nitrogen content of about 7 $t.ha^{-1}$. BUT—when an unmanured plot was taken out of production and allowed to return to 'wilderness', its soil nitrogen rose over the next 80 years to about 7 $t.ha^{-1}$—the same as the manured cultivated plots. On the other hand, application to cropped plots of commercial fertiliser at a rate equivalent to 144 kg $N.ha^{-1}.y^{-1}$ produced a steady state nitrogen content of only 3.5 $t.ha^{-1}$—about 0.5 $t.ha^{-1}$ more than the unmanured cultivated plots. (Legg and Meisinger, 1982).

6 A comparison of nitrogen budgets for West Germany in 1986 and Tanzania over the period 1982-84 is instructive. Overall, West Germany was estimated to experience a nett gain of 47 kg $N.ha^{-1}.y^{-1}$ whereas Tanzania suffered a loss of 27 kg $N.ha^{-1}.y^{-1}$. Germany applied 63 times as much mineral fertiliser, 47 times as much animal manure, experienced 10 times as much atmospheric deposition and had 3 times the rate of biological N fixation as did Tanzania (Smaling *et al.*, 1996).

7 Nitrate, which carries a negative charge, is not adsorbed to clays etc which, within the 'normal' pH range, also carry a negative charge. It is therefore easily leached from soil. Ammonium ion, on the other hand, is positively charged, binds tightly to soil particles, is not very mobile in soils and not easily leached. If conditions become moderately alkaline, however, NH_4^+ is lost to the atmosphere as gaseous ammonia (NH_3).

8 Lund *et al.* (1978), cited by Legg and Meisinger (1982).

9 McLaughlin *et al.* (1992).

10 Various sources cited by Legg and Meisinger (1982).

11 This immediate part of the discussion is based on information tabulated by McLaughlin *et al.* (1992).

12 Some comparative application rates to cropland of all fertilisers (kg.ha^{-1}.y^{-1}) for 1989-91 [and for 1994] are (mean and range): Africa, 20 (0-361), [18 (0-275)]; Asia, 123 (2-440), [? (1-472), excludes Singapore (4800)]; Australia, 26 [35]; Canada, 46 [50]; Europe, 192 (84-725), [? (6-592), excludes Iceland (3433)]; New Zealand, 899 [212]; USA, 99 [103] (World Resources Institute, 1994; World Resources Institute, 1998). Reference to Australian Bureau of Statistics (1992a) indicates that 'cropland' for the purpose of this calculation includes both conventional cropland and sown pasture. The definitions of 'cropland' for the other countries listed is not clear. The relatively low rate of application in Australia can be attributed in part to the restrictions imposed by low rainfall.

13 Cheney *et al.* (1980), as cited by Galbally *et al.* (1992) estimate that, in an 'average year', about 175 Mt of fuel is burnt in Australian fires. Of that, a total of 129 Mt, or 74%, is assigned to grasslands, 'agricultural waste' (sugar cane and cereal stubble), and to clearing forest for agriculture. They also estimate that, in 'extreme' years, up to four times as much fuel can be burnt. Walker (1981), also as cited by Galbally *et al.*, estimated that, on average, a total of 298 Mt fuel is burnt annually of which 13 Mt, or about 4%, is on 'introduced pasture'. No other agricultural environment is listed but, if 'grassy woodland' and 'open forest/grassland' are treated as pasture, then the total fuel consumption in the 'agricultural' category amounts to 231 Mt, or 77.5% of the total.

   According to analyses by Aldrich and Leng (1969) as tabulated by Carter (1988), wheat grain contains about 2.5% N, representing around 71.5% of the nitrogen in the wheat plant, above ground. The remaining 28.5% is in the 'straw'. According to these figures, an average yield of 1360 kg.ha^{-1} (see Note 21) would remove some 34 kg N. ha^{-1} in the grain, leaving about 13 kg N in the straw to be burnt or ploughed in, according to practice. An analogous calculation based on the mean protein content of five strains of wheat (Watt and Merrill, 1963) and a protein nitrogen content of 16%, gives a removal of about 26 kg N.ha^{-1} in the grain and, presumably, about 10 kg.ha^{-1} left in the straw. (In a series of Indian wheats giving yields averaging 1720 and 3330 kg.ha^{-1} when grown under two different sets of conditions, the grain accounted for 38% and 40% respectively of the total mass of the plant above the ground: Sharma, 1992.)

   Similar calculations for phosphorus give, with the first assumptions, 4.6 kg P.ha^{-1} and 1.5 kg P.ha^{-1} in the grain and straw respectively and, with the second assumptions, 5.2 kg P.ha^{-1} and 1.6 kg P.ha^{-1} for grain and straw.

14 There is another factor likely to be involved in the discrepancy between the apparent excess of nitrogen application and the decline in the content of soil organic matter. The deposition of 'excess' nitrogen on native grasslands (in this case in North America, but almost certainly in most other regions as well) is associated with a reduction in species diversity and of the C:N ratio of the biomass, an increase in N mineralisation and the content of NO_3^- in the soil, high rates of loss of total N, and reduced carbon storage within the system (Wedin and Tilman, 1996).

15 Smil (1993).

16 This, of course, ignores other benefits of crop rotation and using legumes as 'green manure'. One of the principal benefits is the maintenance or improvement of overall soil quality.

17 Given the diet of a large proportion of the world's population, this is might be a bit generous. It is not as generous, however, as Smil's (1993) assumption of 70 g. protein.hd^{-1}.d^{-1}.

18 The assumptions used in these calculations included an energy requirement of 45 MJ/kg NH$_3$ for its manufacture from N$_2$ (Slesser, 1984), an energy release of 29 GJ from the complete combustion of one tonne of coal and of 42 GJ for one tonne of crude oil (Smil and Knowland, 1980), together with 50% efficiency in the capture of heat released in their combustion. For the purpose of calculating CO$_2$ emissions, the coal was assumed to be pure carbon and crude oil saturated hydrocarbon.

19 Schlesinger as reported by Moffat (1998). See also 'Various' (2002).

20 Moffat (1998). See also 'Various', (1997); Homer-Dixon (2001); Global Environmental Outlook 2000 (2000).

21 It could increase temporarily if more 'virgin' land were brought into production but, in the absence of miracles, that would accelerate land degradation.

22 Hot fires can also produce some P$_4$O$_{10}$ and P$_4$O$_6$. Both of these compounds are volatile. See Handreck (1997).

23 Rock phosphate (the agricultural term) or phosphate rock is not a pure substance even in terms of the phosphate compounds it contains. It is usually a fluorapatite or fluorapatite carbonate that can be loosely represented by the formula Ca$_5$(PO$_4$)$_3$F,CO$_3$. Depending on the other substances present the phosphorus content is likely to lie within the range 4-17% P, which is roughly equivalent to 10-40% P$_2$O$_5$. The most common form of phosphate fertiliser used, at least in Australia, is 'superphosphate', produced by treating phosphate rock with sulphuric acid. In Australia 'single superphosphate' contains 9-9.1% P. 'Double' and 'triple' superphosphate have higher phosphorus contents. (See, for example, Costin and Williams, 1983.)

24 Lipsett and Dann (1983) gave an assessment of the quantities of calcium, magnesium, nitrogen, phosphorus, potassium, sulphur and a number of trace elements exported in Australian wheat on the basis of 1976 statistics.

25 Concentrations of dissolved phosphorus in the sea range from 10-20 μg P.l^{-1} near the surface to 60-100 μg.l^{-1} at depths of 1000 m or more. Total, (that is dissolved and suspended particulate) phosphorus averages 70-75 μg.l^{-1} (Slansky, 1986). Solubilities also vary with water temperature and hence with latitude.

26 Beckerman (1995).

27 Range 0.393-0.404 g.kg^{-1} (Kennish, 1989).

28 For example, Australian soils have been assigned an average content of 0.03% total P compared with a range of 0.4-0.9% for the USA and of 0.4% average for 'ten typical English soils' (various sources cited by Blair, 1983).

29 Total consumption from Australian Bureau of Statistics (1993a) with phosphorus contents as in Table A14.

30 One of the minor uncertainties lies in the fate of the sewage. The assumption implicit in the tables is that the sewage is 'lost'—if not from the terrestrial ecosystem, then certainly from the agricultural system. The assumption is not entirely true inasmuch as some sewage is discharged onto land although, with the exception of the Werribee sewage farm in Victoria, rarely onto farm land. The proportion so used is gradually increasing. Some sewage sludge is dried and sold as fertiliser, some is burnt and the ashes are sold as fertiliser. Many rural households discharge their wastes through septic tanks or the more modern oxidation tanks. In a few of the former cases and most of the second the effluent is used to irrigate domestic gardens. Composting toilets, which offer the opportunity of using the compost as manure, are also gaining ground in rural situations. All of these options, however, add up to but a minute proportion of the total sewage discharge. I cannot quantify that proportion with any confidence—but see Chapter 14.

31 McLaughlin *et al.* (1992) cite ratios for [P applied in fertiliser]:[Premoved in produce], ranging from 1.9:1 for 'sorghum, maize and rice' to 15:1 for fresh fruit. For pasture (listed as 'Wool, meat, milk and live animals') their ratio was approximately 6.2:1. Blair (1983) gives ratios of 1.4:1 for cereals, 1.2:1 for sugar cane, 11:1 for fruit and vegetables and 2.5:1 for pasture. Another national general estimate of the ratio of application in fertiliser to removal in produce is at least 5:1 (Gifford *et al.*, 1975, cited by Barrow, 1983). Apart from obvious differences associated with soil types, the ratio is obviously affected by the amount of fertiliser applied and the history of its application.

32 The details of what happens are sensitive to the chemistry of the fertiliser. Phosphorus is released more slowly from rock phosphate than from superphosphate, and consequently produces a different set of dynamics within the soil (see, for example, Kumar *et al.*, 1993). Animal manures are different again since their phosphorus content is relatively low—c 0.06-0.1% P, depending on the animal, its diet and the water content of the dung. At least ninety times as much animal manure as superphosphate must therefore be used to provide an equivalent amount of total phosphorus. But since the manure contains a very high proportion of organic material and since much of the phosphorus is bound to this material, its effect on soil structure, chemistry and the dynamics of the phosphate cycle within it, are vastly different from those produced by inorganic phosphate fertilisers.

33 In some circumstances, this generalisation needs qualification—of course. For example, traditional wet rice cultivation in South East Asia commonly sustains yields of the order of 6 t.ha^{-1}.y^{-1} when fertilised almost solely with animal manure and irrigation water. But like Australian dryland farming, 'inputs' of phosphorus exceed the losses in harvesting, in this case by 9-10 fold (Prill-Brett, 1986). The constant availability of liquid water and the relatively uniform high temperature ensure that phosphorylated organic compounds are hydrolysed more than fast enough to meet the needs of a crop. Moreover, the aquatic system is simpler than soil in several respects, including a much lower surface area of soil particles to compete with the crop for inorganic phosphate once it has been released by hydrolysis.

34 The Institute of Ecology cited by Wells (1975).

35 Bennett *et al.* (2001) have attempted to address quantitatively the accumulation of phosphorus in soils together with effects of leaching and drainage into waterways. They estimate an accumulation of 10.5-15.5 Mt P.y^{-1} in "freshwater and terrestrial eco-systems". Some statements are confusing. For example, they state "*Production* of phosphate *rock* in 1995-1996 was 19.8 Tg.yr^{-1}" (19.8 Mt.y^{-1}). (My emphasis.) The data used for Fig. 11.1 (above) translate to a *consumption* for the year 1995/96 equivalent to about 99 Mt rock. *Production* for 1998, 1999 and 2000 amounted respectively to 145, 136, 132 Mt phosphate rock. It would seem, therefore that Bennett *et al.*'s 19.8 Mt re-ferred to elementary P which would be equivalent to about 145 Mt phosphate rock.

36 Readily accessible reserves will normally be mined in preference to those that are not. Mining the latter will demand more energy. Similarly, other things being equal, deposits with a high phosphorus content will be mined in preference to those with less phosphorus. Slansky (1986) defines 'resources' as "all rocks from which profitable in-dustrial production can be envisaged now or in the foreseeable future, taking into ac-count relevant economic and technological factors". 'Reserves' are defined as "identi-fied resources, the mining of which seems profitable or, in some cases, almost profit-able ... under present economic conditions".

37 Respectively from Cook (1983), us Bureau of Mines (1983) as cited by Slansky (1986) and Steen (1998) citing quantities for 1996.

38 Based on rates of phosphate rock production as cited by Cook (1983) and the Inter-national Fertilizer Industry Association (http://www.fertilizer.org). Rates of applica-tion are as represented in Fig. 11.1.

39 Wells (1975). For example, the estimates ranged from a population of 3.6 x 10^9 con-suming 2.1 kg.P.hd^{-1}.y^{-1} to a population of 100 x 10^9 consuming 893 kg.P.hd^{-1}.y^{-1}!

40 See Costin and Williams (1983).

41 Lewis (1983). These are presumably 1966 rates. At the time Australian currency was beginning its transition from pounds, shillings and pence to a decimal system.

42 Gibbs (1996).

43 For example Coghlan (1996), MacKenzie (1994), Pearce (1994c), Bongaarts (1998) and Chapter 14. See also Cohen (1995a) for a comprehensive treatment of human pop-ulation dynamics.

Notes with Chapter 12

1 It can also have its limitations in assessing the performance of an organisation. For example George and Sabelli (1994), in a comprehensive account of the activities and modus operandi of the World Bank, have this to say:

"Preston (President of the Bank) doesn't say either which criteria will measure the Bank's performance. These could be anything from several thousand more televi-sion sets in the favelas of Rio de Janeiro to increased proportions of female children in primary school, to higher *per capita* disposable income, to zero cases of cholera and vitamin-A blindness—with a prodigious variety of quantitative and qualitative measurements in between. The vexed, value-laden question of development indicators is completely glossed over. We aren't told how sustainable poverty reduction will be measured any more than we are told who will measure it."

2   Towards the end of Chapter 2 of his book, *Small is Stupid,* which addresses environmental arguments and polemics from an economic perspective, Beckerman (1995) says "However, the whole population-environment nexus lies outside the scope of this book."

3   Noble and Dirzo (1997). L. R. Brown (1998) comments: "Over the last century, the world has lost close to half of its original forest area". Global Environment Outlook 2000, citing the World Resources Institute (1997) states that 80% "of the forests that originally covered the Earth have been cleared, fragmented or otherwise degraded". Nearly 40% of remaining natural forests are considered to be at risk from "logging, mining and other large-scale development projects". The same report gives a nett global loss of 56 x 10^6 ha forest between 1990 and 1995. About a third of the total is attributed to agriculture.

4   Clearing forests can reduce the organic content of the upper layers of soil by as much as 50% (Sombroek *et al.*, 1993).

5   Turner *et al.* (1997) have estimated CO_2 fluxes to and from the atmosphere for the (former) USSR, the USA and Brazil. They concluded that virtually all terrestrial biological activity (including agriculture and the management of forests) in the first two countries produced a nett uptake of CO_2 equivalent to 377 and 121 Mt C.y^{-1} for the USSR and the USA respectively. In Brazil's case there was an estimated nett efflux to the atmosphere of 248 Mt C.y^{-1}. When the combustion of fossil fuels was included in the assessment, however, there was a nett *efflux* of 762, 1185 and 301 Mt C.y^{-1} from the USSR, the USA and Brazil respectively. For the period 1980-89 Pacala *et al.* (2001) estimated a carbon sink in the "coterminous United States" of 300-580 Mt. C.y^{-1}. Jingyun Fang *et al.* (2001) have reported an assessment of changes in carbon storage capacity of Chinese forests over the period 1949-1998. See also Note 7.

Recent evidence indicates the global carbon balance is approximately as follows: about 8 Gt C.y^{-1} is released to the atmosphere of which some 6.5 Gt is derived from the combustion of fossil fuels and 1.6 Gt from deforestation. About 2 Gt is assimilated in the oceans and 500 Mt in northern forests. The fate of about 1.5 Gt is uncertain but is thought also to be assimilated in northern forests. This leaves a nett annual addition of around 3.5-4 Gt C (approximately 13-15 billion tonnes CO_2) to the atmosphere (IGBP Terrestrial Carbon Working Group, 1998; Williams, 1998).

6   For example, see Zaimeche (1994) for a discussion of the apparent effects of deforestation on local climate in Algeria.

7   Hobbs *et al.* (1997). One estimate of the production of 'greenhouse' gases by forest fires is 1.6 Mt. C.y^{-1}—attributed to tropical 'slash and burn' farmers (Kaiser, 1997a). See also Note 5.

8   Baker (1997). In Indonesia fires are used routinely "as a cheap and quick method of land-clearing during the dry months". Some 300,000 ha were designated for clearing in the second half of 1997.

Fires got out of hand and, by the end of November, 1997 at least 1,700,000 ha were estimated to have been burnt (Vidal, 1997c). Most of Indonesia, in particular Sumatra and Borneo, most of Malaysia and part of Thailand were covered in smoke so dense that visibility was described as 'zero' in the areas most severely affected (Baker, 1997; Williams, 1997b). Another assessment had visibility reduced to about 100 m over a very wide area (Australian Broadcasting Corporation news bulletin, 25th September, 1997). According to Williams (1997a): "At least 20 million Indonesians are exposed to dangerous air pollution levels and tens of millions more to unhealthy smog on the edges of the fire zones and in the big cities." Other reports (*e.g.* Vidal, 1997c) stated that "some 100,000 people across Southeast Asia" have sought medical attention for respiratory problems and that about 500 had died as a result (direct and indirect?) of the smoke. An additional 234 were killed in the crash of an airliner, possibly caused by restricted visibility.

The Indonesian government had previously attributed the annual burning to shifting cultivators, but the distribution of the fires, revealed in recent satellite images, has shown that logging and plantation companies, "many with links to senior officials" are primarily responsible (Williams, 1997a). According to Vidal (1997c), the Indonesian Environment Forum estimated that about 80% of the fires "started in intensive palmoil or timber plantations, a smaller number in forests, and even fewer as a result of local farmers routinely burning off small one- or two-hectare plots". Williams (1997a) adds that, in 1966, 82% of Indonesia was covered with 'primary forest'. By mid-1997 the area had been reduced to about 55%. Baker reports that in 1992 and 1994 more than 2,000,000 ha of Indonesian forest were destroyed by fire before the monsoon periods. Indonesian officials said that the 1997 conditions were much worse.

In 1996 Indonesia was rated as 'the world's biggest' exporter of plywood. More than 30% of all logging concessions within Indonesia are reported to be controlled by "10 companies with close links to the family of (former) President Suharto". Some of these companies are Malaysian (Baker, 1997).

Suharto is reported to have apologised for the effect of the smoke on neighbouring countries but "went on to argue that environmental degradation was sometimes necessary to achieve economic development" (Baker, 1997).

9 Mims *et al.* (1997).

10 For example, Dixon *et al.* (1994), whose tabulated summary I have condensed in Table 12.1, attributed to Australia a clearing rate of about 100,000 ha.y^{-1}. On the other hand, Barson *et al.* (1995), have estimated the average clearing rate of Australian agricultural land over the period 1983-93 at approximately 518,000 (517,714) ha.y^{-1}. Their estimates for 1988 and 1990 are approximately 718,000 and 664,000 ha respectively. A report from the Australian Conservation Foundation in November 1998 attributed to Australia a clearing rate of 'native vegetation' of 400,000 ha.y^{-1}. Sivertsen (1994) has assessed the clearing of 'native woody vegetation' in the NSW wheat belt which covers some 180,000 km^2 (about 22.5% of the state). Over the period 1977-1985 about 70% of the 'remaining native woody vegetation' was cleared. A report by the ABC on 23rd

August, 2000 indicated clearing in the state of Queensland to have averaged 400,000 ha.y^{-1} since 1997.

Tolba *et al.* (1992), as cited by Holdgate (1996), assess total annual deforestation rates in 87 tropical countries over the period 1981-90 at 16.9 x 10^6 ha compared with the 15.4 x 10^6 ha listed by Dixon and associates. A 1997 FAO report (as cited by Noble and Dirzo, 1997) puts global rates of deforestation at an average 13 x 10^6 ha.y^{-1} from 1980-1995. About 10% of this clearance was compensated by new plantations. One of the causes of discrepancies may well lie in inaccuracies, or perceived inaccuracies, in remote sensing measurements from satellites (see, for example, Hecht, 1996). Referring specifically to Amazonia, Nepstad *et al.* (1999) argue that a widespread impoverishment of forests by logging and fire is not adequately recognised in general assessments of rates of deforestation. In particular, they estimate that, annually, logging crews severely damage 10,000-15,000 km^2 of forest and that this area is not included in deforestation mapping programmes. They conclude that current estimates of deforestation rates represent less than half the area degraded in 'normal' years, and an even smaller fraction during drought. This has relevance for figures cited by the World Resources Institute (1998) indicating annual rates of deforestation in the Brazilian Amazon peaking at 30,000 km^2 in 1995 and falling to about 18,000 km^2 in 1996.

11 Examples of different assessments of African forests arising out of perceptions such as these can be found in de Selincourt (1996) and Grove (1996). Conway (1997) defines forest as "land with a minimum cover of 10 percent of crown coverage of trees". Holmgren *et al.* (1994) have produced a fairly optimistic assessment of the number of trees on Kenyan farms, but they estimate trees largely as 'woody biomass'. This would satisfy some of the definitions referred to in the text and the facts are indeed encouraging in their implications for the state of Kenya's agricultural soils. The report, however, is not likely to allay concerns about the broader ecological role of forests.

12 For example, Anderson (1993).

13 In its Table 17.1, however, World Resources Institute (1994) lists an area for 'Forest and Woodland' and a global total in that classification of 3898 million hectares for the years 1989-91. This is 6.4% less than the area of forest derived in Table 12.1 from Dixon *et al.* (1994).

14 Dixon *et al.* (1994).

15 Skole and Tucker (1993). See also Aldhous (1993). During the 1980s, Brazil alone had a clearance rate of 1-2 million ha.y^{-1} (INPE, 1992, cited by Turner *et al.*, 1997).

16 For example, Anderson (1995b), Baird (1996). See also Note 8.

17 FAO as cited by L. R. Brown (1998).

18 A report, *Towards a Sustainable Paper Cycle*, prepared by the International Institute for the Environment and Development; cited by Knight (1996). See also Holdgate (1996). Another estimate is that about 10 million ha of additional land is needed annually to provide the food needed by the increments in population (Palmer and Synnott, 1992, as cited by Noble and Dirzo, 1997). But see also Chapter 14.

Farmers in tropical Africa, having been encouraged or obliged to abandon their traditional practices and grow cash crops, can rarely afford the amount of fertiliser needed to sustain commercial farming. According to one assessment (Abelson, 1995), about 10% of the amount of fertiliser needed is applied and "in desperation, poverty-stricken, hungry rural people destroy moist tropical forests".

19 For example, Homewood (1993), Shankland (1993). In the late 1970s cattle graziers are estimated to have cleared 20,000 km² of South American forests annually (see Margules and Gaston, 1994). Clearing and fragmentation of the Amazon is accelerating with the construction, in progress or planned, of "Roads, railways, gas and oil pipelines, industrial waterways, mining operations and hydro-electric dams" (Rocha, 1997c). These activities are financed internationally with an aim, among others, of improving access to ports. Deforestation in Guyana and associated displacement of the local people is reportedly financed by 'aid' money given by the British Overseas Development Administration to 'reform' Guyana's timber industry (Colchester, 1997).

20 Rifkin (1992). Conway (1997) states that an increase in cattle numbers accounts for about half of the global rate of deforestation and about 90% of deforestation in the Amazon.

21 Norman Myers in a radio interview broadcast by the Australian Broadcasting Corporation in its programme 'Earthbeat' on 19th and 21st November, 1996. See also Walker (1996). Myers (cited by Margules and Gaston, 1994) also estimates that the human population of Ecuadorian Amazonia increased from about 45,000 in 1950 to about 350,000 in 1994. Peasant farmers were displaced in the process.

22 For example, Barraclough and Ghimire (1996).

23 Dudley *et al.* (1998)

24 For example, Kaiser and Gallagher (1997).

25 For example, Conway (1997).

26 See MacIlvain *et al.* (1998).

27 Simard *et al.* (1997).

28 Skole and Tucker (1993) have given quantitative information about this type of effect in Amazonian rainforests. See also Holdgate (1996) for a level-headed discussion of some of the ramifications of biological diversity. The relation between area of habitat and species diversity was first quantified by MacArthur and Wilson (1967) who noted that a tenfold change in area is reflected in a corresponding twofold change in the number of species sustained within the area. The history of the recognition of this relation has been summarised by Wilson (1996). The quantitative relation between area and number of species, however, is more complex than this (*e.g.* Harte *et al.*, 1999; Rosenzweig, 1999).

The details of the dynamic responses of complex ecosystems to challenges can, of course, vary widely with the type of ecosystem—as can the effects of area on the degree of complexity and the dynamics of those responses. See Wardle *et al.* (1997a and b), Tilman *et al.* (1997a and b), Hooper and Vitousek (1997), Diamond, (2001), Terborgh *et al.* (2001).

29 *E.g.* Laurance *et al.* (1997), Gascon *et al.* (2000),

30 Cotton and rice currently account for the greatest part of the world pesticide market (World Resources Institute, 1994, Chapter 6). In 1991-92, Australia used 4,688

tonnes of pesticides of all types (Australian Bureau of Statistics, 1996b). World Resources Institute (1994) lists comparable information for only a few countries: *e.g.* Colombia, 20,019 t; Germany, 36,937 t; Thailand, 36,694 t; Zimbabwe, 4,268 t.

31 Wargo (1996). In June 1997 a research chemist died some 10 months after spilling a few drops of dimethyl mercury onto one of the latex rubber gloves she was wearing as part of her protective clothing. The compound penetrated the gloves within a few seconds and was absorbed directly and rapidly through the skin. Within a few months she had developed neurological symptoms. The concentration of mercury in her blood was found to be 80 times the level regarded as the threshold for toxicity. Dimethyl mercury has been used widely to fumigate crops. (Holden, 1997).

32 Selinger (1986) has provided a useful summary of the essential chemistry of the principal pesticides.

33 For example, Dinham (1991).

34 World Resources Institute (1994), Chapter 6.

35 Probably the best example of the first mechanism is to be found in the antibiotic, penicillin. This specifically blocks the biosynthesis of bacterial peptidoglycan—or 'murein'—the major constituent of the cell walls of Gram-positive bacteria and an important constituent of the cell envelopes of Gram-negative bacteria. So far murein has been found only in bacteria and, because its synthesis is thus apparently unique to bacteria, penicillin produces virtually no biochemical side effects in mammals, although allergic reactions are not uncommon.

Penicillin is bactericidal—not bacteriostatic. Because it prevents synthesis of an essential structural cell component, it is effective only against bacteria that would otherwise be growing. Dormant forms, or cells in an environment that does not permit growth, are not affected. For that reason penicillin cannot be used effectively in conjunction with another antibiotic that is merely bacteriostatic. Resistance to penicillin occurs principally through production of the enzyme, penicillinase, which breaks down the antibiotic. An analogous type of mechanism for an insecticide would involve a highly specific inhibitor of chitin biosynthesis. An insecticide known by the trade name of 'Helix' reportedly functions by inhibiting chitin formation, but its specificity is not clear from the account of its use that I have seen (McHugh, 1996).

36 Soderlund and Bloomquist (1990).

37 Rats suffer liver damage on a diet containing 5 ppm DDT. During the period when DDT was widely used in the industrial countries, it was commonly found at concentrations of 10 ppm in human body fat (Wargo, 1996).

38 World Resources Institute (1994, Chapter 6) reports that "the manufacturer of chlordane, one of the most toxic and persistent pesticides ever formulated, increased its exports tenfold between 1987 and 1990 to over 3.6 tons per day". Over the period, 1970-1990, global pesticide use increased 3-fold while the cash value of agricultural produce increased by about 58%. According to a radio news report by the Australian Broadcasting Corporation on 28th July, 1997, there are some 20,000 tonnes of old pesticides accumulated throughout Africa—and in need of safe disposal. However, some restrictions have been placed on export of pesticides. World Resources Institute

(1998) cites statistics from the Foundation for Advancement in Science and Education to the effect that, for the years 1992-94 inclusive, export of 107,782 tonnes of pesticides from the USA was "banned" or "restricted".

39 A nerve impulse is transmitted from one nerve cell to another, or from a nerve cell to an appropriate cell in striated muscle, by the transfer of a neurotransmitter—a substance with the capacity to attach itself specifically to a receptor site and stimulate that site. If the neurotransmitter remains attached, however, the receptor is blocked against subsequent stimuli. Continued functioning of a nerve therefore depends on enzymic breakdown of the neurotransmitter molecule, once it has done its job.

One of the most important, and probably the most comprehensively studied, of the neurotransmitters is the ester, acetylcholine. After discharge it is hydrolysed by the enzyme, acetylcholinesterase. Nerve gases and organophosphate insecticides (and some other insecticides such as the nitrogenous compounds collectively known as 'carbamates') inhibit this enzyme.

40 There are numerous accounts of illnesses caused in this way. The practice can lead to antagonism, threats of violence, or actual violence between different types of farmers—in Australia perhaps most often between graziers and cotton farmers. A brief account of illness, antagonism and threats of violence in a cotton area of north western New South Wales has been given by Passey (1995a). See also Hogarth (1997a and b) and Note 42.

41 In 1995, 200 itinerant grape pickers in California were sprayed by a 'cropduster' while they were working. The medical consequences were serious and extensive. Apart from any other consideration, one might be forgiven for wondering about the competence of those responsible for spraying fruit that are ready to pick. According to the report (Hirst and Woodley, 1997), the pilot's licence was suspended after 'months of indifference'. McHugh (1996) has given a comprehensive account of similar types of incident in the cotton fields of northern NSW. Workers directly affected by spraying included 'flaggers' (the human equivalent of marker buoys) sprayed in the line of duty, and 'chippers' affected in most instances by drift while weeding nearby rows. On one occasion in 1979 a team of aboriginal chippers was directly sprayed, despite having been assured that they would be able to leave the field before spraying started. Later the manager of the farm justified, or at least explained, this incident in terms of being 'in a hurry'.

42 Bartle (1991) has described experiences of people in Britain accidently exposed to pesticides, mostly organophosphates. There is some serious biochemical misinformation in the article, but there is no reason to doubt the sociological aspects—in particular the indifference, obfuscation and bureaucratic ineptitude encountered by victims of pesticide poisoning when they sought redress. McHugh (1996) has also documented incidents in which householders and people using roads in the NSW cotton districts were sprayed with and affected by pesticides. Complaints about such incidents—and there were many—were at best of limited effectiveness, usually because the 'no scientific proof' argument (see Chapter 10) was invoked. Sometimes complaints generated death threats (see also Note 40).

43 Van den Bosch (1978) has described, *inter alia*, an itinerant Mexican farm labourer in California in the 1970s applying parathion to a crop. He did not read English and did not know what the pesticide was. He was bare chested and his body was coated with parathion.

A publication about farm machinery and agricultural mechanisation (Kaul and Egbo, 1985—and reprinted at least until 1992) is part of a series of 'authoritative core texts for post 'O' level students' and describes itself as "a core textbook covering all the mechanism requirements of the general student of Agriculture". On page 107 there is a photograph illustrating the use of a "shoulder-mounted powered crop duster". The duster is operating and the farm worker using it is exposed to a cloud of the pesticide. He is black, his sleeves are rolled up, he is wearing what might be goggles but there is no sign of any kind of mask. The book has an appendix on "safety for agricultural equipment" dealing entirely with precautions against physical injury. I could find no mention of safety measures against poisoning.

44 Green *et al.* (2001). Arsenic, being an element, cannot be changed into 'something else'—other than by the techniques of nuclear physics, an approach that does not immediately suggest itself for treating large quantities of soil. Arsenic compounds can be treated chemically to change them into different compounds, but the arsenic will remain as arsenic and the new compounds will also be toxic. The ultimate insanity in this general context was the use of lead arsenate as a pesticide applied to, among other things, fruit trees. The only way to get rid of arsenic from soil—or for that matter from anywhere—is to move it physically 'somewhere else'—which is what the Bogong moths were doing.

45 Passey (1995b).

46 Erlichman (1995).

47 Anon (1995b).

48 An increasing concentration of pesticides in groundwater, on which Denmark depends to a large extent on for its domestic water supply, has led to the Danish government's considering the banning of all pesticides (Anon, 1997a). According to another report, Sweden is contemplating the banning of all persistent 'chemicals', whether or not they are toxic (MacKenzie, 1997). It is not clear from the report how a 'chemical' is defined in this case.

49 See Karliner *et al.* (1997) from which most of this information was derived.

50 For some comments on poisoning in India, see Kumar (1995b).

51 A joint report of WHO/UNEP cited by Dinham (1991).

52 Pearce (1996).

53 Both examples cited by World Resources Institute (1994). World Resources Institute (1998) cites another estimate suggesting that "between 50 million and 100 million people in the developing world may receive intensive pesticide exposure, and another 500 million receive lower exposures". These contacts were expected to lead to 3.5-5 million "acute pesticide poisonings per year, with a much larger number suffering from subacute effects".

54 For example, Watterson (1991). 'Safe limits' for pesticides in foods are recommended by the World Health Organization or by national committees of some kind —or they are decreed by government departments. However, these limits rarely recognise the implications of additive or, more seriously, synergistic effects of ingesting different pesticides. Nor do they take sufficient account of the additive effects of ingesting any one pesticide from different types of food.

55 Australian Bureau of Statistics (1996b). In 1994, during a drought, cattle in northern NSW and southern Queensland were fed cotton 'trash', the foliage remaining after the harvested cotton had been processed. The cotton had been sprayed with the insecticide 'Helix' (see Note 35) which was subsequently detected in tissues of the animals, exports were suspended and cattle on a number of farms in both states were quarantined during further testing (see McHugh, 1996). Rifkin (1992) citing the (US) National Research Council states "Beef ranks second only to tomatoes as the food posing the greatest cancer risk due to pesticide contamination ... Beef is the most dangerous food in herbicide contamination and ranks third in insecticide contamination."

56 For example, Wargo (1996) points out that the USA currently imports about 10 billion (10^{10}) bananas annually but samples, on average, less than 200 ('167').

57 Groth (1996). This is a book review of Wargo (1996)—which should also be consulted.

58 Again, see World Resources Institute (1994). Coghlan (1991) has given some British examples. Note that in his brief report, Coghlan uses 'animal' to refer specifically to mammals.

59 Pimentel and Levitan (1988) cited by World Resources Institute (1994).

60 Meade (1997). The report does not identify the insecticide.

61 Resistance to any sort of pesticide can be expected to develop in target organisms if populations of those organisms are exposed to sub-lethal concentrations of the substance. The first case of natural resistance to the herbicide 'glyphosate', sold commercially as 'Roundup' or 'Zero', was reported for Australian rye grass ('Background Briefing', Australian Broadcasting Corporation, 14th September, 1997). This herbicide is widely considered to be highly specific for plants, hence essentially harmless to animals, and to be rapidly inactivated in the ground. McHugh (1996) points out that over the period 1992-1994, the New South Wales Poisons Information Centre averaged annually some 400 enquiries about treating glyphosate poisoning.

Monsanto, the chemical company which developed and patented glyphosate, has genetically modified some crops to resist the herbicide. They are designated 'Roundup Ready' and can withstand direct spraying with glyphosate. One implication of this attribute is, of course, that cultivated land can be sprayed comprehensively while crops are growing. This raises interesting possibilities for pesticide poisoning. In 1997 Monsanto split into two companies, one of which will specialise in the genetic modification ('engineering') of commercial crops. (Monbiot, 1997a and b).

Monbiot (1998) also reports that, under new proposals, the US Department of Agriculture would allow fruit and vegetables that have been "genetically engineered, irradiated, treated with additives and raised on contaminated sewage sludge" to be labelled 'organic'. Livestock could qualify despite being "housed in batteries, fed with the offal of other animals and injected with biotics".

62 Resistance to the natural toxin produced by the bacterium, *Bacillus thuringiensis* ('Bt') is inevitable. The relevant gene from this bacterium has been incorporated into the genomes of some crops, in particular maize and cotton. In 1997 about 2.9×10^6 ha, or 9% of the US maize crop, and 850,000 ha, or 15% of the cotton crop, were modified in this way. In 1996 'Bt' cotton plantations in Texas experienced severe infestation with cotton bollworm. (Wadman, 1997a).

63 Several sources cited by World Resources Institute (1994). An instructive illustration of the effect of a pesticide on production of a crop is given by the application of DDT to cotton crops in India over the period 1970-1978. Over that period the amount of DDT used increased approximately exponentially from 2880 to 7040 tonnes. At the same time cotton production increased by 18.8%. This is my interpolation from the graphical data of Chapin and Wasserman (1981) as cited by Montesano *et al.* (1988).

Notes with Chapter 13

1 This is based first on a total of 233,688 rural farmers, farm managers and farm labourers in 1991 (Australian Bureau of Statistics, 1996b, Table 4.14, p. 75) and a total population of 17.2 million. In Table 4.3 and elsewhere the same publication gives slightly different numbers and it is not always possible to distinguish between farmers and those in agricultural service industries. The calculation was adjusted for the amount of produce exported but not for the amount imported. The result was also subject to the choice of mass of produce or its phosphorus content when calculating the effect of export (see Table A14), but this effect was not sufficient to change the whole number from 99%. (By 2000 the number of farm workers had increased to 319,074 and the total population was estimated to be 19.3 million (Australian Bureau of Statistics, 2001; Australian State of the Environment Committee, 2001). These numbers give a surplus of slighly over 98%.)

2 Information in Table 17.2 of World Resources Institute (1994) about the distribution of the labour force points to a 'basic' surplus for the USA of slightly over 99%. Other assumptions of the type I have made for Australia point an adjusted surplus of around 96-97%.

3 Australian Bureau of Statistics (1994, 2001). Definitions based on the criterion of 'Estimated Value of Agricultural Output' (EVAO!) give different statistics according to that value. In 1985-86 the EVAO was raised from $2,500 to $20,000 annually. In 1993-94 it was lowered to $5,000. The total number of farms so defined fell from 251,982 in 1960-61 to 117,189 in 1993-94 before the change in EVAO, and rose to 150,389 after the change. (Australian Bureau of Statistics, 1996b). The mean area of the relevant establishments was 1858 ha in 1960-61, 3882 ha in 1993-94 before and 3199 ha after the change in EVAO. The total areas corresponding to those averages were 468 million, 455 million and 469 million hectares respectively.

4 According to one report, the largest single cattle station in the world was Anna Creek in South Australia, with a total area of 3,002,800 ha. In late 1997, it was to be broken up to provide a mining company with access to artesian water (Hoy, 1997).

5 As in Fig. 13.2, this calculation used the midpoint of each size range to calculate total area.

6 Using a total of 288,690 as the agricultural work force—that is both 'urban' and 'rural'. (Australian Bureau of Statistics, 1996b).

7 The list of "establishments with agricultural activity" includes "plant nurseries", "cut flowers and flower seed growing" etc.

8 These areas are very different from those that might be inferred from some other publications. For example, Lowry (1991) attributes to Oceania (in 1982), 99.72 ha 'arable or cropland' per agricultural worker. The next highest area in his Table 4 is North America with 91.98 ha. One perpetual difficulty in comparisons of this kind lies in the definitions.

9 Australian Bureau of Statistics (1996a, 2001).

10 Wahlquist (1993).

11 A reduction in the number of species is an integral part of the conversion of a natural ecosystem to any type of farm, but modern agribusiness has added another dimension to the process. The commercialisation of plant breeding with its attendant seed industry has reduced the number of varieties of specific crops that are readily available (see, for example, Vellvé 1993). This has potential dangers analogous to those associated with a reduction in the number of species—but with more immediate social, political and economic ramifications.

There are other disturbing aspects of the industrial control of agricultural seeds. After Monsanto split into two companies (Chapter 12, Note 61), the branch devoted to 'life sciences' developed a number of genetically engineered crops resistant to the herbicide glyphosate (sold under the trade name of 'Roundup'). The idea here, of course, is to enable the herbicide to be applied after a crop is established. But that is not all. Monsanto added another little twist in what has been called 'terminator technology'. Crops that have been modified in this way produce sterile seeds, thereby preventing farmers from saving seeds for next year's planting. To make the package even more attractive, Monsanto has reportedly employed private detectives to report any unofficial exchange of seeds among farmers (Service, 1998; for an account of some of the wider implications of this practice see also Berlan and Lewontin, 1999). According to Vidal (1999), pressure from the President of the Rockfeller Foundation eventually persuaded Monsanto to abandon research into 'terminator technology'.

12 Australian Bureau of Statistics (1996b). Over the past few decades the energy requirements of modern agriculture have received a good deal of attention. See, for example, Stanhill (1984). It is highly probable that, within the first 10-20 years of the 21st century, oil prices will rise sharply as accessible stocks are depleted (Hatfield, 1997a and b; Houthakker, 1997; Campbell and Laherrère, 1998).

13 Australian Bureau of Statistics (1996b, 2001). Of course yields vary in response to rainfall.

14 Tilman *et al.* (1996). Tilman and his associates used 147 experimental plots with 1-24 species in different combinations. See also Hector *et al.* (1999).

15 Tangley (1996).

16 MacVean (1997). See also Katz (1997) and Young (1997).

17 Schroeder and Balassa (1963) cited by McLaughlin *et al.* (1996).

18 McLaughlin *et al.* (1996). According to two separate issues of World Resources, the global average amount of fertiliser (presumably of all types) applied in a year to cropland (kg.ha^{-1}) was, 81 in 1979-81, 107 in 1984, 96 in 1989-91 and 113 in 1994 (World Resources Institute 1994 and 1998). Rates for some selected regions in 1994 were: Africa, 18 (range for African nations 0-275); Asia, ? (range 1-4800) (Asia's average for 1989-91 was 123); North America 92 (Canada 50, USA 103); South America 60 (range 4-95); Europe ? (average for 1989-91, 192), range 6-3433); Australia, 35; New Zealand, 212. The low application rate in Australia is largely an acknowledgment of a limited water supply.

19 World Resources Institute (1994); Worldwatch Institute (2001).

20 Australian Bureau of Statistics (2001).

21 Assuming a mean mass of cattle of 400 kg, a mean human mass of 50 kg globally, of 60 kg in the USA and of 55 kg in Australia. Australian steers normally reach a mass of around 350 kg by the age of one year. The average carcase weight of adult cattle slaughtered in 1990-92 was about 235 kg; for all cattle (*i.e.* including calves) it was about 214 kg (Australian Bureau of Statistics 1992c; Coombs 1993). The corresponding calculation for sheep assumed an average mass of 60 kg.

22 *E.g.* Australian Bureau of Statistics (1996b).

23 For those unfamiliar with this type of country, an article by Carter (1997) is instructive. The article is about a very large highly mechanised shearing property in the north west of NSW. The nature of the country is evident from some aerial photographs. Apart from anything else there is almost no ground cover within the area illustrated.

24 Rifkin (1992).

25 Australian Bureau of Statistics (2001).

26 Statistics for 1990-91 give 34,540 t P.

27 Stocking rates for beef cattle in Australia range from over 0.5.ha^{-1} in coastal regions to less than 0.01.ha^{-1} inland (Australian Bureau of Statistics, 1996b; Coombs, 1993). About a third of Australia's beef cattle are located in the tropical north—mainly on savanna (Stewart, 1996). Australian dairy cattle are confined to coastal districts at stocking rates from about 0.8 to 1.7.ha^{-1} (*e.g.* Bartsch and Mason, undated).

28 McIntosh (1997). The concurrent rates of depletion of some other elements were (kg.ha^{-1}.y^{-1}): N, 27; K, 19; Mg, 1.4; Ca, 30.

29 The distribution of P between grain and straw is based on Pratley (1988).

30 Holt *et al.* (1996).

31 *E.g.* McNaughton *et al.* (1997).

32 But rarely very stringently—they are an influential political force.

33 I understand that South African graziers are required by law to reduce stocking densities to 30% of 'normal' as soon as an area is declared officially to be in a state of drought. I am not sure how a 'normal' stocking rate is defined.

34 Pearce (1997).

35 *E.g.* Costin (1959).

36 According to a summary of 'schoolboy howlers' published in the *Guardian Weekly* some time in 1996, "trees can break wind for up to 200 metres".

37 A friend and former colleague of mine, investigating various aspects of the physiology and ecology of the platypus, spent some of his time at sites on the upper Shoalhaven River in NSW. Over the course of a few years he noted increasing bank erosion and sediment deposition at relevant locations. Accordingly he suggested to the grazier on whose property some sites were located that it might be a good idea to exclude cattle from the bank. The grazier's response was to deny him further access to the sites.

Another friend with a riparian grazing property on the lower Tuross River in NSW fenced off the bank to a distance of about 10 m from the water. In the space of three years there was a dense growth of understorey in the protected area. The bank was stabilised and suffered no damage in a subsequent flood. See also Chapter 7 and comments on the early settlement of the Monaro district in New South Wales.

38 On present evidence, prions are protein molecules that exert their effect by modifying a protein already present in the host's nervous tissue. That protein is chemically the same as the prion, that is to say, it has the same amino acid composition and sequence. The two protein molecules, the 'normal' one and the prion, are functionally different, however, because each has a distinctive conformation or shape. The prion apparently exerts its effect by changing the conformation of the 'healthy' protein to that of the invading prion. This process has two effects; it causes the transformed protein to interfere with the functioning of affected cells, and it provides the invading protein with a means of proliferating—two fundamental characteristics of a pathogen. The word 'prion' is a kind of acronym derived from 'protein only' (see, for example, Prusiner, 1995, 1997; Telling *et al.*, 1996).

39 Much has been written about the BSE outbreak. Some relevant references are: Anderson *et al.* (1996), Butler (1996), Editors (1996), O'Brien (1996). See also Valleron *et al.* (2001), d'Aignaux *et al.* (2001) for information on the epidemiology of Creutzfeld-Jacob disease.

40 According to a news report at the time the cattle were in a holding yard at an abattoir. A telephone enquiry to the Queensland Department of Primary Industries on 12th December, 1992 did not change that understanding, although the location of the cattle was not emphasised during the conversation. According to the Senate Standing Committee on Rural and Regional Affairs (1992), however, 5,500 cattle were affected and were yarded, not at an abattoir, but in a commercial feedlot. 'Chicken litter' has since been banned as a cattle feed by the Queensland government and nationally by the Australian Lot Feeders Association. It continues to be used, however, in North America and in Europe. Cement dust is sometimes included in cattle feeds in the United States (Coats, 1989; Rifkin, 1992).

41 For example, Coats (1989), Johnson (1991), Rifkin (1992).

42 Coghlan (1997a).

43 According to the World Resources Institute (1994), 37% of global grain production in 1992 was fed to livestock. The proportions for some selected regions and countries were as follow: Africa, 16%; Asia, 16%; Denmark, 85%; Canada, 75%; USA, 69%; Australia, 55%. The major components of cattle feed in Australian feedlots are sorghum and barley. In 1991-92 feedlots consumed some 460,000 tonnes sorghum and 420,000 tonnes barley, representing about 39% and 9% respectively of total Australian production of those commodities (Senate Standing Committee, 1992).

44 The Senate Committee's report (Note 43) states that the main health problems in (Australian) feedlots are "acidosis and nutritional disorders associated with a high concentrate diet". It acknowledges that respiratory infections "are significant in other countries". A newspaper article, describing recent developments in Australian feedlots (Hoy, 1996), attributed to two representatives of the Australian Lot Feeders Association statements that can be paraphrased as advising conventional graziers to 'get real'. This provoked a series of angry letters in reply ('Various', 1996a) pointing out, among other things, that (in an abattoir) "Feedlot cattle stink. As their hides are removed, great gobs of pus run down the carcass."

45 E.g. Coats (1989).

46 Unless stated otherwise, quantitative information about Australian feedlots is from Senate Standing Committee (1992).

47 Hoy (1996).

48 The calculation of nitrogen and phosphorus excretion is based on general rates listed in Table 5.17 of Australian Bureau of Statistics (1996b).

49 A submission made to the committee by the President of the Royal Society for Prevention of Cruelty to Animals included the following: " ... whether the animals are overcrowded, whether all of them can sit down, lie down, stretch, turn around and have some degree of movement; in other words, meeting their basic behavioural needs."

50 Comment from Mr Bob McCarthy, Chairman of Australian Meat and Livestock Industry Policy Council.

51 The grain fed to feedlot cattle in 1991-92 would have required approximately 540,000 ha. That amounts to slightly less than 1 animal per hectare (see above in text for cattle densities on pasture). For this calculation I used grain consumption and numbers of cattle as in the Senate Committee's report. Grain yields were derived from Australian Bureau of Statistics (1992b).

52 The Senate Committee was told of damage to a dairy farm from feedlot effluent that flowed near the house and killed pasture grasses.

53 Dickenson *et al.* (1996).

54 Hill (1996) notes that Ireland was England's first plantation, dating back to "long before the seventeenth century". His observations on the colonisation of Ireland and America, and the attitudes of the colonisers to the 'natives', make interesting reading.

55 Goldsmith (1997).

56 As cited by George and Sabelli (1994). Summers left the World Bank in 1993 to become Under-Secretary of the US Treasury.

57 A comparison of the labour requirements of shifting agriculture (see Chapter 7) and traditional wet rice cultivation is informative in this respect. According to Dickenson *et al.* (1996) shifting agriculture in most areas requires at most 120-200 (man)-days.ha^{-1} to produce a crop. Traditional wet rice cultivation, as practised in China, required up to 500 (man)-days.ha^{-1}.y^{-1}. See also Note 62.

58 Swearingen (1994). See also Conway (1997).

59 See, for example, Marten (1986) for a comprehensive account of traditional farming in Southeast Asia. Tiessen *et al.* (1992) have provided some insight into the dynamics of chemical changes in Brazilian soil under shifting cultivation. A comprehensive account of the African environment, its peoples and its agriculture has been given by Lewis and Berry (1988).

60 It can reasonably be argued that the most serious direct ecological threat to the agrarian countries is population pressure. This expresses itself in various ways, all of which reflect the basic principle that an excessive population density will consume food material and produce wastes faster than can be accommodated by other processes within the system. One obvious sign of such an imbalance is the use of animal dung for domestic fuel. This practice alone is likely to lead to soil degradation. Moreover, when population densities are so high, supplies of uncontaminated drinking water are likely to be severely limited. Obviously, problems such as these are also sensitive to economic conditions.

61 *E.g.* Lewis and Berry (1988).

62 A comprehensive FAO report (Alexandratos, 1988) estimates that, during 1982-84, tractors supplied the equivalent of 6% "labour day equivalents" in "93 developing countries" (range 1% for sub-Saharan Africa to 22% for Latin America). It goes on to estimate an annual increase of 4.1% in the contribution of tractors to "total power use" over the period 1984-2000. Conway (1997) has given a concise objective summary of the 'Green Revolution'.

63 Modified from World Resources Institute (1994).

64 For example, by George (1988), Conway (1997), and many issues of the journal, *The Ecologist*.

65 Rural Poland is experiencing a similar type of transition (Nougarède, 1997).

66 *E.g.* Sun (1994).

67 Patel (1997).

68 Cited by Elliott (1998). Elliott also reports that in Niger, a third of children die under the age of five. See also Note 80.

69 Chalmers (1997).

70 O'Toole (1996), Dhombres (1996), Sévilla (1997), Vidal (1997a), Rouard (1998).

71 As cited by Dickenson *et al.* (1996).

72 I have not consulted the original FAO report.

73 George and Sabelli (1994) have discussed in some detail the 'ethos' of the World Bank and its effects on the Bank's objectivity and policies. Similar questions can be raised about FAO (see, for example 'Various' 1991) and the IMF.

74 *E.g.* Smil (1993).

75 In the figure from which these wheat yields were interpolated, Smil also includes a graph of rice yields in China over the same period. This graph, which does not state its source, gives the rice yield in 1990 as slightly over 5600 kg.ha^{-1}. A similar graph from the United States Department of Agriculture as shown by L. R. Brown (1995) gives the 1990 yield of rice for China as approximately 3.9 tons.ha^{-1}. If 'ton' means a 'short ton' (2000 lb), this is equivalent to approximately 3500 kg.ha^{-1}. If it means a 'long ton' the yield is approximately 4000 kg.ha^{-1}. If it means 'metric ton' then, of course, is 3,900 kg.ha^{-1}. Elsewhere Smil compares stocking densities of sheep in China (0.8.ha^{-1}) with those in other countries, including New Zealand to which he attributes a density of "more than" 110.ha^{-1}. If he is talking about grazing and not some kind of intensive system, such a stocking density is impossible. A quick look at numbers listed by the World Resources Institute (1994) gives an average stocking density of sheep + goats + cattle on 'permanent pasture' in New Zealand of 4.7 ha^{-1}.

Shiva (1991) has also raised questions about the meaning of the apparent high yields given by the new strains and has outlined other consequences of their introduction into the Punjab.

76 These topics are largely outside the scope of this book, but see Nader *et al.* (1993); Hildyard, 1996; George (1988); George and Sabelli (1994); Mittelman (1997)—and many others.

77 Some implications for human health are graphically summarised by Smith (1997). This is a letter to the editor of the *Guardian Weekly* from a doctor running a district hospital in a province of Zambia. See also Anon (1996a) and Blustein (1997) for some supplementary comment on international approaches to Third World economics.

78 Lewis (1992).

79 Sweeney (1997).

80 According to one report, Africa annually pays its creditors—mainly governments of industrial countries, the World Bank and the IMF—about $10 billion. This is more than the total annual expenditure on health and education. Arrears in repayments have doubled since 1990 (Watkins, 1996).

81 Construction of the Chixoy dam in Guatemala necessitated the 'resettlement' of some 2500 Mayan Indians. According to a World Bank report, the resettlement was 'mismanaged'. This apparently means that "369 Mayan Indians—mainly women and children—(were) tortured, shot, stabbed, garotted and bludgeoned to death by the Guatemalan military in punishment for their community demanding they be properly compensated for the loss of their homes" (Rocha, 1997a). Asian logging companies are responsible for much of the deforestation in the Amazon *(e.g.* Rocha, 1997b). The literal survival of the U'wa people in Columbia is threatened by oil companies (Vidal, 1997b). See also Note 70.

In Chile, a wealthy American conservationist—and philanthropist—progressively bought some 300,000 ha of land, largely rainforest, which he intended to give to the Chilean government on condition that the land be declared a natural sanctuary. His proposal invoked widespread protest which was apparently organised. His offer was rejected. The most likely explanation of the government's refusal to accept the gift seems to be that "powerful business groups and their allies in government fear large-scale land preservation" (Escobar, 1997).

Indonesia, under the Suharto régime, had a reputation as perhaps the most corrupt nation in the world. Chomsky (1998) makes the following comment about the financial problems which that country experienced early in 1998:

"The current International Monetary Fund 'rescue package' for Indonesia approximates the estimated wealth of the Suharto family. One Indonesian economist estimates that 95 per cent of the country's foreign debt of some $80 billion is owed by 50 individuals, not the 200 million who end up suffering the costs."

82 See, for example, an assessment of the results of development in Lesotho (Ferguson and Lohmann (1994).

Notes with Chapter 14

1 Annual consumption increased from 10^9 MW-h (3.6×10^9 GJ) in 1860 to 93×10^9 MW-h (3.3×10^{11} GJ) in 1991 (Cohen 1995b). From 1985 to 1998 consumption of inanimate energy in Latin America increased by about 30%, in Africa by 40% and in Asia by 50% (Campbell and Laherrère, 1998).

2 I quoted some estimates of erosion rates in Chapter 10. Conway (1997), citing various sources, gives global land degradation since 1945 as about 2×10^9 ha, or 22.5% "of the world's agricultural, pasture, forest and woodland". Eighty per cent of Third World cropland (more than 4×10^8 ha) is given as degraded. Neither the types nor degrees of degradation are specified. The annual 'loss' of land is estimated to be 5-10 million hectares.

3 One can identify anabolic and catabolic reactions in metabolic systems but, although such reactions might be classified loosely as 'constructive' and 'destructive' in relation to specific substances, they do not normally warrant such evocative descriptions. But they can if they get out of balance. In a purely social or economic context, however, one can identify processes that are predominantly, if not wholly, destructive. The manufacture of and international trade in armaments is one example.

4 Diamond (1997) has discussed this topic at some length with due consideration of the alternative explanation that the extinctions were caused by climate change. For more recent specific evidence of human involvement in the extinction of megafauna see Miller *et al.* (1999) and, for supplementary comment, Flannery (1999). At a broader level see Lawton and May (1995); Harvey and May (1997).

5 Scheffer *et al.* (2001).

6 L. R. Brown (1997). The value for 2000 is interpolated from L. R. Brown (2001). One of the factors in this decline has been the deliberate withdrawal of arable land from the cultivation of grain. Conway (1997) reports that, in 1987-88, the USA withdrew for conservation programmes about 20 million hectares of cereal cropland. (There is however a major discrepancy between Conway's statement of 20 million hectares in the text (p. 117), attributed to the Economic Research Service of the US

Department of Agriculture, and his Fig. 7.10 which indicates a loss of about 9 million hectares in North America during the 1980s.) Conway adds that the biggest withdrawal was in the (former) Soviet Union and was undertaken because of soil erosion.

Brown (1997) comments that in some ways "carryover stocks are the most sensitive indicator of food security". When these stocks are equivalent to less than 60 days' supply, prices become 'highly volatile'. He notes that stocks in 1996 amounted to 50 days' supply.

7 L. R. Brown (1997). Given the amount of recreational and illegal commercial fishing that occurs throughout the world, the statistics must have a high level of uncertainty. It is unlikely, however, that the overall trend is wrong. For some other observations on global fishing see, for example, Dayton (1995); Holmes (1994b, 1997); Masood (1996); Safina (1995); 'Various' (1995b). FAO (2001) states that "world production of fish, shellfish and other aquatic animals" was 125 million tonnes in 1999, 74% of which was obtained from "capture fisheries" (presumably aquaculture). Canada's response to a declining fish population has included blaming seals and killing them.

8 Myers (1998).

9 Pauley *et al.* (1998).

10 From time to time the explosive growth of the human population is referred to as a 'cancer'. The description is emotional and melodramatic but the analogy does have some merit in relation to mechanisms. To put it simply, the development of a cancer commonly has two principal stages. The first is the 'transformation'—essentially a change or mutation in the genetic material of a cell. The second stage, which might not occur for some years, is the 'promotional' stage in which the altered gene begins to express itself, the cell begins to multiply and resist the normal regulatory mechanisms that, in a healthy organism, ensure a dynamic balance among cells and organs. In a number of ways this balance is a functional counterpart of the balance of organisms within a healthy ecosystem. If unchecked, the multiplication of cancer cells will eventually destroy the host organism—their habitat. The initial appearance of Man on the planet might be compared functionally to the original mutation or transformation. The counterpart of the promotional stage would be the adoption of agriculture with its attendant positive feedback mechanism.

11 I use the verb 'sustain' here in the same sense as the adjective 'sustainable'—which has become a cliché in this context. In other words, I have not attempted to define the length of time the population is expected to be sustained. But see later in the text.

12 Cited by Bongaarts (1994a).

13 Cited by MacKenzie (1994).

14 Coghlan (1996). The median projection for 2050 in a later assessment by Lutz *et al.* (1997) is 9968 million. Projections have since been revised downwards because of the AIDS epidemic in Africa.

15 There are frequent articles along such lines in the daily press and in popular science journals such as *New Scientist*. See Bongaarts (1994b) for an accessible comprehensive article on this topic. Among other things, Bongaarts concludes that:

"The expansion of agriculture will be achieved by boosting crop yields and by using existing farmland more intensively, as well as by bringing more arable land into cultivation where such action proves economical. Such events will transpire more slowly than in the past, however, because of environmental constraints. In addition, the demand for food in the developed world is approaching saturation."

Conway (1997) has subtitled his book "Food for all in the 21st century".

16 Some may argue that virtually all ecological damage is eventually repairable while geophysical conditions are favourable. I have added the adverb 'essentially' as a reminder that, for any such repair to be relevant to human interests, it has to occur within quite strict time limits.

17 Other approaches to this problem are a High Quality Crystal Ball (HQCB) and a Direct Line to God (DLG).

18 Through the IMF and the World Bank, the 'West's' policy towards the agrarian countries (perhaps especially Africa) has been to emphasise technological 'development' to raise farm yields in order to meet the demands of growing populations. Population control has not been ignored but my impression is that it has not been given the emphasis of 'development.' As we have seen, global *per capita* grain production has fallen.

19 In Europe there has recently been a belated acknowledgment of the importance of hedgerows in the general ecology of farmland. *E.g.* Anon (1997b).

20 Pimentel *et al.* (1995a) list from various sources examples of the extent of reduction in erosion rates achieved in the USA by some of these approaches. The most dramatic was a drop from 91 t.ha^{-1}.y^{-1} for cotton grown 'conventionally' to 3 t.ha^{-1}.y^{-1} under a system of 'no-till' cultivation. This method, however, depends on a substantial application of herbicide, an approach with its own set of disadvantages. The method, which involves sowing by direct drilling, generally reduces the rate of depletion of organic matter in soil but does not appear to reverse it (Hamblin and Kyneur, 1993, cited by Australian Bureau of Statistics, 1996b).

21 For example, Alexandratos (1988); Conway (1997).

22 For example, Folke *et al.* (1997), 'Various' (1996b). See also Boyden *et al.* (1981).

23 *E.g.* 1 hectare per 10 persons (Imhoff *et al.*, 1971). At an average flow rate of 300 l.hd^{-1}.d^{-1}, this is equivalent to a daily discharge of 3000 l.ha^{-1}, or an annual rate of about 1.1 Ml.ha^{-1} (*cf* some irrigation rates in Chapter 9). At the height of a southern NSW summer, actively growing tree plantations (mainly eucalypts and pines) can transpire water at rates of 80,000-100,000 l.ha^{-1}.d^{-1} (Myers *et al.*, 1995). See also Note 24.

24 For example, Pimentel *et al.* (1995a) cite results from other sources indicating that a maize crop, at a yield of 7 kg.ha^{-1}, transpires about 4 Ml.ha^{-1} while an additional 2 Ml.ha^{-1} evaporates from the soil. Such rates will obviously be affected by climate and weather.

25 *E.g.* Yeates (1995).

26 According to one report (Coghlan, 1993) half of the sewage sludge produced in Britain is applied to farmland of some kind—including pasture. According to another report, by 1996 about 80% of Sydney's sewage sludge was 'reused, mostly as fertiliser in agriculture and horticulture' (Anon, 1996b).

27 Increases have been reported in the cadmium content of the livers and kidneys of lambs that had grazed on English pasture fertilised with sewage sludge (Coghlan 1997b). Heavy metals and refractory organic substances are also prone to concentration as they move up the 'food chain' (*e.g.* Mentasi and Ramel, 1995).

It has recently been reported that a number of companies within Australia have been disposing of (some) industrial waste as fertiliser—but labelled 'soil conditioner'. The waste contains a range of heavy metals as well as other toxic elements including arsenic. At the time of writing it was expected that such activity "may be subjected to tough new state regulations" (Ryle, 2002).

28 E.g. Aschmann *et al.* (1992). See also Note 23.

29 E.g. Day (1998).

30 Rae and Gruen (1997).

31 Conway (1997) defines sustainability as "The ability of the agroecosystem to maintain productivity when subject to stress or shock".

32 The nett proportion of biomass that can be taken annually from an area of land without impoverishing or degrading the soil will depend on many factors. It will be affected by the physical environment—topography, types of rock and soil, rainfall, frequency of flooding etc. It will be affected by the major types of biota, by the total biomass, the growth rates of the vegetation, rates of nitrogen fixation, by the amount of droppings from passing birds and any other wild animals. And, of course, it will be affected by the absolute yield of a crop. A nett loss of 5% of 3000 kg.ha^{-1}.y^{-1} will obviously raise more problems than 5% of 500 kg.ha^{-1}.y^{-1}. In most situations a continuing loss of 5% of harvested nutrient elements is likely to be too high to be sustainable. See also Chapters 4 and 7.

33 Nominating 'neighbouring' farms is an acknowledgment of the logistical difficulties of transporting sewage to distant farms. The acknowledgment also assumes that, in the interests of nutritional variety, there would be exchange of produce among different regions. Genuine sustainability would require that the amount of material exchanged was essentially equivalent in all directions so that, overall, there was no nett transfer from one region to another.

34 I noted in Chapter 9 that the incorporation of phosphorus into human bodies is very small in relation to the amount carried in foodstuffs and its derivative wastes. Nevertheless, in the longer term (if there is one) the disposal of human dead will need to be assessed more seriously by ecological criteria than it is at present. With my earlier assumption of an average individual mass of about 50 kg, a human population of about 6 billion contains some 3 million tonnes of phosphorus. This is roughly equivalent to 59 million tonnes of high quality phosphate rock.

35 In parts of India, reaction against the environmental, economic and social effects of the so-called 'Green Revolution' and policies imposed by the IMF etc has stimulated a popular move back to traditional farming (*e.g.* Nellithanam *et al.*, 1998; Ainger, 1999). An analogous movement is in train in Finland. In this case the primary motivation is evidently to restore traditional village life (Pietilä, 1997).

36 There are intermediate positions between conventional decentralisation and wholesale return of the population to live and work on farms that might warrant consideration. For example, the siting of an 'institute' such as a school, a university, a shopping centre etc essentially on farmland could facilitate the recycling of (we hope) uncontaminated sewage and some food wastes within surrounding farms.

37 I say 'if necessary' because in some regions there is excessive deposition from the atmosphere (*e.g.* Amann and Klaassen, 1995; Hungate *et al.*, 1997; 'Various', 1997—and Chapter 11). That would probably change, however, under the régime that I am proposing.

38 This is a very complex problem. I am reluctant to suggest what that fraction should be since, among other things, it would be sensitive to the total demand from all sources—all farms as well as urban requirements. The total allocation would need to be set at a level that could be met during a severe drought. On the relatively small, labour-intensive farms that would arise from the previous argument, demand for dedicated irrigation water would presumably be less than for present-day large commercial farms because, among other things, of the availability of a significant volume of septic tank effluent. It is conceivable that a return to mixed farming with more trees might increase local rainfall in some regions. In the interests of avoiding salination, a limit (as a proportion of rainfall) should also be placed on the volume of sewage used for irrigation—although, under the idealised conditions that I describe, sewage or septic tank effluent should contain approximately only those salts that were removed in produce from the farms receiving the effluent.

39 There is sometimes talk of harvesting native animals, in particular kangaroos, from such properties. In one important respect, this would be worse than running cattle. In general, Australian native animals grow faster than cattle. Harvesting kangaroos at a rate to maintain a constant total mass of the animals within a specified area of land would therefore remove nutrients from the soil faster than maintaining the same mass of cattle. If it should be necessary to cull the kangaroo population, soil conservation would require that their carcases remain on the land.

40 World Resources Institute (1994). See also Note 14.

41 For example Balter (1999). Statistics for 9 major nations in 1997 attributed to South Africa the highest incidence of HIV infection among adults (12.91%). The next highest was India with 0.82% (Flavin, 2001).

Notes with Chapter 15

1 The demographic transition is conventionally seen to have four phases. In the past, the populations of agrarian countries changed slowly because both birth and death rates, although high, were of similar magnitude. That is Phase 1. Phase 2 arises when improved living conditions extend life expectancy, but birth rates remain high. Accordingly the population increases. Phase 3 involves a further improvement in living conditions to the extent that children are no longer so essential for support in old age.

In consequence the birth rate declines and so, of course, does the growth rate of the population. In Phase 4 birth and death rates are again similar, but are now relatively low. The population is again approximately stationary but is substantially larger than it was in Phase 1. This is more or less the current state of Western Europe. The topic has been discussed at length. See, for example, Cohen (1995a), Day (1983), Gould and Lawton (1986), Moffett (1994).

2   Elliott (1998); O'Kane (1998); Watkins (1996). See also Chapter 13.

3   Moffett (1994).

4   McCrone (1993). Elsewhere McCrone objectively summarises Freud's propensity for fabrication.

5   Brown (1986).

6   Diamond (1997) discusses the relation between the size of a community and the level of organisation within religions.

7   Lorenz (1966).

8   See Wrangham and Peterson (1997) for an informative account of chimpanzee raiding parties. The success of Argentine ants in their invasion of North America has been attributed largely to their avoidance of intraspecific warfare (Holway *et al.*, 1998).

9   *E.g.* by Morris (1981).

10   Perhaps a modern equivalent of burning witches is the shooting by the American 'religious right' of obstetricians who undertake abortions.

11   See Ali (2002) for a very lucid, objective and informative account of the history and contemporary attitudes of these cultures to one another.

12   Cited by Koestler (1980).

13   As cited by Elliott and Brittain (1997).

14   Podmore *et al.* (1998).

15   Tickell (1998). See also Seymour (1996).

16   Anon (1997c).

17   For example, Jung Chang (1991); Kristoff and Wudunn (1994).

18   Of course I am not alone in referring to the myths of economics. Mishan (1986), for example, devotes a book specifically to that topic.

19   Cassidy (1997).

20   For example, Chote (1992). This is an account of the errors in computer models used by the British Treasury.

21   For example Cobb *et al.* (1995), Eckersley (1998). Cobb *et al.* give a concise history of the GDP in the course of which they make the following observation:

> "It is not accidental that both the habitat and social structure have suffered severe erosion in recent decades; these are precisely the realms that eighteenth- and nineteenth-century assumptions precluded from the reckoning of national well-being —in capitalist and socialist economies alike. This erosion has been mainly invisible in terms of economic policy because our index of progress ignores it; as a result, the nation's policies have made it worse."

The authors proposed replacement of the GDP with a more comprehensive measure, the 'Genuine Progress Indicator' (GPI).

22 Eckersley (1998a).

23 Pearce *et al.* (1989).

24 As summarised by Pearce (1995f).

25 In 1998 an election in the Australian state of Queensland saw 'One Nation', a new populist party of the extreme 'right', do so well (it won 11 seats) that the previous government, a coalition of National and Liberal parties—both 'right wing'—lost office. This disconcerted the Federal Government, a similar coalition with the Liberals the major party, which was due to hold an election within a year. It took to criticising One Nation's policies which, until then, it had largely ignored. The government's criticisms extended to One Nation's racist attitudes and, in particular, its comments about Australia's becoming 'Asianised'. The main thrust of the government's attack on these attitudes was that they would endanger trade with Australia's Asian neighbours and discourage Asian tourism—and thereby affect 'the economy'.

26 Constanza *et al.* (1997). See also Fullerton and Stavins (1998), Masood and Garwin (1998), and Pimm (1997a). Occasionally financial considerations can moderate environmental damage. For example, several large 'dirty' corporations (including Monsanto, British Petroleum and DuPont) have improved their financial performances by 'cleaning and greening' their operations (Chichilnisky and Heal, 1998).

27 E.g. Editorial (1995); Hadfield (1994); Pearce (1995g); Schofield and Shaoul (1997). Poupeau (2002) has summarised economic, environmental, social and political consequences of 'privatising' public water supplies in Argentina and Bolivia. Sheil (2000) has discussed the general effects of economic rationalism on water supplies in Australia. See also Walker and Walker (2001) for a comprehensive appraisal of 'privatisation' in Australia.

28 Warde (1999). Cassidy (1997) gives ratios of approximately 60:1 for 1978 and 170:1 for 1995.

29 See Pimm and Lawton (1998).

30 E.g. Janzen (1998); Ando *et al.* (1998). Ando and associates give considerable weight to economic criteria. See also DiSilvestro (1993). A general question that suggests itself here is the competence of otherwise of the management, and whether a management policy is based primarily on economic or ecological criteria.

31 See Soulé and Sanjayan (1998).

32 One of the better known challenges came from Mishan (1967). See also Coombs (1990).

33 See Mittelman (1997).

34 Both Britain (Arlidge, 1999) and the USA (ABC Radio National, 25th March, 1999) have been trying to persuade Australia to establish dumps for their respective high level radioactive wastes. Although the Federal Government declared that this will not happen, some individual politicians argued on economic grounds for the proposal, especially that from the USA.

35 For example, Hertz (2001), Keen (2001), Nader *et al.* (1993), Wallach and Sforza (1999).

36 John Browne, Chief Executive of British Petroleum, is "perhaps the only fossil-fuel executive to speak decisively and meaningfully for corporate responsibility". He is also accredited with arguing that the possibility of human influence on climate is 'too significant to be ignored'. (Anon, 1998). Another exception is the intention of the British pharmaceutical company, SmithKline Beecham to donate supplies of 'albendazole' to treat and prevent the parasitic disease, lymphatic filariasis or 'elephantiasis' in much of the 'Third World'. The value of the donation will amount to $1.6 billion over 20 years. (Mihill, 1998). It could be argued, of course, that this makes for good public relations.

37 Saul (1997). See also McCarthy (1998). This is a review of J. Tirman, *Spoils of War. The Human Costs of America's Arms Trade*, Free Press (1997?), which I have not read. Doyle (1998) gives semi-quantitative information of the global arms trade.

38 See Pilger (1998).

39 Rowan (1998).

40 Atkinson (1998); Denny (1998); Frodon *et al.* (1998); Hoedeman *et al.* (1998)—and many others. Subsequent reports, however, suggest that 'GATS' (General Agreement on Trade and Services) is trying to move in the same direction ('Background Briefing', ABC 23rd June, 2002).

41 Soros argues that democracy and the current global free market are fundamentally incompatible. In the course of an address to the World Economic Forum in Davos, Switzerland early in 1998, he commented that if "left to its own devices the global market would undoubtedly destroy itself". (Reported in the ABC's 'Background Briefing', broadcast on 5th and 7th July, 1998.) See also Soros (1998).

42 For example, London (1984) comments that "Environmental, resource and population stresses are diminishing, and with the passage of time will have less influence than now upon the quality of human life on our planet" (p. 180). See also Kaufman (1994) and Rubin (1994). These two authors adopt similar approaches in their generalisations, their quotations and criticisms of some of the more sweeping statements by 'environmentalists'. In essence, Kaufman argues that the human population must manage the biosphere rather than leave it to its own devices. That might be fair enough—depending on the competence of the management. His view of the future, however, is coloured by perceptions of economic and technological development. He is also prone to making the odd generalisation of great profundity—such as "From the White House to Greenpeace, environmentalists have not yet learned that we do not adapt and survive by imitating plants and animals, but by becoming more human"—whatever that might mean (p. 179).

43 E.g. Barraclough (1998); Homburger (1996); Motluk (1996a and b); Sugarman (1996); Williams (1996); Concar and Day (1998); Concar (1998); Editorial (1998); Gibbs (1998).

44 Wadman (1997b).

45 My supporting information for this statement consists of correspondence dating back to 1980, other personal communications, newspaper reports and some government reports. See also Thomas and Orlova (2001).

46 Kaiser (1996); Lehrman (1996).

47 Butler (1996, 98).

48 E.g. Azar and Rodhe (1997); Kaiser (1997b); May (1997).

49 A television programme, *Against Nature*, was a two-part programme first broadcast on Britain's Channel 4 and subsequently shown in Australia by the ABC on 21st and 28th July, 1998. It was a very crude piece of propaganda setting out to deny, among other things, the 'enhanced greenhouse effect'. Another programme, entitled *Staking the Globe*, was made in Denmark and broadcast in Australia by SBS (the Special Broadcasting Service) on 12th May, 1998. It was a revealing investigation into an organised campaign seeking to deny global warming. For factual accounts of recent temperature changes see, for example, Kerr (1998), Vogel and Lawler (1998), Barnett *et al.* (2001), Levitus *et al.* (2001).

50 See Diamond (1991).

Biliography

Abelson, P. H. (1995). International agriculture. *Science*, **268**, 11.

Adamson, D. A. and Fox, M. D. (1982). Change in Australasian vegetation since European Settlement. In J. M. B. Smith (Ed.), *A History of Australian Vegetation*. McGraw Hill, Sydney, pp 109-146.

Ager, D. (1992). The myth of the caveman. *New Scientist*. **133** (1806) 48.

Ahmad, A. (1999). The Narmada water resources project, India: implementing sustainable development. *Ambio* **28**, 398-403.

d'Aignaux, J. N. H. *et al.* (2001). Predictability of the UK variant Creutzfeldt-Jacob disease epidemic. *Science*, **294** 1729-1730.

Ainger, K. I. (1999). The meek fight for their inheritance. *Guardian Weekly*, 21st February, p.23.

Aldhous, P. (1993). Tropical deforestation: not just a problem in Amazonia. *Science*, **259**, 1390.

Alexander, M. (1971). *Microbial Ecology*. Wiley, New York.

Alexander, G. C. and Stevens, R. J. (1976). Per capita phosphorus loading from domestic sewage. *Water Res.* **10**, 757-764.

Alexandratos, N. (Ed.). (1988). *World Agriculture: Toward 2000*. An FAO Study. Belhaven Press, London.

Ali, T. (2002). *The Clash of Fundamentalisms*. Verso, London and New York.

Alwen, M. (1995). Damming the Gorges. *The Ecologist* **25**, 33-34.

Amann, M. and Klaassen, G. (1995). Cost-effective strategies for reducing nitrogen depositions in Europe. *J. Env. Management* **43**, 289-311.

Amato, M. and Ladd, J. N. (1994). Application of the ninhydrin-reactive N assay for microbial biomass in acid soils. *Soil Biol. Biochem.* **26**, 1109-1115.

Anderson, D. M. (1994). Red tides. *Scientific American* **271** (2), 52-58.

Anderson, I. (1993). Western Australia 'censored' forest reports. *New Scientist* **137** (1863), 4.

Anderson, I. (1995a). Australia's growing disaster. *New Scientist* **147** (1988), 12-13.

Anderson, I. (1995b). Lawless loggers 'will go on plundering paradise'. *New Scientist* **147** (1995), 6.

Anderson, R. M. *et al.* (1996). Transmission dynamics and epidemiology of BSE transmission in British cattle. *Nature* **382**, 779-788.

Ando, A. *et al.* (1998). Species distributions, land values and efficient conservation. *Science* **279**, 2126-2128.

Anon (1990). Prehistoric people 'ruined their own environment'. *New Scientist* **125** (1705), 10.

Anon (1993). We still like to be beside the sewage. *New Scientist* **140** (1903), 9.

Anon (1995a). The sea that turned to dust. *Guardian Weekly*, 24th September, p.12.

Anon (1995b). Garden cancer. *New Scientist* **145** (1968), 13.

Anon (1996a). Britain 'squeezing Third World debtors'. *Guardian Weekly* 29th September, p.12.

Anon (1996b). Sewage sludge, better in forests than seas. *Ecos* (CSIRO) **88** (Winter) 4.

Anon (1997a). Denmark considering banning all pesticides. *New Scientist* **154** (2085), 13.

Anon (1997b). Small furry animals bounce back. *New Scientist* **153** (2067), 11.

Anon (1997c). Euro-gridlock. *New Scientist* **149** (2015), 11.

Anon (1998). Oil boss honoured for backing sustainability. *Nature* **392**, 430.

Arlidge, J. (1999). N-waste may go to Outback. *Guardian Weekly*, 28th February, p.8.

Arrhenius O. (1938). The phosphate content of the soils of the Isle of Gotland. *Sveriges Geol. Undersökn.*, Ser. C, No.413 *Årsbok* **32**, 1-15.

Aschmann, S.G. *et al.* (1992). Nitrogen movement under a hardwood forest amended with liquid wastewater sludge. *Ag. Ecosystems, Env.* **38**, 249-263.

Ashton, P. and Blackmore, K. (1987). *On the Land. A Photographic History of Farming in Australia.* Kangaroo Press, Kenturst, NSW.

Asswad, R. M. E. (1995). Agricultural prospects and water resources in Libya. *Ambio*, **24**, 324-327.

Atkinson, M. (1998). Rich man's club makes poor offer. *Guardian Weekly*, 3rd May, p.18.

Attwood, H., Forster, F. and Gandevia, B. (Eds), (1984). *Medical History Australia (Occasional Papers).* Medical History Unit, University of Melbourne, Parkville.

Attwood, H. and Kenny, G. (Eds). (1987). *Reflections on Medical History and Health in Australia.* Medical History Unit, University of Melbourne, Parkville.

Australia: State of the Environment 1996. CSIRO Publishing, Collingwood, Victoria.

Australian Bureau of Statistics (1991a). *Australian Farming in Brief.*

Australian Bureau of Statistics (1991b). *Foreign Trade Statistics.* Report June 1991.

Australian Bureau of Statistics (1992a). *Australia's Environment. Issues and Facts.* (Catalogue No. 4140.0).

Australian Bureau of Statistics (1992b). *1990-91 Summary of Crops. Australia.* (Catalogue No. 7330.0).

Australian Bureau of Statistics (1992c). *1990-91 Livestock and Livestock Products. Australia.* (Catalogue No. 7221.0).

Australian Bureau of Statistics (1993a). *Apparent Consumption of Foodstuffs and Nutrients. Australia 1990-91.* (Catalogue No. 4306.0).

Australian Bureau of Statistics (1993b). *Characteristics of Australian Farms.* (Catalogue No. 7102.0).

Australian Bureau of Statistics (1994). *Characteristics of Australian Farms.* (Catalogue No. 7102.0).

Australian Bureau of Statistics (1996a). *Australians and the Environment.* (Catalogue No. 4601.0).

Australian Bureau of Statistics (1996b). *Australian Agriculture and the Environment.* (Catalogue 4606.0).

Australian Bureau of Statistics (2001). *Agriculture 1999-2000.* (Catalogue 7113.0).

Australian State of the Environment Committee (2001). *State of the Environment 2001.* CSIRO Publishing, Collingwood, Victoria.

Azar, C. and Rodhe, H. (1997). Targets for stabilization of atmospheric CO_2. *Science*, **276**, 1818-1819.

Bagla, P. and Kaiser, J. (1996). India's spreading health crisis draws global arsenic experts. *Science*, **274**, 174-175.

Baird, N. (1996). Unwisdom of the Solomons. *New Scientist* **149** , 30-33.

Baker, M. (1997). Burn-off choking region's economies. *Sydney Morning Herald*, 27th September, 1997, p.25.

Balter, M. (1999). AIDS now world's fourth biggest killer. *Science* **284**, 1101.

Barber, M. and Ryder, G. (1993). *Damming the Three Gorges.* Earthscan, London.

Barker, G. (1985). *Prehistoric Farming in Europe.* University Press, Cambridge.

Barnett, T. P. *et al.* (2001). Detection of anthropogenic climate change in the world's oceans. *Science* **292**, 270-274.

Barraclough, J. (1998). Indian addiction. *Guardian Weekly*, 22nd March, p.15.

Barraclough, S. L. and Ghimire, K. B. (1996). Deforestation in Tanzania. *The Ecologist*, **26**, 104-109.

Barrow, N. J. (1983). Soil fertility changes. In Costin and Williams, pp 197-220.

Barson, M., Lloyd, J. and Farquhar, G. (1995). Land use changes. *Search*, **26**, 122-125.

Bartle, H. (1991). Quiet sufferers of the silent spring. *New Scientist* **130** (1769), 26-31.

Bartsch, B. and Mason, W. (Eds). (undated). *Feedbase 2000.* CSIRO Publications, Melbourne. (ISBN 0 643 05233 x).

Beale, B. (1995a). Has King Cotton killed the water? *Sydney Morning Herald*, 15th April, 17.

Beale, B. (1995b). Mismanagement, drought ruining wetlands. *Sydney Morning Herald*, 15th April, 4.

Beaumont, P. (1981). Water resources and their management in the Middle East, in Clarke and Bowen-Jones, pp 40-72.

Beckerman, W. (1995). *Small is Stupid.* Duckworth, London.

Beder, S. (1997). *Global Spin.* Scribe Publications, Melbourne.

Behrenfeld, M. J. *et al.* (1996). Confirmation of iron limitation of phytoplankton photosynthesis in the equatorial Pacific Ocean. *Nature* **383**, 508-510.

Beirne, K. G. and Pratley, J. E. (1988). Irrigation for crop production. In Pratley, pp 306-340.

Bek, P., and Robinson, G. (1991). *Sweet water or Bitter Legacy.* NSW Department of Water Resources (ISBN 0 7305 7860 7).

Bellwood, P. (1985). *Prehistory of the Indo-Malaysian Archipelago.* Academic Press, Sydney.

Bellwood, P. (1991). The Austronesian dispersal and the origin of languages. *Scientific American* **265** (1) 70-75.

Bellwood, P. *et al.* (1992). New dates for prehistoric Asian rice. *Asian Perspectives*, **31**, 161-170.

Berlan, J.-P. and Lewontin, R. C. (1999). Menace of the genetic-industrial complex. *Le Monde Diplomatique* (English edition), January, pp 8-9.

Bernal, M. P., Roig, A. and García, D. (1993). Nutrient balances in calcareous soils after application of different rates of pig slurry. *Soil Use and Management*, **9**, 9-14.

Bilham, R. and Barrientos, S. (1991). Sea-level rise and earthquakes. *Nature* **350**, 386.

Bird, M. I., Chivas, A. R. and Head, J. (1996). A latitudinal gradient in carbon turnover times in forest soils. *Nature* **381**, 143-146.

Biswas, A. K. (1998). Deafness to global water crisis: causes and risks. *Ambio* **27**, 492-493.

Black, F. L. (1992). Why did they die? *Science*, **258**, 1739-1740.

Blaikie, P. and Brookfield, H. (Eds) (1987). *Land Degradation and Society.* Methuen, London.

Blaikie, P. and Brookfield, H. (1987a). Questions from history in the Mediterranean and western Europe. In Blaikie and Brookfield, pp 122-142.

Blair, G. J. (1980). The biogeochemical cycle of sulfur. In J. R. Freney and A. J. Nicholson (Eds), *Sulfur in Australia*, Aust. Acad. Sci., Canberra, pp 3-13.

Blair, G. J. (1983). The phosphorus cycle in Australian agriculture. In Costin and Williams, pp 92-111.

Blustein, P. (1997). Foreign aid 'has no impact'. *Guardian Weekly* 1st June, p.16.

Bogucki, P. (1988). *Forest Farmers and Stockherders.* University Press, Cambridge.

Bogucki, P. and Grygiel, R. (1983). Early farmers of the North European Plain. *Scientific American* **248** (4), 96-104.

Bolton, G. (1989). The spread of colonization. In Hardy and Frost, pp 183-193.

Bongaarts, J. (1994a). Population policy options in the Developing World. *Science*, **263**, 771-776.

Bongaarts, J. (1994b). Can the growing human population feed itself? *Scientific American* **270** (3), 18-24.

Bongaarts, J. (1998). Demographic consequences of declining fertility. *Science*, **282**, 419-420.

Bonyhady, T. (2000). *The Colonial Earth.* Melbourne University Press, Melbourne.

van den Bosch, R. (1978). *The Pesticide Conspiracy.* University of California Press, Berkeley.

Boserup, E. (1965). *The Conditions of Agricultural Growth.* Aldine, Chicago.

Bowler, J. (1990). The last 500,000 years. In Mackay and Eastburn, pp 94-109.

Boyden, S. (1987). *Western Civilization in Biological Perspective.* Oxford University Press, New York.

Boyden, S. *et al.* (1981). *The Ecology of a City and its People.* Australian National University Press, Canberra.

Bray, W. and Trump, D. (1982) *Dictionary of Archaeology*. 2nd edn, Penguin, London.

Brown, A. D. (1986). The brain and the bomb. The ultimate problem in applied biology? *Search* **17** (3-4), 95-99.

Brown, A. D. (1990). *Microbial Water Stress Physiology*. Wiley, Chichester.

Brown, L. R. *et al.* (1993). *State of the World* 1993. Earthscan, London.

Brown, L. R. (1995). Nature's limits. In L. R. Brown *et al.*, *State of the World* 1995. Earthscan, London pp 3-20.

Brown, L. R. (1997). Facing the prospect of food scarcity. In L. R. Brown *et al.*, *State of the World* 1997. W. W. Norton, N.Y., pp 23-41.

Brown, L. R. (1998). The future of growth. In L. R. Brown *et al.*, *State of the World* 1998. Earthscan, London, pp 3-20.

Brown, L. R. (2001). Eradicating hunger: a growing challenge. In L. Starke (Ed.), *State of the World* 2001. Earthscan, London, pp 43-62.

Bunney, S. (1994). Did modern culture begin in prehistoric caves? *New Scientist* **141** (1908), 16.

Burke, M. (1994). Phosphorus fingered as coral killer. *Science*, **263**, 1086.

Butler, D. (1996). BSE researchers bemoan 'ministry secrecy'. *Nature* **383**, 467-468.

Butler, D. (1998). British BSE reckoning tells a dismal tale. *Nature* **392**, 532-533.

Calder, N. (1984). *Timescale*. Chatto & Windus, London.

Campbell, C. J. and Laherrère, J. H. (1998). The end of cheap oil. *Scientific American* **278** (3), 60-65.

Caraco, N. F. (1995). Influence of human populations on P transfers to aquatic systems: a regional scale study using large rivers. In H. Tiessen (Ed.), *Phosphorus in the Global Environment*. Wiley, N.Y. pp 235-244.

Carmichael, W. W. (1994). The toxins of cyanobacteria. *Scientific American* **270** (1), 64-71.

Carter, J. (1997). The big shed. *Australian Geographic* (46), 32-39.

Carter, O. G. (1988). The nutrition of crops. In J. E. Pratley, pp 233-261.

Cassidy, J. (1997). Why Karl Marx was right. *The Australian*, 20-21st December, pp 21, 24. (Originally published in the *New Yorker*.)

Cavalli-Sforza, L. L., Menozzi, P. and Piazza, A. (1993). Demic expansions and human evolution. *Science* **259**, 639-646.

Chalmers, J. (1997). Golf ousts rice fields in Vietnam village. *Guardian Weekly* 19th January, p.7.

Chamberlain, J. (1992). Oxygen loss threatens Lake Victoria. *New Scientist* **135** (1840), 8.

Charman, P. E. V. and Murphy, B. W. (Eds). (1991). *Soils. Their Properties and Management*. Sydney University Press/Oxford University Press, Melbourne.

Charman, P. E. V. and Roper, M. M. (1991). Soil organic matter. In Charman and Murphy, pp 206-214.

Chartres, C. J. *et al.* (1992). Land degradation as a result of European settlement of Australia and its influence on soil properties. In Gifford and Barson, pp 3-33.

Chau Kwai-cheong. (1995). The Three Gorges Project in China: resettlement prospects and problems. *Ambio* **24**, 98-102.

Chichilnisky, G. and Heal, G. (1998). Economic returns from the biosphere. *Nature* **391**, 629-630.

Chomsky, N. (1998). The poor always pay debts of the rich. *Guardian Weekly*, 24th May, p.15.

Chote, R. (1992). Why the Chancellor is always wrong. *New Scientist* **136** (1845), 26-31.

Chowdhury, A. M. R. (1999). Testing of water for arsenic in Bangladesh. *Science* **284**, 1622.

Cipolla, C. M. (1992). *Miasmas and Disease. Public Health and the Environment in the Pre-Industrial Age.* English translation by Elizabeth Potter. Yale University Press, London.

Clapp, B. W. (1994). *An Environmental History of Britain.* Longman, Harlow, Essex.

Clarke, J. I. and Bowen-Jones H. (1981). *Change and Development in the Middle East.* Methuen, London.

Clarke, P. (1986). *A Colonial Woman. The Life and Times of Mary Braidwood Mowle, 1827-1857.* Allen and Unwin, Sydney.

Clarke, W. C. (1971). *Place and People. An Ecology of a New Guinean Community.* University of California Press, Berkeley.

Close, A. (1990). River salinity. In Mackay and Eastburn, pp 126-144.

Coale, K. H. *et al.* (1996) A massive phytoplankton bloom induced by an ecosystem-scale iron fertilization experiment in the equatorial Pacific Ocean. *Nature* **383**, 495-501.

Coats, C. D. (1989). *Old MacDonald's Factory Farm*, Continuum, New York.

Cobb, C., Halstead, T. and Rowe, J. (1995). If the GDP is up, why is America down? *Atlantic Monthly,* **276** (4), 59-78.

Coghlan, A. (1991). Animals bear brunt of pesticide poisonings. *New Scientist* **131** (1782), 9.

Coghlan, A. (1993). The toxic sludge that just won't die. *New Scientist* **139** (1884), 15.

Coghlan, A. (1996). Doomsday has been postponed. *New Scientist* **152** (2050), 8.

Coghlan, A. (1997a). Foul farm air sickens workers and livestock. *New Scientist* **154** (2078), 10.

Coghlan, A. (1997b). Lamb's liver with cadmium garnish. *New Scientist* **153** (2074), 4.

Cohen, J. E. (1995a). *How Many People Can the Earth Support?* Norton, New York.

Cohen, J. E. (1995b). Population growth and Earth's human carrying capacity. *Science,* **269**, 341-346.

Cohen, M. N. (1977). *The Food Crisis in Prehistory.* Yale University Press, New Haven.

Cohen, M. N. (1989). *Health and the Rise of Civilization.* Yale University Press, New Haven.

Colchester, M. (1997). Aid money is helping the rape of Guyana. *Guardian Weekly*, 17th August, p.12.

Colls, K. and Whitaker, R. (1990). *The Australian Weather Book.* Child & Associates, French's Forest, NSW.

Concar, D. (1998). Their learned friends. *New Scientist* **158** (2134), 5.

Concar, D.: Day, M. (1998). Undercover operation. *New Scientist* **158** (2134), 4.

Connolly, B. and Anderson, R. (1987). *First Contact.* Viking Penguin, New York.

Constanza, R. *et al.* (1997). The value of the world's ecosystem services and natural capital. *Nature* **387**, 253-259.

Conway, G. (1997). *The Doubly Green Revolution.* Penguin, London.

Cook, P. (1983). World availability of phosphorus: an Australian perspective. In Costin and Williams, pp 3-41.

Coombs, B. (Ed.). (1993). *Australian Beef.* Morescope, Camberwell, Victoria.

Coombs, H. C. (1990). *The Return of Scarcity.* Cambridge University Press, Melbourne.

Cooper, D. J., Watson, A.J. and Nightingale, P.D. (1996). Large decrease in ocean-surface CO_2 fugacity in response to *in situ* iron fertilization. *Nature* **383**, 511-513.

Cooter, W. S. (1978). Ecological dimensions of mediæval agrarian systems. *Agricultural History*, **52**, 458-477.

Cornell, S., Rendell, A. and Jickells, T. (1995). Atmospheric inputs of dissolved organic nitrogen to the oceans. *Nature* **376**, 243-246.

Costin, A. B. (1959). Replaceable and irreplaceable resources and land use. *J. Aust. Inst. Ag. Sci.* 3-9.

Costin, A. B. and Williams, C. H. (Eds). (1983). *Phosphorus in Australia.* Centre for Resource and Environmental Studies, Australian National University, Canberra.

Cowan, C. W. and Watson, P. O. (1992). *The Origins of Agriculture. An International Perspective.* Smithsonian Institution Press, Washington DC.

Coward, D. H. (1988). *Out of Sight. Sydney's Environmental History* 1851-1981. Department of Economic History, Australian National University, Canberra.

Creagh, C. (1992). Understanding arid Australia. *Ecos* (CSIRO) (73) 14-20.

Crosson, P. (1995). Soil erosion estimates and costs. *Science*, **269**, 461-464.

Culotta, E. (1993). Seesawing on syphilis. *Science*, **260**, 892.

Cunningham, C. (1996). *The Blue Mountains Rediscovered.* Kangaroo Press, Kenthurst, NSW.

Curson, P. and McCracken, K. (undated). *Plague in Sydney. The Anatomy of an Epidemic.* University of New South Wales Press, Kensington, NSW.

Dai Qing (1994). *Yangtze! Yangtze!* English Edition, Earthscan, London.

Daily, G. C. (1995). Restoring value to the world's degraded lands. *Science*, **269**, 350-354.

Dalal, R. C. and Mayer, R. J. (1986a). Long-term trends in fertility of soils under continuous cultivation and cereal cropping in southern Queensland. 1. Overall changes in soil properties and trends in winter cereal yields *Aust. J. Soil. Res.* **24**, 265-279.

Dalal, R. C. and Mayer, R. J (1986b). Long-term trends in fertility of soils under continuous cultivation and cereal cropping in southern Queensland. 11. Total organic carbon and its rate of loss from the soil profile. *Aust. J. Soil. Res.* **24**, 281-292.

Dargavel, J. (1995). *Fashioning Australia's Forests.* Oxford University Press, Melbourne.

Davis, K. (1991). Population and resources: fact and interpretation. In Davis and Bernstam, pp 1-21.

Davis, K. and Bernstam, M. S. (Eds). (1991). *Resources, Environment and Population.* Oxford University Press, N.Y.

Dawson, J. (1881). *Australian Aborigines.* George Robertson, Melbourne. Facsimile Edition (1981), Australian Institute of Aboriginal Studies, Canberra.

Day, L. H. (1983). *Analysing Population Trends*. Croom Helm, Beckenham, Kent.

Day, M. (1998). Fields of filth. *New Scientist* **157** (2120), 4.

Dayton, L. (1995). The killing reefs. *New Scientist* **148** (2003), 14-15.

Denny, C. (1998). Talks on cross-border investment treaty collapse. *Guardian Weekly*, 1st March, p.19.

Derry, T. K. and Williams, T. I. (1960). *A Short History of Technology*. Oxford University Press, New York.

Dhombres, D. (1996). Brazil's landless face long and hard battle. *Guardian Weekly*, 5th May, p.13.

Diamond, J. (1991). *The Rise and Fall of the Third Chimpanzee*. Radius, London.

Diamond, J. (1997). *Guns, Germs and Steel*. Jonathan Cape, London.

Diamond, J. (2001). Damned experiments! *Science* **294**, 1847-1848.

Dickenson, J. *et al.* (1996). *A Geography of the Third World*. Routledge, London.

Dingle, T. and Rasmussen, C. (1991). *Vital Connections. Melbourne and its Board of Works* 1891-1991. McPhee Gribble/Penguin, Ringwood, Victoria.

Dinham, B. (1991). FAO and pesticides: promotion or proscription? *The Ecologist* **21**, 61-65.

DiSilvestro, R. L. (1993). *Reclaiming the Last Wild Places*. Wiley, N.Y.

Dixon, B. (1994). Genes of yesteryear. *New Scientist* **142** (1920), 40-41.

Dixon, R. K. *et al.* (1994), Carbon pools and flux of global forest ecosystems. *Science*, **263**, 185-190.

Dodgshon, R. A. (1987). *The European Past. Social Evolution and Spatial Order*. Macmillan, Basingstoke and London.

Donald, C. M. (1964/5). The progress of Australian agriculture and the role of pastures in environmental change. *Aust. J. Sci.*, **27**, 187-198.

Dovers, S. (Ed.). (1994). *Australian Environmental History. Essays and Cases*. Oxford University Press, Melbourne.

Doyle, R. (1996). Soil erosion of cropland in the U.S., 1982 to 1992. *Scientific American* **275** (4), 23.

Doyle, R. (1998). The arms trade. *Scientific American* **279** (1), 17.

Dudley, N. *et al.* (1998). The timber trade and global forest loss. *Ambio* **27**, 248-250.

Dury, G. H. (1981). *An Introduction to Environmental Systems*. Heinemann, New Hampshire.

Dynesius, M. and Nilsson, C. (1994). Fragmentation and flow regulation of river systems in the northern third of the world. *Science*, **266**, 753-761.

Dyos, H. J. and Wolff, M. (1973). (Eds). *The Victorian City. Images and Realities*. Routledge and Kegan Paul, London.

Dyson, F. J. (1996). Reality bites. *Nature* **380**, 296.

Eastburn, D. (1990). The River. In Mackay and Eastburn, pp 2-15.

Eckersley, R. (Ed.). (1998). *Measuring Progress*. CSIRO Publishing, Collingwood, Victoria.

Eckersley, R. (1998a). Perspectives on progress: economic growth, quality of life and ecological sustainability. In Eckersley, pp 3-34.

Editorial (1995). Privatization in tears. *Nature* **376**, 624.

Editorial (1998). The whole truth. *New Scientist* **158** (2134), 3.

Editors (1993). Next step, Mr Preston: step down. *The Ecologist*, **23**, 83.

Editors (1994). Lies, dam lies and ballistics. *The Ecologist* **24**, 44.

Editors (1996). BSE: madness in the method. *The Ecologist*, **26**, 45.

Edwards, K. (1991). Soil formation and erosion rates. In Charman and Murphy, pp 36-47.

Elliott, G. L. and Leys, J. F. (1991). Soil erodibility. In Charman and Murphy, pp 181-192.

Elliott, L. (1998). Why the poor are picking up the tab. *Guardian Weekly*, 17th May, p. 14.

Elliott, L. and Brittain, V. (1997). Seven richest could end world poverty. *Guardian Weekly*, 22nd June, p. 3.

Elvin, M. and Liu, T. (1998). *Sediments of Time. Environment and Society in Chinese History*. Cambridge University Press, Cambridge.

Erlichman, J. (1995). Alarm grows over sheep dip injury toll. *Guardian Weekly*, 11th June, p. 9.

Escobar, G. (1997). Conservationist angers Chile. *Guardian Weekly* 112th January, p. 18.

Evans, R., Brown, C., and Kellett, J. (1990). Geology and groundwater. In Mackay and Eastburn, pp 76-93.

Evans, S. (1983). *Historic Sydney as Seen by its Early Artists*. Doubleday, Lane Cove, NSW.

Evans, W. (Ed.). (1975). *Diary of a Welsh Swagman*. 1869-1894. Macmillan, Sydney.

FAO (2001). *The State of Food and Agriculture* 2001. Food and Agriculture Organization, Rome.

Ewald, P. W. (1993). The evolution of virulence. *Scientific American* **268** (4), 56-62.

Ferguson, J. and Lohmann, L. (1994). The anti-politics machine. *The Ecologist* **24**, 176-181.

Finley, M. I. (1980). *Ancient Slavery and Modern Ideology*. Chatto and Windus, London.

Flannery, T. F. (1999). Debating extinction. *Science* **283**, 182-183.

Flavin, C. (2001). Rich planet, poor planet. In L. Starke (Ed.), *State of the World* 2001. Earthscan, London, pp 3-20.

Flood, J. (1980). *The Moth Hunters*. Australian Inst. of Aboriginal Studies, Canberra.

Flood, J. (1989). *Archaeology of the Dreamtime*. Collins, Sydney.

Flood, J. (1990). *The Riches of Ancient Australia*. Queensland University Press, St Lucia.

Folke, C. *et al.* (1997). Ecosystem appropriation by cities. *Ambio* **26**, 167-172.

Freiberg, C. *et al.* (1997). Molecular basis of symbiosis between *Rhizobium* and legumes. *Nature* **387**, 394-401.

Frisch, R. E. (1988). Fatness and fertility. *Scientific American* **258** (3), 70-77.

Fritz, H. and Duncan, P. (1993). Large herbivores in rangelands. *Nature* **364**, 292-293.

Frodon, J.-M., Labe, Y.-M. and Vulser, N. (1998). Europe 'will defend cultural exceptions'. *Guardian Weekly*, 1st March, p. 18.

Fry, K. (1994). *Beyond the Barrier. Class Formation in a Pastoral Society.* Crawford House Press, Bathurst.

Fullerton, D. and Stavins, R. (1998). How economists see the environment. *Nature* **395**, 433-434.

Galbally, I. E. *et al.* (1992). Biosphere-atmosphere exchange of trace gases over Australia. In Gifford and Barson, pp 117-149.

Gascon, C., Williamson, G. B., da Fonseca, G. A. B. (2000). Receding forest edges and vanishing reserves. *Science* **288**, 1356-1358.

George, S. (1988). *A Fate Worse than Debt.* Penguin, London.

George, S. and Sabelli, F. (1994). *Faith and Credit.* Penguin, London.

Gibbs, W. W. (1996). Gaining on fat. *Scientific American* **275** (2), 70-76.

Gibbs, W. W. (1998). Big tobacco's worst nightmare. *Scientific American* **279** (1) 16, 19-20.

Gifford, R. M. and Barson, M. M. (Eds), (1992). *Australia's Renewable Resources: Sustainability and Global Change.* BRR Proceedings No. 14. Bureau of Rural Resources and CSIRO Division of Plant Industry, Canberra.

Gimpel, J. (1992). *The Medieval Machine.* Pimlico, London.

Gladkih, M. I., Kornietz, N. L. and Soffer, O. (1984). Mammoth-bone dwellings on the Russian Plain. *Scientific American* **251**, (5) 136-143.

Glantz, M. H. (Ed.). (1994). *Drought Follows the Plow.* Cambridge University Press, New York.

Glantz, M. H. (1994a). Drought, desertification and food production. In Glantz, pp 9-30.

Glanz, J. (1994). New soil erosion model erodes farmers' patience. *Science* **264**, 1661-1662.

Gleick, P. H. (2000). *The World's Water 2000-2001.* Biennial Report on Freshwater Resources. Island Press, Washington D.C.

Gleick, P. H. (2001). Making every drop count. *Scientific American* **284** (2) 28-33.

Global Environment Outlook 2000. (2000). UNEP, Earthscan, London.

Godden, D. *et al.*: Gifford, R. *et al.*: Nicholls, N. (1998). Climate change and Australian wheat yield. *Nature* **391**, 447-449.

Goldsmid, J. (1988). *The Deadly Legacy. Australian History and Transmissible Disease.* New South Wales University Press, Sydney.

Goldsmith, E. (1997). Development as colonialism. *The Ecologist* **27**, 69-76.

Goldschmidt, T. (1996). *Darwin's Dreampond. Drama in Lake Victoria.* M.I.T. Press, Cambridge MA.

Gornitz, V., Rosenzweig, C. and Hillel, D. (1994). Is sea level rising or falling? *Nature* **371**, 481.

Goss, H. (1995). Flying detectives help farmers save land. *New Scientist* **148** (1998), 25.

Goudie, A. (1986). *The Human Impact on the Natural Environment.* Blackwell, Oxford.

Gould, W. T. S. and Lawton, R. (Eds). (1986). *Planning for Population Change.* Croom Helm, Beckenham, Kent.

Gras, N. S. B. (1925). *A History of Agriculture.* F. S. Crofts & Co., N.Y.

Green, K. *et al.* Long distance transport of arsenic by migrating Bogong moths from agricultural lands to mountain ecosystems. *The Victorian Naturalist* 118, 112-115.

Gregg, S. A. (1988). *Foragers and Farmers*. University of Chicago Press, Chicago.

Grigg, D. B. (1974). *The Agricultural Systems of the World*. University Press, Cambridge.

Grigg, D. (1982). *The Dynamics of Agricultural Change*. St Martin's Press, New York.

Groth, E. (1996). The pesticide regulation treadmill. *Science*, 274, 61-62.

Grove, S. (1996). The wrong trees. *New Scientist* 150 (2033), 32.

Le Guenno, B. (1995). Emerging viruses. *Scientific American* 273 (4), 30-37.

Hadfield, P. (1994). Revenge of the rain gods. *New Scientist* 143 (1939), 14-15.

Hammond, N. (1986). The emergence of Maya Civilization. *Scientific American* 255 (2), 98-107.

Han, T.-M. and Runnegar, B. (1992). Megascopic eukaryotic algae from the 2.1 billion-year-old Negaunee Iron Formation, Michigan. *Science*, 257, 232-235.

Hancock, W. K. (1972). *Discovering the Monaro. A Study of Man's Impact on his Environment*. The University Press, Cambridge.

Handreck, K. A. (1997). Phosphorus requirements of Australian plants. *Aust. J. Soil. Res.* 35, 241-289.

Harding, L. (2002). 'It is a dangerous time to be a tall poppy in India.' *Guardian Weekly*, 14-20th March, p.24.

Hardy, J., and Frost, A. (Eds). (1989). *Studies from Terra Australis to Australia*. Aust. Acad. Humanities, Canberra.

Harlan, J. (1971). Agricultural Origins: Centers and Noncenters. *Science* 174, 468-473.

Harrison, G. A. *et al.* (1977). *Human Biology*. Oxford University Press, Oxford.

Harte, J., Kinzig, A. and Green, J. (1999). Self-similarity in the distribution and abundance of species. *Science* 284, 334-336.

Harvell, C. D. *et al.* (1999). Emerging marine diseases—climate links and anthropogenic factors. *Science* 285, 1505-1510.

Harvey, P.H. and May, R. M. (1997). Case studies of extinction. *Nature* 385, 776-778.

Hatfield, C. B. (1997a). Oil back on the global agenda. *Nature* 387, 121.

Hatfield, C. B. (1997b). A permanent decline in oil production? *Nature* 388, 618.

Heathcote, R. L. (1994). Australia. In Glantz, pp 91-102.

Hecht, J. (1996). Eye in the sky needs help on the ground. *New Scientist* 149 (2018), 13.

Hector, A. *et al.* (1999). Plant biodiversity and productivity experiments in European grasslands. *Science* 286, 1123-1127.

Hertz, N. (2001). *The Silent Takeover. Global Capitalism and the Death of Democracy*. Random House, London.

Hildyard, N. (1996). Migrant labour in the global economy. *The Ecologist* 4, 133-134.

Hill, C. (1996). *Liberty Against the Law*. Penguin, London.

Hillel, D. (1992). *Out of the Earth. Civilization and the Life of the Soil*. Aurum Press, London.

Hillel, D. (1994). *Rivers of Eden: The Struggle for Water and the Quest for Peace in the Middle East*. Oxford University Press, N.Y.

Hingston, F. J. and Raison, R. J. (1982). Consequences of biochemical interactions for adjustment to changing land use practices in forest systems. In I. E. Galbally and J. R. Freney (Eds), *The Cycling of Carbon, Nitrogen, Sulphur and Phosphorus in Terrestrial and Aquatic Ecosystems*. Australian Academy of Science, Canberra, pp 11-24.

Hirst, D. and Woodley, B. (1997). Steinbeck's wrath reincarnated. *The Australian*, Weekend Review 15-16th March, p. 3.

Hobbs, P. V. *et al.* (1997). Direct radiative forcing by smoke from biomass burning. *Science*, **275**, 1776-1778.

Hoedeman, O. *et al.* (1998). MAIgalomania: the new corporate agenda. *The Ecologist*, **28**, 154-161.

Hogarth, M. (1997a). Minister in a spin over cotton pesticide war. *Sydney Morning Herald*, 10th May, p.10.

Hogarth, M. (1997b). Pesticide findings put blame on cotton. *Sydney Morning Herald*, 19th May, p.5.

Holden, C. (1997). Death from lab poisoning. *Science*, **276**, 1797.

Holdgate, M. (1996). The ecological significance of biological diversity. *Ambio* **25**, 409-416.

Holmes, B. (1994a). Peruvian mummy clears colonists of TB charge. *New Scientist* **141** (1917), 5.

Holmes, B. (1994b). Biologists sort the lessons of fisheries collapse. *Science* **264**, 1252-1253.

Holmes, B. (1997). Destruction follows in trawlers' wake. *New Scientist* **154** (2086), 4.

Holmgren, P., Masakha, E. J. and Sjöholm, H. (1994). Not all African land is being degraded: a recent survey of trees on farms in Kenya reveals rapidly increasing forest resources. *Ambio*, **23**, 390-395.

Holt, J. A., Bristow, K. L. and McIvor, J. G. (1996). The effects of grazing pressure on soil animals and hydraulic properties of two soils in semi-arid tropical Queensland. *Aust. J. Soil. Res.* **34**, 69-79.

Holway, D. A., Suarez, A. V. and Case, T. J. (1998). Loss of intraspecific aggression in the success of a widespread invasive social insect. *Science* **282**, 949-952.

Homburger, F. (1996) Tobacco research: one researcher's experience. *Science*, **273**, 1322-1323.

Homer-Dixon, T. (2001). *The Ingenuity Gap*. Vintage. London.

Homewood, B. (1993). Killers escape. *New Scientist* **137** (1862), 5.

Hooper, D. U. and Vitousek, P. M. (1997). The effects of plant composition and diversity on ecosystem processes. *Science*, **277**, 1302-1305.

Houthakker, K. S. (1997). A permanent decline in oil production? *Nature* **388**, 618.

Hoy, A. (1996). Feed-lot versus grass-fed: a cattle war brews. *Sydney Morning Herald*, 14th September, p. 7.

Hoy, A. (1997). Mining giant chops up world's bigest cattle station. *Sydney Morning Herald*, 11th October, p. 9.

Hughes, R. (1993). *The Culture of Complaint*. Oxford University Press, N.Y.

Hungate, B. A. *et al.*: Henebry, G. M.: Wedin, D. A., Tilman, D. (1997). Atmospheric nitrogen deposition. *Science*, **275** (5301), 739-741. 3 letters to editor.

Hunt, P. (1993). Hong Kong rounds up toxic waste. *New Scientist* **138** (1876), 6.

IGBP Terrestrial Carbon Working Group (1998). The terrestrial carbon cycle: implications for the Kyoto protocol. *Science* **280**, 1393-1394.

Iinuma, J. (1973). The introduction of American and European agricultural science into Japan in the Meiji era. In R. T. Shand (Ed.), *Technical Change in Asian Agriculture.* Australian National University Press, Canberra, pp 1-8.

Imhoff, K., Müller, W. J. and Thistlethwayte, D. K. B. (1971). *Disposal of Sewage and Other Water-Borne Wastes.* Ann Arbor Science Publishers Inc., Ann Arbor.

Jackson, R. V. (1988). *Population History of Australia.* McPhee Gribble/Penguin, Melbourne.

Jacobs, J. (1969). *The Economy of Cities.* Random House, New York.

Jacobs, T. (1990). River regulation. In Mackay and Eastburn, pp 38-58.

Jacobsen, T. and Adams, R. M. (1958). Salt and silt in ancient Mesopotamian agriculture, *Science* **128**, 1251-1258.

Janzen, D. (1998). Gardenification of wildland nature and the human footprint. *Science* **279**, 1312-1313.

Jingyun Fang *et al.* (2001). Changes in forest biomass carbon storage in China between 1949 and 1998. *Science* **292**, 2320-2322.

Johnson, A. (1991). *Factory Farming,* Blackwell, Oxford.

Johnson, K. (1994). Creating place and landscape. In Dovers, pp 37-54.

Johnson, D. and Lewis, L.A. (1995). *Land Degradation: Creation and Destruction.* Blackwell, Oxford.

Jung Chang (1991). *Wild Swans.* Flamingo, London.

Jones, E., and Raby, G. (1989). The fatal shortage. Establishing a European economy in New South Wales, 1788-1805. In Hardy and Frost pp 153-167.

Kaiser, J. (1996). Paper-trail cleanup memo sparks furor. *Science* **274**, 173.

Kaiser, J. (1997a). A nasty brew from forest fires. *Science,* **277**, 1205.

Kaiser, J. (1997b). Panels lead the way on the road to Kyoto conference. *Science,* **278**, 216-217.

Kaiser, J. and Gallagher, R. (1997). How humans and nature influence ecosystems. *Science,* **277**, 1204-1205.

Karliner, J., Morales, A. and O'Rourke, D. (1997). The barons of bromide. *The Ecologist* **27**, 90-98.

Kasting, J. F. (1993). Earth's early atmosphere. *Science* **259** 920-925.

Katz, D. B. (1997). Eco-friendly coffee farming. *Science,* **275**, 12.

Kaufman, W. (1994). *No Turning Back.* BasicBooks, New York.

Kaul, R. N. and Egbo, C. O. (1985). *Introduction to Agricultural Mechanisation.* Macmillan, London.

Keen, S. (2001). *Debunking Economics.* Pluto Press, Anandale, NSW; Zed Books, London and New York.

Keller, M. *et al.* (1996), If a tree falls in the forest ... *Science,* **273**, 201.

Kendall, H. W., Pimentel, D. (1994). Constraints on the expansion of the global food supply. *Ambio* **23** (3), 198-205.

Kennish, M. J. (Ed.). (1989). *Practical Handbook of Marine Science*. CRC Press, Boca Raton, Florida.

Kerr, R. A. (1998). The hottest year, by a hair. *Science*, **279**, 315-316.

Kerr, R. A. (2001). It's official: humans are behind most of global warming. *Science* **291**, 586.

Keys, D. (1993). Snapshot of a medieval city. *New Scientist* **140** (1895), 10.

King, F. H. (1911). *Farmers of Forty Centuries*. Rodale Press, Emmaus, Pennsylvania. (Reprint distributed by St Martin's Press.)

Kinnersley, D. (1994). *Coming Clean*. Penguin, London.

Kirkby, M. J. (1980). The problem. In M. J. Kirkby and R. P. C. Morgan (Eds), *Soil Erosion*, Wiley, Chichester, pp 1-16.

Knapp, A. B. (1988). *The History and Culture of Ancient Western Asia and Egypt*. Dorsey Press, Chicago.

Knight, P. (1996). Better than it looks on paper. *New Scientist* **151** (2049), 16-17.

Koestler, A. (1980). *Bricks to Babel*. Random House, N.Y.

Kolber, Z. S. *et al*. (2001). Contribution of aerobic photoheterotrophic bacteria to the carbon cycle in the ocean. *Science* **292**, 2492-2495.

Korn, M. (1996). The dike-pond concept: sustainable agriculture and nutrient recycling in China. *Ambio* **25**, 6-13.

Kristoff, N. D. and Wudunn, S. (1994). *China Wakes*. Times Books, N.Y.

Kumar, S. (1993). India backs down on Narmada dam. *New Scientist* **139** (1886), 8.

Kumar, S. (1995a). Indian dam builders flout environmental rules. *New Scientist* **147** (1990), 10.

Kumar, S. (1995b). Everyday poisons take their toll in Third World. *New Scientist* **148** (2007), 8.

Kumar, V., Gilkes, R. J., and Bolland, M.D.A. (1993). Forms of phosphate in soils fertilized with rock phosphate and superphosphate as measured by chemical fractionation. *Aust. J. Soil Res.* **31**, 465-480.

Kyle, M. A. and McClintock, S. A. (1995). The availability of phosphorus in municipal wastewater sludge as a function of the phosphorus removal process and sludge treatment method. *Water Environment Res.* **67**, 282-289.

Ladurie, E. L. R. (1978). *Montaillou. Cathars and Catholics in a French Village*. 1294-1324. English translation by Barbara Bray. Penguin, Harmondsworth, Middlesex.

Lampard, E. E. (1973). The urbanizing world, In Dyos and Wolff, Vol. 1, pp 3-57.

Langdon, J. (1982). The economics of horses and oxen in medieval England. *Ag. Hist. Rev.* **30**, 31-40.

LaRoche, J., *et al*. (1996). Flavodoxin as an *in situ* marker for iron stress in phytoplankton. *Nature* **382** (6594) 802-805.

Laurance, W. F. *et al*. (1997). Biomass collapse in Amazonian forest fragments. *Science* **270**, 1117-1118.

Lawton, J. H. and May, R. M. (Eds). (1995). *Extinction Rates*. Oxford University Press, Oxford.

Lean, G. and Hinrichsen, D. (1992). *Atlas of the Environment*, Helicon, Oxford.

Leeper, G. W. and Uren, N. C. (1993). *Soil Science. An Introduction.* Melbourne University Press, Melbourne.

Legg, J. O. and Meisinger, J. J. (1982). Soil nitrogen budgets. In F. J. Stevenson (Ed.), *Nitrogen in Agricultural Soils.* Publication No. 22, Agronomy, American Society of Agronomy *et al.*, Madison. pp 503-566.

Legge, A. J. and Rowley-Conwy, P. A. (1987). Gazelle killing in Stone Age Syria. *Scientific American* **257** (2), 76-83.

Lehrman, S. (1996). California backs off destruction of records. *Nature* **383**, 470.

Levitus, S. *et al.* (2001). Anthropogenic warming of Earth's climate system. *Science* **292**, 267-270.

Lewis, J. N. (1983). Australia's phosphate policy. In Costin and Williams, pp 267-284.

Lewis, L. A. and Berry, L. (1988). *African Environments and Resources.* Unwin Hyman, London.

Lewis, S. A. (1992). Banana bonanza. *The Ecologist* **22**, 289-290.

Lindahl-Kessing, K. (1998) Conference on the Aral Sea—women, children, health and environment. *Ambio* **27**, 560-564.

Lines, W. J. (1991). *Taming the Great South Land.* Allen and Unwin, North Sydney.

Lipsett, J. and Dann, P. R. (1983). Wheat: Australia's hidden mineral export. *J. Aust. Inst. Ag. Sci.* 81-89.

Livi-Bacci, M. (1992). *A Concise History of World Population.* (English translation by C. Ipsen) Blackwell, Oxford.

London, H. (1984). *Why Are They Lying to Our Children?* Stein and Day, New York.

Loomis, R. S. (1978). Ecological dimensions of medieval agrarian systems: an ecologist responds. *Agricultural History* **52**, 478-483.

Lorenz, K. (1966). *On Aggression.* (English translation by Marjorie Latzke) Methuen, London.

Lowi, M. R. (1993). *Water and Power.* Cambridge University Press, Cambridge.

Lowry, I. S. (1991). World urbanization in perspective. In Davis and Bernstam, pp 148-176.

Lutz, W., Sanderson, W. and Scherbov, S. (1997). Doubling of world population unlikely. *Nature* **387**, 803-804.

MacArthur, R. H. and Wilson, E. O. (1967). *The Theory of Island Biogeography*, Princeton University Press, Princeton.

McCarthy, C. (1998). Peddlers of violence and death. *Guardian Weekly*, 15th February, p. 20.

McCrone, J. (1993). *The Myth of Irrationality.* Carroll & Graf, New York.

McEvedy, C. (1992). *The New Penguin Atlas of Medieval History.* Penguin, Harmondsworth, Middlesex.

McEvedy, C. and Jones, R. (1978). *Atlas of World Population History.* Allen Lane, London.

McHugh, S. (1996). *Cottoning On.* Hale and Ironmonger, Sydney.

MacIlvain, C. *et al.* (1998). When rhetoric hits reality in debate on bioprospecting. *Nature* **392**, 535-540.

McIntosh, P. (1997). Nutrient changes in tussock grasslands, South Island, New Zealand. *Ambio* **26**, 147-151.

Mackay, N. (1990). Understanding the Murray. In Mackay and Eastburn, pp viii-xix.

Mackay, N. and Eastburn, D. (Eds). (1990). *The Murray*. Murray Darling Basin Commission, Canberra.

Mackay, N. and Landsberg, J. (1992). The 'health' of the Murray-Darling river system. *Search* **23**, 34-35.

MacKenzie, D. (1994). World unites to fight soaring population. *New Scientist* **142** (1922), 5.

MacKenzie, D. (1997). If it lasts, ban it, says Sweden. *New Scientist* **155** (2089), 22.

McLaughlin, M. J., Fillery, I. R. and Till, A. R. (1992). Operation of the phosphorus, sulphur and nitrogen cycles. In Gifford and Barson, pp 67-116.

McLaughlin, M. J. *et al.* (1996). Review: the behaviour and environmental impact of contaminants in fertilizers. *Aust. J. Soil Res.* **34**, 1-54.

McNaughton, S. J., Banyikwa, F. F. and McNaughton, M. M. (1997). Promotion of the cycling of diet-enhancing nutrients by African grazers. *Science*, **278**, 1798-1800.

MacNeish, R. S. (1991). *The Origins of Agriculture and Settled Life*. University of Oklahoma Press, Norman, Oklahoma.

McTainsh, G. H. and Boughton, W. C. (1993). *Land Degradation Processes in Australia*. Longman Cheshire, Melbourne.

McTainsh, G. and Leys, J. (1993). Soil erosion by wind. In McTainsh and Boughton, pp 188-233.

Macumber, P. (1990). The salinity problem. In Mackay and Eastburn, pp 110-125.

MacVean, C. (1997). Coffee growing: sun or shade? *Science*, **275**, 1552.

Maisels, C. K. (1990). *The Emergence of Civilization*. Routledge, London.

Margules, C. R. and Gaston, K. J. (1994). Biological diversity and agriculture. *Science*, **265**, 457.

Marten, G. G. (Ed.). (1986). *Traditional Agriculture in Southeast Asia. A Human Ecology Perspective*. Westview Press, Boulder.

Masibay, K. (2000). Drinking without harm. *Scientific American* **283**, 17.

Masood, E. and Garwin, L. (1998). Costing the Earth: when ecology meets economics. *Nature* **395**, 426-430.

Masood, E. (1996). Scientific caution 'blunts efforts' to conserve fish stocks. *Nature* **379**, 481.

May, R. M. (1997). Kyoto and beyond. *Science*, **278**, 1691.

Meade, K. (1997). Fruit-fly spraying ruins rich coffee crop. *The Weekend Australian*, 31st May, p. 7.

Melbourne, B. (1990). *The Phosphorus Cycle in Australia: Implications for Sustainability*. B.Sc. (Hons) Thesis, Australian National University.

Meltzer, D. J. (1992). How Columbus sickened the New World. *New Scientist* **136** (1842), 38-41.

Mentasi, E., Ramel, C. (1995). Spread of toxic substances and environmental pollution. *Ambio* **24** (4), 250-251.

Mihill, C. (1998). Drug company donates $1.6 bn to defeat tropical disease. *Guardian Weekly*, 1st February, p. 1.

Miller, G. H. *et al.* (1999). Pleistocene extinction of *Genyornis newtoni*: human impact on Australian megafauna. *Science* **283**, 205-208.

Miller, S. K. and Kumar, S. (1993). Narmada dam fails World Bank's final test. *New Scientist* **138** (1868), 5.

Milthorpe, F. L. (1982). Interaction of biogeochemical cycles in nutrient-limited environments: wheat-pasture and forest systems. In I. E. Galbally and J. R. Freney (Eds), *The Cycling of Carbon, Nitrogen, Sulphur and Phosphorus in Terrestrial and Aquatic Ecosystems.* Australian Academy of Science, Canberra pp 35-45.

Mims, F. M. *et al.* (1997). Smoky skies, mosquitoes and disease. *Science*, **276**, 1774-1775.

Mishan, E. J. (1967). *The Costs of Economic Growth.* Pelican, Harmondsworth, Middlesex.

Mishan, E. J. (1986). *Economic Myths and the Mythology of Economics.* Humanities Press International, Atlantic Highlands, N.J.

Mittelman, J. H. (Ed.). (1997). *Globalization.* Lynne Rienner, Boulder, Colorado.

Moffat, A. S. (1996). Ecologists look at the big picture. *Science*, **273**, 1490.

Moffat, A. S. (1998). Global nitrogen overload problem grows critical. *Science* **279**, 988-989.

Moffett, G. D. (1994). *Critical Masses.* Penguin, New York.

Monbiot, G. (1997a). Watch those beans. *Guardian Weekly*, 28th September, p. 23.

Monbiot, G. (1997b). The seeds of a conspiracy. *Sydney Morning Herald*, 20th September, p.26. (Originally published in *The Guardian*.)

Monbiot, G. (1998). Give us this day our toxic bread. *Guardian Weekly*, 22nd March, p. 14.

Monnin, E. *et al.* (2001). Atmospheric CO_2 concentrations over the last glacial termination. *Science* **291**, 112-114.

Montesano, R., Cabral, J. R. P. and Wildbourn, J. D. (1988). Environmental carcinogens. *Ann. N.Y. Acad. Sci.* **534**, 67-73.

Moore, P. D. (1995). Too much of a good thing. *Nature* **374**, 117-118.

Morgan, R. P. C. and Davidson, D. A. (1986). *Soil Erosion and Conservation.* Longman, London.

Morris, D. (1981). *The Soccer Tribe.* Jonathan Cape, London.

Morton, S. R. (1990). The impact of European settlement on the vertebrate animals of arid Australia: a conceptual model. *Proc. Ecol. Soc. Aust.* **16**, 201-213.

Morton, S. R. (1994). European settlement and the mammals of arid Australia. In Dovers, pp 141-166.

Moss, A. J. *et al.* (1993). A preliminary assessment of sediment and nutrient exports from Queensland coastal catchments. *Environment Technical Report No. 5*, Queensland Department of Environment and Heritage.

Moss, P. (1972). *Town Life Through the Ages.* Harrap, London.

Motluk, A. (1996a) The dirtiest dilemma of all. *New Scientist* **151** (2045), 12-13.

Motluk, A. (1996b). Cancer charities tighten rules on 'dirty money'. *New Scientist* **152** (2050), 6.

Mumford, L. (1961). *The City in History.* Secker & Warburg, London.

Murray Darling Basin Commission (1992). *Investigation of Nutrient Pollution in the Murray-Darling River System.* (Ref. No. 311/1048/0504: ISBN 1 875209 13 1).

Myers, B. *et al.* (1995). *Effluent Irrigated Plantations.* CSIRO Technical Paper No. 2.

Myers, N. (1998). Lifting the veil on perverse subsidies. *Nature* **392**, 327-328.

Nader, R. *et al.* (1993). *The Case Against Free Trade.* Earth Island Press, San Francisco.

National Farmers' Federation (Australia). (1992). *Australian Agiculture* 1991/92.

Nellithanam, R., Nellithanam, J. and Samiti, S. S. (1998). Return of native seeds. *The Ecologist,* **28**, 29-33.

Nepstad, D. C. *et al.* (1999). Large-scale impoverishment of Amazonian forests by logging and fire. *Nature* **398**, 505-508.

New South Wales Water Resources Council. (1994a). *Surface Water* (ISBN 0 7310 2267 x).

New South Wales Water Resources Council. (1994b). *Groundwater.* (ISBN 0 7310 2257 2).

Newby, H. (1987). *Country Life. A Social History of Rural England.* Cardinal Edition (1988), Sphere Books, London.

Nickson, R. *et al.* (1998). Arsenic poisoning of Bangladesh groundwater. *Nature* **395**, 338.

Noble, I. R. and Dirzo, R. (1997). Forests as human-dominated ecosystems. *Science,* **277**, 522-525.

Nougarède, N. (1997). Rural Poland faces uncertain future. *Guardian Weekly* 25th May, p. 14.

Nutrition Search Inc. (1979). *Nutrition Almanac.* McGraw-Hill, N.Y.

Oades, J. M. (1995). Krasnozems—organic matter. *Aust. J. Soil Res.* **33**, 43-57.

O'Brien, C. (1996). Protein test favors BSE-CJD link. *Science,* **274**, 721.

Odum, E. P. (1971). *Fundamentals of Ecology.* W. B. Saunders, Philadelphia.

O'Kane, M. (1998). She is three and suffers from a plague that kills millions—the plague of debt. *Guardian Weekly,* 17th May, p. 1.

O'Loughlin, E. M. (1980). An overview of Australia's water resources. In O'Loughlin (Ed.), *Irrigation and Water Use in Australia.* Australian Acad. Sci., Canberra, pp 1-23.

Olsen, S. R. and Dean, L. A. (1965). Phosphorus. In C. A. Black (Ed.), *Methods of Soil Analysis. Part* 2. Chemical and Microbiological Properties. American Soc. Agronomy. pp 1035-1041.

Ortloff, C. R. (1988). Canal Builders of Pre-Inca Peru. *Scientific American* **259** (6), 74-80.

O'Toole, G. (1996). Brazil's poor pay bloody price in battle for land. *Guardian Weekly,* 23rd June, p. 7.

Pacala, S. W. *et al.* (2001). Consistent land- and atmosphere-based U.S. carbon sink estimates. *Science* **292**, 2316-2320.

Pakulski, J. D. *et al.*. (1996). Iron stimulation of Antarctic bacteria. *Nature* **383**, 133-134.

Passey, D. (1995a). On the land, they fear another crop of sickness. *Sydney Morning Herald* 30th September, p. 7.

Passey, D. (1995b). Chemical delivers a life sentence to shearers. *Sydney Morning Herald*, 16th September, p. 3.

Patel, T. (1997). Rampant urban pollution blights Asia's crops. *New Scientist* 154 (2086), 11.

Pauley, D. *et al.* (1998). Fishing down marine food webs. *Science*, 279, 860-863.

Pearce, D., Markandya, A. and Barbier, E. B. (1989). *Blueprint for a Green Economy.* Earthscan, London.

Pearce, F. (1992). A shanty town that's here to stay. *New Scientist* 135 (1837), 22-25.

Pearce, F. (1994a). Neighbours sign deal to save Aral Sea. *New Scientist* 141 (1909), 10.

Pearce, F. (1994b). How disappearing lakes are swelling the oceans. *New Scientist* 141 (1909), 17.

Pearce, F. (1994c). Fifty years from now the world population may be falling. *New Scientist* 143 (1943), 6.

Pearce, F. (1995a). Dead in the water. *New Scientist* 145 (1963), 26-31.

Pearce, F. (1995b). Sea life sickened by urban pollution. *New Scientist* 146 (1982), 4.

Pearce, F. (1995c). Poisoned waters. *New Scientist* 148 (2000), 29-33.

Pearce, F. (1995d). Death and the devil's water. *New Scientist* 147 (1995), 14-15.

Pearce, F. (1995e). The biggest dam in the world. *New Scientist* 145 (1962), 25-29.

Pearce, F. (1995f). Global row over human life. *New Scientist* 147 (1991), 7.

Pearce, F. (1996). Deadly sprays worse than useless. *New Scientist* 152 (2057), 7.

Pearce, F. (1997). Norway's tundra is trampled underfoot. *New Scientist* 154 (2086), 6.

Peierls, B. L., *et al.* (1991). Human influence on river nitrogen. *Nature* 350, 386-387.

Philip, J. R. (1978). Water on Earth. In A. K. McIntyre, (Ed.), *Water: Planets, Plants and People.* Australian Academy of Science, Canberra pp 35-59.

Pierce, J. T. (1990). *The Food Resource.* Longmans/Wiley, New York.

Pietilä, H. (1997). The villages in Finland refuse to die. *The Ecologist* 27, 178-181.

Pilger, J. (1998). *Hidden Agendas.* Vintage (Random House), London.

Pimentel, D. *et al.* (1995a). Environmental and economic costs of soil erosion and conservation benefits. *Science*, 267, 1117-1123.

Pimentel, D. *et al.* (1995b). (Untitled; reply to Crosson, 1995). *Science* 269, 464-465.

Pimm, S. L. (1997a). The value of everything. *Nature* 387, 231-232.

Pimm, S. L. (1997b). In search of perennial solutions. *Nature* 389, 126-127.

Pimm, S. L. and Lawton, J. H. (1998). Planning for biodiversity. *Science* 279, 2068-2069.

Pitty, A. F. (1979). *Geography and Soil Properties.* Methuen, London.

Podmore, I. D. *et al.* (1998). Vitamin C exhibits pro-oxidant properties. *Nature* 392, 559.

Post, J. D. (1985). *Food Shortage, Climatic Variability and Epidemic Disease in Preindustrial Europe.* Cornell University Press, Ithaca.

Postel, S. L., Daily, G. C. and Ehrlich, P. R. (1996). Human appropriation of renewable fresh water. *Science*, 271, 785-788.

Poupeau, F. (2002). Commodifying rain. *Le Monde Diplomatique* (English Edition), May, p. 13.

Prasad, R. (1986). Fertilizer nitrogen: requirements and management. In M. S. Swaminathan and S. K. Sinha (Eds), *Global Aspects of Food Production*, Tycooly, London. pp 199-226.

Pratley, J. E. (Ed.). (1988). *Principles of Field Crop Production.* Sydney University Press/Oxford University Press, Melbourne.

Pretty, J. N. (1990). Sustainable agriculture in the middle ages: the English manor. *Ag. Hist. Rev.* 38, 1-19.

Price, T. D. and Petersen, E. B. (1987). A Mesolithic Camp in Denmark. *Scientific American* **256** (3), 90-99.

Prill-Brett, J. (1986). The Bontok: traditional wet rice and swidden cultivators of the Philippines. In Marten, pp 54-84.

Provan, D. M. J. (1973) The soils of an Iron Age farm site—Bjellandsynæ, SW Norway. *Norwegian Archaeological Review,* 6, 30-41.

Prusiner, S. B. (1995). The prion diseases. *Scientific American* **272** (1), 30-37.

Prusiner, S. B. (1997). Prion diseases and the BSE crisis. *Science,* **278**, 245-251.

Pyne, S. J. (1991). *Burning Bush.* Henry Holt, New York.

Rae, J. and Gruen, N. (1997). *A Full Repairing Lease.* Draft Report of the Australian Industry Commission of Inquiry into Ecologically Sustainable Land Management.

Randhawa, M. S. (1980). *A History of Agriculture in India.* Vol. 1. Indian Council of Agricultural Research, New Delhi.

Rayment, G. E. and Higginson, F. R. (1992). *Australian Laboratory Handbook of Soil and Water Chemical Methods.* Inkata Press, Melbourne.

Reay, B. (1990). *The Last Rising of the Agricultural Labourers.* Oxford University Press, Oxford.

Reisner, M. (1986). *Cadillac Desert.* Viking Penguin, New York.

Riemer, A. (1998). *Sandstone Gothic.* Allen and Unwin, Sydney.

Rifkin, J. (1992). *Beyond Beef.* Penguin, New York.

Robinson, M. E. (1976). *The New South Wales Wheat Frontier 1851 to 1911.* Australian National University Press, Canberra.

Rocha, J. (1997a). Asian loggers strip the Amazon's assetts. *Guardian Weekly* 19th January, p.24.

Rocha, J. (1997b). Rivers of money, rivers of tears. *Guardian Weekly,* 23rd May, p. 28.

Rocha, J. (1997c). Amazon road to ruin. *Guardian Weekly,* 31st August, p. 23.

Rolls, E. (1994). More a new planet than a new continent. In Dovers, pp 22-36.

Rose, F. G. G. (1987). *The Traditional Mode of Production of the Australian Aborigines.* Angus & Robertson, North Ryde, NSW.

Rosen. G. (1973). *Disease, Debility and Death.* In Dyos and Wolff, pp 625-667.

Rosenzweig, M. L. (1999). Heeding the warning in biodiversity's basic law. *Science* **284**, 276-278.

Rosewell, C. J. *et al.* (1991). Forms of erosion. In Charman and Murphy, pp 12-35.

Ross, P. E. (1992). Eloquent Remains. *Scientific American.* **266** (5), 73-81.

Rouard, D. (1998). Brazil ranchers grow rich on 'slave labour'. *Guardian Weekly,* 10th May, p. 16.

Rowan, D. (1998). Meet the new world government. *Guardian Weekly,* 22nd April, p. 14.

Rubin, R. T. (1994). *The Green Crusade.* The Free Press, New York.

Ruddle, K., and Zhong, G. (1988). *Integrated Agriculture-Aquaculture in South China. The Dike-Pond System of the Zhujiang Delta.* Cambridge University Press, New York.

Runnels, C. N. (1995). Environmental degradation in ancient Greece. *Scientific American* **272** (3), 72-75.

Rutherford, I. (1990). Ancient river, young nation. In Mackay and Eastburn, pp 16-36.

Ryle, G. (2002). Action on toxic waste fertiliser. *Sydney Morning Herald*, 25th April, p. 5.

Safina, C. (1995). The world's imperiled fish. *Scientific American* **273** (5), 30-37.

Sahaglan, D. L., Schwartz, F. W. and Jacobs, D. K. (1994). Direct anthropogenic contributions to sea level rise in the twentieth century. *Nature* **367**, 54-57.

Saul, J. R. (1997). *The Unconscious Civilization.* Penguin, Ringwood, Victoria.

Saunders, N. (1992). Rich Existence in Pre-Columbian America. *New Scientist*, **134** (1818) 42-43.

Schofield, R. and Shaoul, J. (1997). Regulating the water industry. *The Ecologist* **27**, 6-13.

Schopf, J. W. (1993). Microfossils of the Early Archean apex chert: new evidence of the antiquity of life. *Science* **260**, 640-646.

Scott, J. C. (1952). *Health and Agriculture in China.* Faber and Faber, London.

Seckler, D. (1987). Economic costs and benefits of degradation and its repair. In Blaikie and Brookfield, pp 84-96.

de Selincourt, K. (1996). Demon farmers and other myths. *New Scientist* **150** (2027), 36-39.

Selinger, B. (1989). *Chemistry in the Marketplace.* 4th edn. Harcourt Brace Jovanovich, Marrickville, NSW.

Sen, A. (1995). We shall not be moved ... *New Scientist* **147** (1989), 51.

Senate Standing Committee On Rural And Regional Affairs (1992). *Beef Cattle Feedlots in Australia*, Senate Printing Unit, Canberra.

Service, R. F. (1998). Seed-sterilizing 'terminator technology' sows discord. *Science* **282**, 850-851.

Sévilla, J.-J. (1997). Hired guns menace Brazil's landless peasants. *Guardian Weekly*, 2nd February, p.13.

Seymour, J. (1996). Trafficking in death. *New Scientist* **151** (2047), 34-37.

Shankland, A. (1993). Brazil's BR-364 highway. *The Ecologist*, **23**, 141-147.

Sharma, R. C. (1992). Analysis of phytomass yield in wheat. *Agron. J.* **84**, 926-929.

Scheffer, M. (2001). Catastrophic shifts in ecosystems. *Nature* **413**, 591-596.

Sheil, C. (2000) *Water's Fall.* Pluto Press, Anandale, NSW.

Shiva, V. (1991). The green revolution in the Punjab. *The Ecologist* **21**, 57-65.

Silvester, W. B. and Musgrave, D. R. (1991). Free living diazotrophs. In M. J. Dilworth and A. R. Glenn (Eds), *Biology and Biochemistry of Nitrogen Fixation.* Elsevier, Amsterdam, pp 162-186.

Simard, S. W. *et al.* (1997). Net transfer of carbon between ectomycorrhizal tree species in the field. *Nature* **388**, 579-582.

Simon, T. (1994). *The River Stops Here.* Random House, New York.

Singer, P. (1993). *How Are We to Live?* Text Publishing, East Melbourne.

Sivertsen, D. (1994). The native vegetation crisis in the wheat belt of NSW. *Search* **25**, 5-8.

Skinner, G. W. (Ed). (1977). *The City in Late Imperial China.* University Press, Stanford.

Skinner, G. W. (1977a). Introduction: urban development in Imperial China. In Skinner, pp 3-32.

Skinner, G. W. (1977b). Regional urbanization in nineteenth century China. In Skinner, pp 211-252.

Skjemstad, J. O. *et al.* (1996). The chemistry and nature of protected carbon in soils. *Aust. J. Soil Res.* **34**, 251-257.

Skole, D. and Tucker, C. (1993). Tropical deforestation and habitat fragmentation in the Amazon: satellite data from 1978 to 1988. *Science*, **260**, 1905-1910.

Slansky, M. (1986). *Geology of Sedimentary Phosphates.* (English Edition), North Oxford Academic, London.

Slesser, M. (1984). Energy use in the food-producing sector of the European Economic Community. In G. Stanhill (Ed.), *Energy and Agriculture.* Springer-Verlag, Berlin. pp 132-153.

Slicher van Bath, B. H. (1963). *The Agrarian History of Western Europe. A.D.* 500-1850. English translation by Olive Ordish. Edward Arnold, London.

Smaling, E., Fresco, L. and de Jager, A. (1996). Classifying, minitoring and improving soil nutrient stocks and flows in African agriculture. *Ambio* **25**, 492-496.

Smil, V. (1984). *The Bad Earth.* M. E. Sharpe, Armonk, N.Y.

Smil, V. (1987). Land degradation in China: an ancient problem getting worse. In Blaikie and Brookfield, pp. 214-222.

Smil, V. (1993). *China's Environmental Crisis.* M. E. Sharpe, Armonk, N.Y.

Smil, V. and Knowland, W. E. (1980). *Energy in the Developing World*, Oxford University Press, N.Y.

Smith, D. I. (1998). *Water in Australia.* Oxford University Press, Melbourne.

Smith, T. (1997). Zambia counts its debt in death. *Guardian Weekly*, 18th May, p.2.

Sneath, D. (1998). State policy and pasture degradation in Inner Asia. *Science* **281**, 1147-1148.

Sombroek, W. G., Nachtergaele, F. O. and Hebel, A. (1993). Amounts, dynamics and sequestering of carbon in tropical and subtropical soils. *Ambio* **22**, 417-426.

Soderlund, D. M. and Bloomquist, J. R. (1990). Molecular mechanisms of insecticide resistance. In R. T. Roush and B. E. Tabashnik (Eds), *Pesticide Resistance in Arthropods*, Chapman and Hall, N.Y. pp 58-96.

Soros, G. (1998). *The Crisis of Global Capitalism.* Little, Brown; London.

Soulé, M. E. and Sanjayan, M. A. (1998). Conservation targets: do they help? *Science* **279**, 2060-2061.

Spender, D. (Ed.). (1991). *The Diary of Elizabeth Pepys.* Grafton, Hammersmith.

Stanhill, G. (Ed.). (1984). *Energy and Agriculture.* Springer-Verlag, New York.

Stanley , D. J. and Warne, A. G. (1993a). Nile Delta: recent geological evolution and human impact. *Science* **260**, 628-634.

Stanley , D. J. and Warne, A. G. (1993b). Sea level and initiation of Predynastic culture in the Nile delta. *Nature* **363**, 435-438.

Stavrianos, L. S. (1988). *The World Since* 1500. A Global History. 5th Edition. Prentice Hall, Englewood Cliffs, New Jersey.

Steen, I. (1998). Phosphorus availability in the 21st century. Management of a non-renewable resource. *Phosphorus and Potasssium*, Issue no. 217. *http://www.nhm.ac.uk/mineralogy/phos/p&k217/*steen.htm

Stephens, J. (1849). *Sanitary Reform. Its General Aspect and Local Importance.* John Stephens, Adelaide. Xerographic copy by The Public Library of South Australia, 1962.

Stevens, J. H. (1981). Irrigation in the Arab countries of the Middle East. In Clarke and Bowen Jones, pp 73-81.

Stewart, J. W. (1996). Savanna users and their perspectives: grazing industry. In A. Ash (Ed.), *The Future of Tropical Savannas. An Australian Perspective.* CSIRO Publishing, Collingwood, Victoria.

Stocking, M. (1987). Measuring land degradation. In Blaikie and Brookfield pp 49-63.

Stone, R. (1999). Coming to grips with the Aral Sea's grim legacy. *Science* **204**, 30-33.

Sugarman, S. D. (1996). Smoking guns. *Science*, **273**, 744-745.

Sun, L. H. (1994). The dragon within begins to stir. *Guardian Weekly*, 30th October, p. 15.

Swearingen, W. (1994). Northwest Africa, In Glantz, pp 117-133.

Sweeney, J. (1997). Rotten bananas. *Guardian Weekly*, 5th October, p. 24.

Tangley, L. (1996). The case of the missing migrants. *Science*, **274**, 1299-1300.

Tegen, I., Lacis, A. A. and Fung, I. (1996). The influence on climate forcing of mineral aerosols from disturbed soils. *Nature* **380** 419-422.

Telling, G. C. *et al.* (1996). Evidence for the conformation of the pathological isoform of the prion protein enciphering and propagating prion diversity. *Science*, **274**, 2079-2082.

Terborgh, J. *et al.* (2001). Ecological meltdown in predator-free forest fragments. *Science* **294**, 1923-1926.

Thomas, V. M., Orlova, A. O. (2001). Soviet and Post-Soviet environmental management: lessons from a case study on lead pollution. *Ambio* **30**, 104-111.

Tickell, O. (1992). Too little, too late for India's dam? *New Scientist* **136** (1849), 8.

Tickell, O. (1998). Death duties. *Guardian Weekly* 5th July, p.23.

Tiessen, H., Salcedo, I. H. and Sampaio, E. V. S. B. (1992). Nutrient and soil organic matter dynamics under shifting cultivation in semi-arid northeastern Brazil. *Agriculture, Ecosystems and Environment*, **38**, 139-151.

Tilman, D., Wedin, D. and Knops, J. (1996). Productivity and sustainability influenced by biodiversity in grassland ecosystems. *Nature* **379**, 718-720.

Tilman, D. *et al.* (1997a). The influence of functional diversity and composition on ecosystem processes. *Science*, **277**, 1300-1302.

Tilman, D. *et al.* (1997b). Biodiversity and ecosystem properties. *Science*, **278**, 1866-1867.

Tilman, D. *et al.* (2001). Forecasting agriculturally driven global environmental change. *Science* **292**, 281-284.

Tisdall, A. L. (1980). The accomplishments of irrigation. In O'Loughlin, pp 24-38.

Trimble, S. W. and Crosson, P. (2000). U.S. soil erosion rates—myth and reality. *Science* **289**, 248-250.

Turner, D. P. *et al.* (1997). Accounting for biological and anthropogenic factors in national land-base carbon budgets. *Ambio* **26**, 220-226.

Valleron, A.-J. *et al.* (2001). Estimation of the epidemic size and incubation time based on age characteristics of VCJD in the United Kingdom. *Science*, **294**, 1726-1728.

van Slyke, L. P. (1988). *Yangtze.* Addison-Wesley, Reading, Mass.

Various (1991) Promoting world hunger. *The Ecologist*, **21** (2). (entire issue).

Various (1995a). Human coastal settlements. *Ambio* **24**, (7, 8) (double issue).

Various (1995b). Overfishing. *The Ecologist*, **25**, (2/3) (entire issue).

Various (1996a). Cattle's feedlot not a happy one. (Several Letters to editor). *Sydney Morning Herald*, 21st September, p. 36.

Various (1996b). The sustainable city. *Ambio* **25** (2) (entire issue).

Various (1997) Nitrogen: a present and future threat to the environment. *Ambio* **26** (5) (entire issue).

Various (2001). Man and the Baltic Sea. *Ambio* **30** (4, 5) (double issue).

Various (2002). Optimizing nitrogen management in food and energy productions, and environmental change. *Ambio* **31** (2) (entire issue).

Vellvé, R. (1993). The decline of diversity in European agriculture. *The Ecologist*, **23**, 64-69.

Vidal, J. (1997a). Landless on the long march home. *Guardian Weekly* 11th May, pp 8-9.

Vidal, J. (1997b). A tribe's suicide pact. *Guardian Weekly*, 12th October, pp 8-9.

Vidal, J. (1997c). A smouldering catastrophe. *Guardian Weekly*, 23rd November, pp 30-31.

Vidal, J. (1999). How Monsanto's mind was changed. *Guardian Weekly*, 20th October, p. 12.

de Villiers, M. (1999). *Water Wars.* Weidenfeld and Nicolson, London.

Vogel, G. and Lawler, A. (1998). Hot year, but cool response in Congress. *Science* **280**, 1684.

Vörösmarty, C. J. *et al.* (2000), Global water resources: vulnerability from climate change and population growth. *Science* **289**, 284-288.

Wadman, M. (1997a). Dispute over insect resistance to crops. *Nature* **388**, 817.

Wadman, M. (1997b). $100m payout after drug data withheld. *Nature* **388**, 703.

Wahlquist, Å. (1993). Advance of the combined corps. *Sydney Morning Herald*, 20th November.

Wahlquist, Å. (1994). Water torture. *Sydney Morning Herald*, 22nd October, 5A.

Wahlquist, Å. (1995). The river that runs through us. *Sydney Morning Herald*, 11th March, 4A-5A.

Wahlqvist, M. L. (Ed). (1982). *Food and Nutrition in Australia*. Methuen, North Ryde, NSW.

Walker, B. and Walker, B. C. (2000). *Privatisation: sell off or sell out?* ABC Books, Sydney.

Walker, G. (1996). Slash and grow. *New Scientist* 151 (2048), 28-33.

Walker, J. (1988). *Jondaryan Station*. University of Queensland Press, St Lucia.

Walker, M. (1978). *Pioneer Crafts of Early Australia*. MacMillan, South Melbourne.

Wallach, L. and Sforza, M. (1999). *Whose Trade Organization?* Public Citizen Foundation, Washington D.C.

Walvin, J. (1992). *Black Ivory*. Harper Collins, London.

Warde, I. (1999). The rise and rise of the DOW. *Le Monde Diplomatique* (English edition) October, p. 13.

Wardle, D. A. *et al.* (1997a). The influence of island area on ecosystem properties. *Science*, 277, 1296-1299.

Wardle, D. A. *et al.* (1997b). Biodiversity and ecosystem properties. *Science*, 278, 1867-1869.

Wargo, J. (1996). *Our Children's Toxic Legacy*. Yale University Press, New Haven.

Warwick, H. (1996). Come hell and high water. *New Scientist* 149 (2023), 38-42.

Wasson, R. J., Olive, L. J. and Rosewell, C. J. (1996). Rates of erosion and sediment transport in Australia. In D. E. Walling and R. Webb (Eds), *Erosion and Sediment Yield: Global and Regional Perspectives*. Proceedings of the Exeter Symposium, IAHS Publication No. 236, pp 139-148.

Watkins, K. (1996). IMF holds gold key for the Third World. *Guardian Weekly*, 16th June, p. 14.

Watt, B. K. and Merrill, A. L. (1963). *Composition of Foods*. Agriculture Handbook No. 8. US Department of Agriculture, Washington D.C.

Watt, J. (1989). The Colony's Health, In J.Hardy and A. Frost (Eds), *Studies From Terra Australis to Australia*. Australian Acad. Humanities, Canberra, pp 137-151.

Watterson, A. (1991). *Pesticides and Your Food*. Merlin Press, London.

Weast, R. C. (Ed.). (1982). *Handbook of Chemistry and Physics*. CRC Press, Boca Raton, Florida.

Wedin, D. A. and Tilman, D. (1996). Influence of nitrogen loading and species composition on the carbon balance of grasslands. *Science*, 274, 1720-1723.

Weiss, H., *et al.* (1993). The genesis and collapse of third millennium North Mesopotamian civilization. *Science* 261, 995-1004.

Wells, F. J. (1975). *The Long-Run Availability of Phosphorus*. Johns Hopkins University Press, Baltimore.

Wells, P. S. (1983). An Early Iron Age Farm Community in Central Europe. *Scientific American* 249 (6), 72-78.

Whelan, R. J. (1995). *The Ecology of Fire*. Cambridge University Press, Cambridge, England.

Whitlock, R. (1990). *A Victorian Village*. Robert Hale, London.

Wiles, J., Turner, G. (1996). Marsh mellow. *Aust Geog.* (41), 68-87.

Wilkinson, R. G. (1973). *Poverty and Progress*. Methuen, London.

Wilks, A. and Hildyard, N. (1994). Evicted! The World Bank and forced resettlement. *The Ecologist*, **24**, 225-229.

Williams, L. (1997a). Asia is burning. *Sydney Morning Herald*, 20th September, p. 41.

Williams, L. (1997b). Under eerie light, a scene from hell. *Sydney Morning Herald*, 27th September, p.27.

Williams, N. (1996). Tobacco funding debate smolders. *Science*, **274**, 28.

Williams, N. (1998). Global change fights off a chill. *Science*, **280**, 1682-1684.

Wilson, E. O. (1996). *Naturalist*. Penguin, London.

Wohl, A. S. (1973). Unfit for human consumption. In Dyos and Wolff, pp 603-624.

Woodford, J. (1995). Report details Warragamba's slow poisoning. *Sydney Morning Herald,* 30th May.

World Commission on Dams (2000). *Dams and Development. A New Framework for Decision-Making*. Final Report.

World Resources Institute (1992). *World Resources* 1992-93. Oxford University Press, New York.

World Resources Institute (1994). *World Resources* 1994-95. Oxford University Press, New York.

World Resources Institute (1998). *World Resources* 1998-99. Oxford University Press, New York.

Worldwatch Institute (2001). *Vital Signs* 2001. W. W. Norton; New York, London.

Wrangham, R. and Peterson, D. (1997). *Demonic Males. Apes and the Origins of Human Violence*. Bloomsbury, London.

Yeates, G. W. (1995). Effect of sewage effluent on soil fauna in a *Pinus radiata* plantation. *Aust. J. Soil. Res.* **33**, 555-564.

Young, A. M. (1997). Eco-friendly coffee farming. *Science*, **275**, 12-13.

Zaimeche, S. E. (1994). The consequences of rapid deforestation: a North African example. *Ambio* **23**, 136-140.

Zimmer, C., (2001). Genetic trees reveal disease origins. *Science,* **292**, 1090-1093.

Zonn, I., Glantz, M. H. and Rubinstein, A. (1994). The virgin lands scheme in the former Soviet Union. In Glantz, pp 135-150.

About the Author

Austin Duncan Brown was born in New Zealand in 1925. He moved to Australia in 1928, taking his parents with him. He holds degrees of B.Sc and M.Sc from the University of Sydney and Ph.D. from the University of Manchester. His professional work has been as a Research Chemist with Australian Paper Manufacturers (1947-1949); Research Officer/Senior Research Officer with CSIRO (Commonwealth Scientific and Industrial Research Organisation) Divisions of Food Research (1950-1956) and Fisheries and Oceanography (1958-1960); Senior Lecturer/Associate Professor in Microbiology at the University of New South Wales (1961-1974); Foundation Professor of Biology at the University of Wollongong (1974-1985); Visiting Fellow, Centre for Resource and Environmental Studies, Australian National University (1992). Currently, Professor Emeritus, Department of Biological Sciences, University of Wollongong.

Work (i.e. excluding conferences etc) outside Australia has included a year at the Low Temperature Research Station, Cambridge, 1954/55; Department of Bacteriology, University of Manchester, 1957/58 (ICI Fellow); San Francisco Medical School, University of California, 1968 (Fulbright Fellow); Department of Biochemistry, Cambridge University, 1973 (Wellcome Foundation; Potter Foundation); Chemistry Department, Yale University, 1977 (Fulbright Fellow); Department of Biochemistry, Norwegian Institute of Technology, 1983 (Norwegian Fisheries Research Council); Department of Chemical Plant Physiology, University of Tübingen, Germany, 1985 (DAAD, German Academic Exchange Service).

Brown has published extensively in the discipline of microbiology with emphasis on the physiology of water stress. Publications include *Microbial Water Stress Physiology*, (Wiley 1990). In the late 1960s he became actively involved with organisations such as Social Responsibility in Science (SRS) and Scientists Against Nuclear Arms. Partly because of that involvement, his concern about the seriousness of environmental threats grew and, unintentionally, led to the establishment of a new government department.

In the early 1970s, under the flag of SRS, Duncan Brown and John Pollak (Department of Biochemistry, University of Sydney), asked relevant municipal councils if they would be willing to fund a chemical and microbiological assessment of the Parramatta River. They were, and the survey was undertaken with the indispensable participation of Margaret Heath and Bill Coady. The report subsequently sent to the councils prompted them to say to the NSW Government, in effect, 'If we can do this, why can't you?' The government established the State Pollution Control Commission which evolved into the Environmental Protection Authority.

Brown is currently a member of The Australian Conservation Foundation and The Wilderness Society.

Index

Cont. from p. 4

There are also ethical questions involved which Duncan Brown does not yet address: should we perceive remaining phosphate reserves as 'Commons', and is every region entitled to claim its own part of the world's phosphate supplies? If so, should the condition be that each region must make its national and regional production systems sustainable and keep its phosphates within the system? This line of thought implies a new kind of development cooperation in which the North and the South will be interdependent: a shared combination of technology and ethics across frontiers.

In short, a challenging book that must be read and discussed, inviting new policy." — **Henk Kieft**, agronomist ETC Ecoculture, Leusden, the Netherlands. ETC is an independent foundation, advisor in sustainable agriculture and rural development worldwide, active in both Europe and the South

"How encouraging is Duncan Brown's conclusion that, if man is to mend his ways, organic agriculture grants him more time than any other method of food production. Brown is realistic enough to acknowledge that the processes that have been set in motion for the sake of civilization—which he pictures as a vicious circle—also have the characteristics of a flywheel that cannot be brought to a halt easily. Is modern man headstrong enough to persist in ignoring serious warnings? Have we become so alienated from nature that we put personal gain before planetary survival? Does human effort still have an input or is the process impelled by its own dynamics?

Feed or Feedback is a very readable description of ecological history and of mankind's expectations for the future, focusing on the relationship between food production and world population." — **Jaap Melgers**, organic farmer and consultant